ENGINEERING/ TECHNOLOGY MANAGEMENT – 2003:

SAFETY ENGINEERING AND RISK ANALYSIS
TECHNOLOGY AND SOCIETY
ENGINEERING BUSINESS MANAGEMENT
HOMELAND SECURITY

presented at

THE 2003 ASME INTERNATIONAL MECHANICAL ENGINEERING CONGRESS
NOVEMBER 15–21, 2003
WASHINGTON, D.C.

sponsored by

THE SAFETY ENGINEERING AND RISK ANALYSIS DIVISION, ASME
THE TECHNOLOGY AND SOCIETY DIVISON, ASME
THE MANAGEMENT DIVISION, ASME

organized by

JONATHAN YOUNG
PACIFIC NORTHWEST NATIONAL LABORATORY

HIROSHI HONDA
TOKYO METROPOLITAN INSTITUTE OF TECHNOLOGY AND
INSTITUTE FOR THE INTERNATIONAL EDUCATION OF STUDENTS

TED A. AANSTOOS
UNIVERSITY OF TEXAS AT AUSTIN

DONALD C. PRICE
RAYTHEON NETWORK CENTRIC SYSTEMS

THE AMERICAN SOCIETY OF MECHANICAL ENGINEERS
Three Park Avenue / New York, N.Y. 10016

ISBN No. 0-7918-3728-9

FOREWORD

The current proceedings for the ASME Engineering and Technology Management (E&TM) Group, consisting of the Technology and Society (T&S) Division, the Safety Engineering and Risk Analysis Division (SERAD) and the Management Division, contain a collection of papers presented in sessions held under the sponsorship of E&TM Group at the 2003 ASME International Mechanical Engineering Congress (IMECE) November 15–21, 2003. The E&TM Group has offered a variety of sessions, shown in Table 1, for their division technical tracks and Engineering Business Management Track, and this section of the Proceedings contains a modest number of peer-reviewed papers, shown in Table 2, for the sessions sponsored by the Technology and Society Division.

We are privileged and honored to have distinguished speakers for our special events: Dr. Richard Hinder, an IEE Fellow and formerly an official of the Department of Trade and Industry, the United Kingdom, for Joint Technology and Society and Management Dinner and Session T&S-1 Management of Technology for Business Development; Dr. John A. Brighton, an ASME Fellow and Assistant Director for Engineering at National Science Foundation, for Henry Robinson Towne Lecture; and Dr. Michael Stamatelatos, Manager, Agency Risk Assessment Program, NASA Headquarters, for SERAD Dinner. Dr. Hinder speaks on "Changing Roles for the Engineer in the 21st Century — A European View," and Dr. Brighton, on "The Role of the National Science Foundation in Advancing the National Research and Education Agenda in Science and Engineering," respectively. Their URLs are shown in Table 1.

I owe a great deal to Dr. Winfred Phillips, ASME President (1998-99), Dr. Clifford Brubaker, Dean of Health and Rehabilitation Sciences at the University of Pittsburgh, and Mr. Marshall Graves of California Governor's Office, among other organizers and members of the T&S Program Committees shown in Table 3, for distinguished T&S sessions. We are blessed with distinguished speakers, Dr. John Watson of National Heart, Lung and Blood Institute and Dr. Gerson Rosenberg of the Pennsylvania State University for Session T&S-2, Dr. Michael Heinrich of National Institute of Child Health and Human Development, Dr. Donna Dean of National Institute of Biomedical and Bioengineering Research, Dr. Joel Myklebust of National Institute on Disability and Rehabilitation Research, and Dr. Rory Cooper, a bronze medalist at the 1998 Seoul Olympics for the 1500 m wheelchair competition and Professor at the University of Pittsburgh for Sessions T&S-5A and –6A, and Fred Allen, Editor for American Heritage of Invention and Technology, Dr. Thomas Hughes of the University of Pennsylvania, Ben Wu, Deputy Under Secretary for Technology at US Department of Commerce, and Dr. Heater Wicke, Senior Environmental Scientist at Office of Senator John McCain for Session T&S-8 among other speakers.

I also owe a great deal to members of the E&TM Group Operating Board and the division committees during the course of preparation for publication of this volume. Particular thanks go to Robert Simmons, E&TM Group Vice President, and Theodore Aanstoos, Engineering Business Management Track Manager, for their coordination work, Frank Spanitz, Jonathan Young, and Stephen Long, Program Representatives of the Technology and Society Division, SERAD, and Management Division, respectively, to Kenneth Horne, William Doerr, and Wolf Yeigh, IMECE 2003 Chairs for the three respective divisions, for their support during the course of organizing our E&TM Group sessions, and Flavio Franco, Chair of the ASME United Kingdom Section, for accepting my invitation to participate in a Technology and Society Division session. It is a pleasure for me to serve as the editor for the four consecutive E&TM Group symposium proceedings from IMECE 2000 to 2003, which were made possible through cooperation of those noted and many others.

Hiroshi Honda, Ph.D., P.E.
E&T M Group Representative

Table 1 List of Sessions Organized by Engineering and Technology Management Group

Day	***Date***	***Time***	***E&TM Session No.***	***EBM Session No.***	***Program Title***	***Organizer***	***E&TM Division and Cosponsoring Body***
Tuesday	18-Nov.	6:30P.M-9:30P.M.	**Special Event**		Joint T&S and Management Dinner (Invited Speaker: Dr. Richard Hinder (URL: www.richardhinder.com), U. K.)	Hiroshi Honda	**T&S/ MGMT**
Wednesday	19-Nov.	7:45A.M.- 9:15A.M.	—	EBMT-1	Engineering Entrepreneurship	Theodore Aanstoos	**MGMT**
Wednesday	19-Nov	9:30A.M.-11:00A.M.	—	EBMT-2	Business Skills Workshop for Young Engineers	Theodore Aanstoos	**MGMT**
Wednesday	19-Nov	9:30A.M.-11:00A.M.	**SERAD-1**		Patient Safety Improvement	Jonathan Young	**SERAD**
Wednesday	19-Nov	11:15A.M-12:45P.M.		EBMT-3	The "Hows" of Small Business Innovation	Wolf Yeigh	**MGMT**
Wednesday	19-Nov	11:15A.M-12:45P.M.	T&S-1		Management of Technology for Business Development	Hiroshi Honda	**T&S**
Wednesday	19-Nov	1 :00P.M.- 2:00P.M.	**Special Event**		Henry Robinson Towne Lecture (Invited Speaker: Dr. John A. Brighton (URL: www.nsf.gov), NSF)	Wolf Yeigh	**MGMT**
Wednesday	19-Nov	2:00P.M.- 3:30P.M.		EBMT-4	Young Managers Speaker	Yohei Yamamuro	**MGMT**
Wednesday	19-Nov	2:00P.M.- 3:30P.M.	T&S-2		The Artificial Heart: Progress and Premise	Winfred Phillips	**T&S**
Thursday	20-Nov.	7:45A.M.- 9:15A.M.	—	EBMT-5	Industry, University and Government Partnerships	Wolf Yeigh	**MGMT**
Thursday	20-Nov	9:30A.M.-11:00A.M.		EBMT-6	Managing Strategic Technology Partnering and Transfer	Susanne Jones/ Pat O'Reilley	**T&S**
Thursday	20-Nov	11:15A.M-12:45P.M.	T&S-3		Trends of Managing Technology and Engineering Education	Hiroshi Honda	**T&S**
Thursday	20-Nov	2:00P.M.- 3:30P.M.		EBMT-7	Current Trends and Strategies for Managing and Protecting Intellectual Property	John Paul and Susie Jones	**T&S**
Thursday	20-Nov	2:00P.M.- 3:30P.M.	T&S-4		The Influence of Maritime Technology on Human Society	Robert Woods	**T&S**
Thursday	20-Nov	3:45P.M.- 5:15P.M.	T&S-5A		The History of Health and Rehabilitation Sciences and Engineering Part I	Cliff Brubaker	**T&S**
Thursday	20-Nov	3:45P.M.- 5:15P.M.	T&S-5B		The Role of Ethics and Ethical Behavior in Engineering	Edward Ostrowski	**T&S/BPPE**
Thursday	20-Nov	5:30P.M.- 7:00P.M.	T&S-6A		The History of Health and Rehabilitation Sciences and	Cliff Brubaker	**T&S**

					Engineering Part II		
Thursday	20-Nov	5:30P.M.- 7:00P.M.	T&S-6B		Professional Conduct and Credentialing of Engineers in a World Market	Charles Proctor	**T&S/BPPE**
Thursday	20-Nov	7:00P.M.- 10:00P.M.	Special Event		SERAD Dinner (Invited Speaker: Dr. Michael G. Stamatelatos (URL: www.nasa.gov), NASA)	William Doerr	**SERAD**
Friday	21-Nov	7:45A.M.- 9:15A.M.		EBMT-8	The Business and Tools of Product Development	Stephen Long	**MGMT**
Friday	21-Nov	7:45A.M.- 9:15A.M.	SERAD-2		Safety Improvement through Design	William Doerr	**SERAD**
Friday	21-Nov	7:45A.M.- 9:15A.M.	T&S-7		Sustainable Engineering	Frank Spanitz	**T&S**
Friday	21-Nov	9:30A.M.-11:00A.M.		EBMT-9	International Issues of Engineering Business Management	Hiroshi Honda	**T&S/MGMT**
Friday	21-Nov	9:30A.M.-11:00A.M.	SERAD-3		Homeland Security	Bob Simms	**SERAD**
Friday	21-Nov	9:30A.M.-11:00A.M.	T&S-8		Engineering Public Policy	Marshall Graves	**T&S**
Friday	21-Nov	11:15A.M.-12:45A.M.	SERAD-4		Process Industry Safety Management	Geoff Kaiser	**SERAD**
Friday	21-Nov	2:00P.M.- 3:30P.M.		EBMT-10	Increasing Market Share and Profits with An Effective Patent Portfolio	Frank Spanitz	**T&S**
Friday	21-Nov	2:00P.M.- 3:30P.M.	SERAD-5		Space, Aerospace, and DOD Risk and Reliability Analysis	Michael Stamatelatos	**SERAD**
Friday	21-Nov	3:45P.M.-5:15P.M.	SERAD-6		Safety, Risk, and Uncertainty Analysis	Ben Ale	**SERAD**
Friday	21-Nov	3:45P.M.-5:15P.M.	T&S-9		Overview of the Engineering and Technology Management Group Sessions	Bob Simmons	**3 Divisions**

Table 2 List of the Technology & Society Division Peer-Reviewed Papers Included in the Proceedings

Paper No.	Session No.	Authors and Affiliations	Paper Title
IMECE2003-43515	T&S-3	Philip S. Schmidt and Joseph J. Beaman, The University of Texas at Austin	Managing a Major Curriculum Reform Program in a Large Research University
IMECE2003-43648	T&S-3	Venkat Allada and Jose F. Vivas, The University of Missouri-Rolla	Case-Based Problem Solving Exercises for Engineering Education
IMECE2003-43533	T&S-3	Bruce A. Harding, Purdue University	The Treasure in Technical Publication Information: A Research Project for All Disciplines
IMECE2003-55237	T&S-3	Hiroshi Honda, Institute for the International Education of Students, and Shuichi Fukuda, Tokyo Metropolitan Institute of Technology	Industry Focused Courses Utilizing Collaborative Learning Approach
IMECE2003-43781	T&S-4	Jose Femenia, U. S. Merchant Marine Academy	Future Developments in Maritime Technology
IMECE2003-43767	T&S-4	Matthew Carr, U. S. Naval Academy	The Impact of Steam Innovations on Ship Design

Table 3 List of Program Committees and Their Respective Chairs of Technology & Society Division

Program Committee	Chair		
	Name	Affiliation	E-Mail Address
History and Society	Hiroshi Honda	Institute for the International Education of Students, and Tokyo Metropolitan Institute of Technology	Hondah9876@aol.com
Intellectual Property	Robert Burns	Finnegan, Henderson, Farabow, Garrett & Dunner, L.L.P.	robert.burns@finnegan.com
Sustainable Engineering	**TBD**	-	-
Ethics & Technology	Donald A. Westbrook Jr.	Honeywell Aerospace	DWestbrook@mail.com
Engineering Public Policy	Marshall W. Graves Jr.	California Governor's Office of Planning and Research	marshall.graves@opr.ca.gov

Note: The T&S executive committee, general committee and program committee meetings are scheduled on the afternoon of November 18 (Tuesday).

CONTENTS

ENGINEERING BUSINESS MANAGEMENT

HOMELAND SECURITY

RISK ANALYSIS AND MANAGEMENT

Proceedings of IMECE'03
2003 ASME International Mechanical Engineering Congress
Washington, D.C., November 15–21, 2003

IMECE2003-41252

Empowering Effective Teamwork for Machine Risk Reduction in the Workplace

J.R. Etherton
National Institute for Occupational Safety and Health, Morgantown, WV

ABSTRACT
The need for teamwork creates new requirements for engineers who use the new ANSI guidelines (B11 TR3-2000) for reducing risks associated with machines. Specifically, the guidelines state that "...a team of involved personnel (e.g., operators, maintenance, engineering) should participate in the risk assessment and reduction effort." The goal of such teams is to achieve tolerable risk. If they are to be effective team members, engineers will need skills in team leadership; identifying tasks and their associated hazards; using risk assessment tools; developing consensus about degree of risk; and selling the team's proposed risk reduction measures to management. This paper describes the stages of the TR3 process and focuses on engineering leadership skill development. Leadership must face the challenges of change and instill a sense of trust in the process so that a machine risk assessment team's injury reduction work can be effective.

INTRODUCTION
From an injury severity perspective, the risk of fatal and nonfatal injury from machinery in US workplaces is high. Two significant measures of injury severity; 1) number of cases of fatal injury and 2) rate of days-away-from-work for nonfatal injury cases, substantiate this assertion. Between 1996 and 1998, there were 464 occupational fatalities (an average of 155 per year) reported to the U.S. Department of Labor's Bureau of Labor Statistics involving caught-in-running-machinery. For the 1995-97 period, there were 92,932 cases of nonfatal injury of this type involving lost workdays. The majority of these (65%) were in the manufacturing industry. The 1996 total incidence rate (cases involving days-away-from-work/10,000 full-time workers) for the manufacturing industry was 238.3 compared to 212.3 for all private industry. The rate for machinery injury in manufacturing was 27.7 (11% of the total rate.) Using National Safety Council (1996) estimates, the caught-in-running-machinery deaths alone cost the nation $122 million. Design safety experts in the U.S. and Europe agree that machine injury rates are high because protective measures that are only appropriate for low level risks are being applied in situations where the risk involves frequent exposures to fatal or serious traumatic injury hazards (Adams, 1997).

In 2002, ASME issued a position statement on "The Role of Risk Analysis in Decision-making." In the statement, ASME International indicated that it "supports advancing the understanding, use, and acceptance of risk analysis, and encourages the larger community to join with us in advancing this critical process." Promulgation and use of consistent methods of risk analysis, including cost/benefit analyses, are recommended throughout government (e.g. regulatory agencies) and the private sector, worldwide. Leaders in business and industry must make many decisions that have risk implications for their employees. ASME's position is that it is important that engineers be able to use risk analysis to help make the best decisions. The first step in facilitating this is to develop a broad consensus on appropriate approaches to risk analysis. It will then be important to provide appropriate training in the risk analysis process. Basic principles set forth in the position statement are:

- No course of action, including taking no action, is risk free.
- Stakeholders should be involved in technical risk studies.
- Risk analysis must be open and transparent.
- Consistent methods of risk analysis should be used.
- Free and open discussion of uncertainties must occur.

The ASME position paper on risk assessment reflects current activities in machine injury prevention. In 1995, a Technical Report subcommittee (TR3) was formed under the auspices of the American National Standards Institute (ANSI) B11 Machine Tool Safety Standards Committee. The subcommittee has labor, machine builder, machine user, government and safety consulting representatives. The subcommittee's purpose was to develop a technical guidance document to bring machine tool risk assessment practice in the United States up to or above the level now required by ISO standard 14121 "Safety of Machinery: Risk Assessment". The ISO standard mandates that a process be followed that ensures that safety measures are appropriate to the risks in machine operation and servicing tasks. The ANSI B11 TR3 (2000) document became available for general use in November 2000, adding the benefits of risk assessment to the array of methods already available for preventing injury at machine tools in the United States.

The TR3 process

The TR3 process follows the flow chart (Figure 1) that prompts analysts to use the risk reduction hierarchy of: 1) eliminate the hazard by design; 2) apply safeguards and protective devices; and 3) apply administrative controls and other protective measures (e.g., warn, train, use personal protective equipment). The general approach is: task and hazard identification; risk estimation; determine if the risk is tolerable; risk reduction; verification of risk reduction via re-estimation of risk; and documentation. The aim of risk assessment on new machinery is to see that hazards are mitigated before they are introduced into the workplace. Employed judiciously, TR3 can help to locate critical elements in a design, identify associated hazards, evaluate their risks, and prompt decisions to reduce risks to acceptable levels. The risk for a task depends on a two-way relationship between the potential severity of harm (catastrophic, serious moderate, or minor) and probability of occurrence of harm (very likely, likely, unlikely, and remote). These factors are entered on a risk estimation matrix (Table 1) and a risk level is determined.

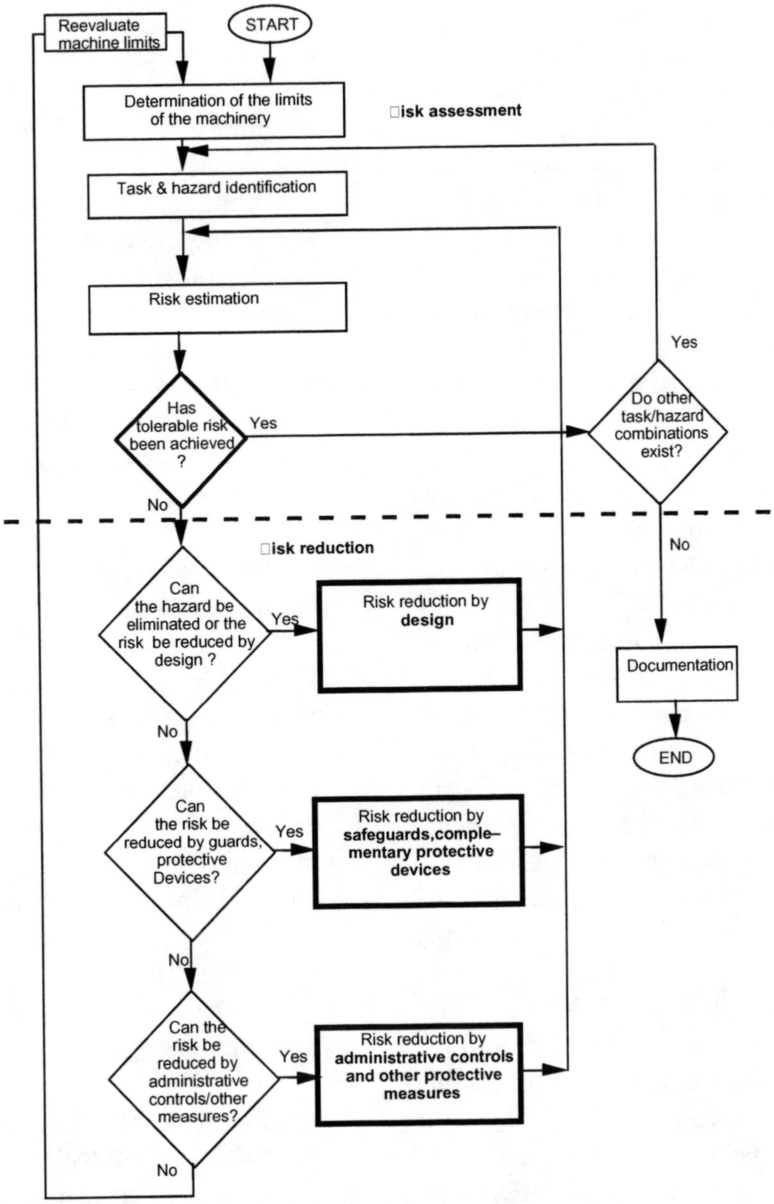

Figure 1

Probability of Occurrence of Harm	□everity of Harm			
	□atastrophic	□erious	□oderate	□inor
Very Likely	High	High	High	□edium
Likely	High	High	□edium	Low
Unlikely	□edium	□edium	Low	Negligible
□emote	Low	Low	Negligible	Negligible

Table 1: □isk □stimation □atrix

The evaluation study

The National Institute for Occupational Safety and Health (NIOSH) has successfully trained five teams in a 2-½ day workshop titled "Machine Risk Reduction" based on the new ANSI B11 TR3 "A Guide to Estimate, Evaluate, & Reduce Risks Associated with Machine Tools." The workshop was the first step in a NIOSH research project to evaluate the effectiveness of this new consensus guideline that tailors the concepts of ISO 14121, "Safety of Machinery - Principles of Risk Assessment" to the U.S. workplace. The workshop was designed to educate team participants from a variety of industries (automotive, industrial and consumer products, metal stamping, and medical devices) on the TR3 risk assessment process. Software tailored to the TR3 process was used to facilitate and document assessments. As the TR3 study proceeds, participating companies focus on two machinery systems for study; one will have the TR3 risk assessment process applied, and the other will remain in use without utilizing the TR3 risk assessment process. At the end of one year, these two machinery systems will be evaluated by NIOSH for safety and productivity improvements. Upon completion of the workshop, participant teams were able to communicate the organization and use of the TR3 document and risk assessment process; understand maintenance risk assessment applications; lead a risk assessment team through a TR3 risk assessment; and comprehend risk assessment implementation issues.

The study is currently in the intervention stage as the 2-person teams return to their workplace to form and train their machine risk reduction team with additional engineering, maintenance, and operating personnel. Facilitating the five teams requires frequent communication and keeping management informed and supportive. Team engineers must get constant feedback to develop the motivation to continue (Mitchell, 1998). And, constant evaluation not only lets supervisors know where they are, but it also does the same for team engineers. NIOSH anticipates that engineers will be vital to the success of these machine risk reduction teams.

Using TR3 can lead to change

In the teamwork phase of the activity, special attention must be given to the human side of the risk assessment process. People may be reluctant to join in on a new process that can bring about changes in the workplace. The team engineers must be ready to handle the challenges of initiating change. Senge (1999) has identified four challenges of initiating change:

- ***Time flexibility***. Do people have sufficient control of their own time? If they don't, they will not have enough time for significant change-initiatives.
  - *TR3 example.* In the early stages of the TR3 evaluation it has become very apparent that this activity is being done within other competing time demands. Weak business conditions have led to focus on other concerns. Teams have had to be flexible in when to start and complete tasks.
- ***Help***. Do people have the coaching support they need? If not, they will

flounder in initiatives that require new skills and capabilities.

- *TR3 example.* Calls to the NIOSH support center provide feedback on using software and tips for avoiding excess analysis.

❖ ***Relevance.*** Is a compelling case made for change? If not, people will conclude that the change initiative is not relevant.

- *TR3 example.* The champions in the TR3 study were able to engage their management to be supportive of the activity. In spite of weak business conditions, the teams have been able to maintain focus on TR3 goals. Having workers themselves on the teams has ensured relevance.

❖ ***Personal alignment.*** Are management's values and goals perceived as authentic and aligned with its actions? If not, people will see that management is not "walking the talk."

- *TR3 example.* The safety programs at these sites are strong programs and the teams do get good support. Given the slow business climate and production demands, it is clear that the success of the project to-date has been due to the strong safety values of the champions and management.

Building trust in the TR3 process

Building an atmosphere of trust in the process is a leadership skill that engineers must bring to their team activities (Covey, 1994). And trust in the process is a precondition to team success. Empowerment of the team and individual team members must be cultivated, much like providing favorable conditions for the growth of a good agricultural crop. It is a matter of nurturing the conditions that create empowerment. In particular, establishing a culture of trust requires the engineering leader to nurture both character and competence among team members (Deming, 1993).

❖ ***Character*:**

- *Integrity* – Integrity means that safety principles are adhered to continually.
- *Maturity* – Maturity implies a balance of courage and consideration that enables you to say what needs to be said about a hazardous condition, to address issues in a straightforward way, but with consideration and respect for the feelings, thoughts , and opinions of others.
- *Abundance mentality* – Leaders should seek alternatives that avoid the zero-sum game approach that the pie is only so large and if a safety improvement gets resources that is less for something else.

❖ ***Competence:***

- *Technical competence* – the safety knowledge and technical skill to complete a thorough assessment and reduced risk.
- *Conceptual competence* – the ability to see the big picture, to examine the we-have-always-done-it-that-way modality; and to see things from the perspective of others.
- *Interdependent competence* – the ability to interact effectively with others. That is, to communicate, listen, and create win-win agreements.

Covey (1994) proposes that occasional questioning by leadership can raise team meetings, such as TR3 meetings, to a higher level of awareness and insight.

- What led us to our mission?
  - A vision of safer conditions for the machine operators and maintenance personnel who work here would be a good answer.
- What led us away?
  - Time constraints are often a hindrance.
- What processes got in the way of accomplishing our mission?
  - The viability of the TR3 approach must not be hindered by factors that are external to it.
- Were we true to principles centered on people, trust, learning, industry?
  - Feedback on these factors by team members, whether formally or informally, can keep the team's TR3 process practical.
- How can we create better alignment with principles?
  - Focusing on finding the way out of dilemmas with principles can lead to valuable safety solutions.

Conclusion

This research on the effectiveness of the ANSI B11-TR3 process is contributing to the ASME goal to promote risk analysis as a technically sound and socially responsible way to help in decision-making by industry, government, and the general public. Teamwork and leadership within teams are essential to the success of TR3 implementation. Partnerships between NIOSH and industry are ensuring that a businesslike approach is followed in conducting the current machine system risk assessment evaluation. In the early stages of this evaluation of the TR3 process, it is evident that engineering leadership skill development is vital for meeting the challenges of change and instilling a sense of trust in the process needed to make a machine risk assessment team's work products effective in injury reduction.

REFERENCES

Adams, P [1997] Establishing a Safe Design Process in a Global Company, in Proceedings Symposium, Benchmarks for World Class Safety Through Design, National Safety Council, Itasca, IL, pp. 11-18.

ANSI B11-TR3 [2000] Risk Assessment and Risk Reduction – A Guide to Estimate, Evaluate and Reduce Risks Associated with Machine Tools, New York, NY.

ASME [2002] Position Statement #02-15, Statement on Role of Risk Analysis in Decision Making, American Society of Mechanical Engineers, New York, NY, www.asme.org

Covey, S [1994] First Things First, Simon and Schuster, New York, NY.

Deming, WE [1993] The New Economics, MIT Press, Cambridge, MA.

Mitchel, G [1998] The Trainer's Handbook, Third Edition, American Management Association, New York, NY.

National Safety Council [1996] Accident Facts, Itasca, IL.

Senge, P [1999] Leading Beyond the Walls, Jossey-Bass Inc., San Fransisco, CA.

Windau J [1999] Worker fatalities from being caught in machinery. In: Fatal Workplace Injuries in 1997, U.S. Department of Labor, Bureau of Labor Statistics, Report 934, Washington, DC

Proceedings of IMECE'03
2003 ASME International Mechanical Engineering Congress
Washington, D.C., November 15–21, 2003

IMECE2003-41962

THE IMPORTANCE OF INHERENTLY SAFER PROCESSES TO SITE SECURITY

Pamelyn G. Lindsey
E. I. du Pont de Nemours & Co., Inc.
DuPont Engineering Technology - Process Safety & Fire Protection
Wilmington, DE, USA
Email: Pamelyn.G.Lindsey@usa.dupont.com

Dennis C. Hendershot
Rohm and Haas Company
Engineering Division
Croydon, Pennsylvania, USA
Email: DHendershot@rohmhaas.com

ABSTRACT

Inherently safer processes (ISP) have become increasingly more important in recent years, specifically since the terrorist events of September 2001. New emphasis on site security and vulnerability has made it imperative for the chemical industry to view new and existing processes in a creative way. The tools available in designing an inherently safer process will allow industry to make these changes. "We ought, when possible, to be removing hazards rather than controlling them" (Kletz [1]).

The concepts of inherent safety will be reviewed with specific examples with site security implications. For example the in-situ intermediate generation of methyl isocyanate (MIC) is a striking example of the concept of minimization and is of particular interest, since after the Bhopal incident, large quantities of MIC would be a target for terrorism. Other examples such as in-situ generation of phosgene use of aqueous ammonia instead of anhydrous, and elimination of chlorine cylinders in water treatment will be discussed with site security implications. The concepts used to create these inherently safer designs can then be integrated into existing process safety management programs to improve the overall safety of the chemical industry.

INTRODUCTION

An Inherently Safer Process (ISP) has been defined by Kletz [2] as "the avoidance of hazards rather than their control by added protective equipment." For purposes of our discussion, we will define hazard as "an inherent physical or chemical characteristic that has the potential for causing harm to people, the environment, or property." [3] Examples of hazards include the inhalation toxicity of phosgene, the flammability of gasoline, and the high temperature and pressure of 600 psig steam. Hazards are intrinsic to a material, or its conditions of use. They cannot be avoided without changing the material, the quantity of material, or the conditions of use. Therefore ISP requires a fundamental redesign of a manufacturing process to avoid hazards.

While the term "inherently safer process" is relatively recent, the concept is actually quite old. Engineers and technologists have always worked to avoid hazards rather than controlling them, understanding that this represents a more robust approach to risk management. Some examples include:

- Robert Stevenson, one of the pioneers of railway development, argued for simplification of the controls on early steam locomotives in 1828, recognizing that existing complex controls made it much more likely that the operators of the locomotives would make mistakes which could lead to accidents [4]. (ISP Principle – Simplify)

- During construction of the Central Pacific Railroad in the 1860s, a British chemist, James Howden, set up a plant to manufacture nitroglycerine, used for tunneling in the Sierra Nevada, in small quantities, as needed, at the construction site. This limited inventory to the amount needed for the day's work, and avoided the hazard of transporting nitroglycerine to the construction site [5]. (ISP Principle – Minimize)
- In 1867, Alfred Nobel invented dynamite, a safer way of using nitroglycerine by absorbing it onto an inert carrier (ISP Principle – Moderate)

In the 1970s a series of major accidents in large petrochemical plants increased concern about the potential for chemical plant accidents. Much of the immediate response to this concern was the addition of complex "add on" safety features to the plants – alarms, interlocks, safety shutdowns, accident mitigation systems. These safety features can be very expensive, both in initial investment and ongoing operating costs. In 1977, Trevor Kletz of ICI, recognizing the benefits of avoiding hazards rather than controlling them, gave the concept the name "inherently safer design," and proposed a set of characteristics which designers could use to identify ISP options for chemical plants. Since 1977, ISP has continued to develop and evolve, and many companies have incorporated ISP into their process risk management programs.

WHAT IS ISP?

A characteristic is said to be inherent if it exists as a permanent and inseparable element, quality or attribute. A chemical process is "inherently safer" if it reduces or eliminates the hazards associated with the materials and operations used in the process, and this reduction or elimination is permanent and inseparable [3].

Trevor Kletz, the renowned author and process safety advisor, said it best. "The most effective methods, therefore, of preventing leaks of hazardous materials are to use so little that it hardly matters if it all leaks out (intensification), or to use a safer material instead (substitution). If we cannot do this and have to store or handle large amounts of hazardous material, we should store or handle it in the least hazardous form (attenuation). Plants in which this is done are said to be inherently safer because they are not dependent on added-on equipment or procedures which might fail; the hazard is avoided rather than controlled and the safety is inherent in the design" [6]. There is no other way to so simply state the important features and concepts of ISP.

Often there may be ISP conflicts. All technological artifacts, including chemical manufacturing plants, have multiple hazards. A change in a process, which eliminates or reduces one hazard may create or increase another hazard. For example, supercritical carbon dioxide may be a substitute for a toxic, flammable solvent in an extraction process. However, the supercritical extraction process must operate at high temperature and pressure, introducing new hazards. Any change to a process, even a change intended to improve safety, has the potential to introduce new hazards. The designer must always be aware of this, and must consider all hazards in identifying the best design.

APPROACHES TO ISP

Approaches to ISP have been grouped into four major strategies by IChemE and IPSG [7] and Kletz [1, 2].

Minimize	Use smaller quantities of hazardous substances (also called Intensification). For example see Figure 1, which shows the impact of using smaller process piping for a potential phosgene release.
Substitute	Replace a material with a less hazardous substance. For example substitution of water based coatings and paints for solvent based alternatives. This reduces fire hazards, toxicity, odor, hazards to the manufacturer and end user, and is generally more environmentally friendly.
Moderate	Use less hazardous conditions, a less hazardous form of a material, or facilities that minimize the impact of a release of hazardous material or energy (also called Attenuation and Limitation of Effects). For example, the impact on consequences of a pipe rupture comparing anhydrous ammonia to 28%

aqueous ammonia for a neutralization process is shown in Figure 2.

Simplify — Design facilities which eliminate unnecessary complexity and make operating errors less likely, and which are forgiving of errors that are made (also called Error Tolerance). Any time human factors are considered and the design is user friendly, the process will be made simpler.

These concepts of ISP are primarily aimed at risk reduction. In this way ISP fits into the overall Process Safety Management program. Risk is defined as a measure of economic loss, human injury, or environmental damage in terms of both the event likelihood and the magnitude or consequences of the event. [8] Therefore, risk reduction efforts aimed at reducing either the frequency or consequences of an incident can be classified into the following categories:

Inherent — Eliminate the hazard by using materials and processes that are not hazardous - for example, by using water for a flammable solvent.

Passive — Minimize the hazard with design features that do not require an active functioning of any device - for example, by using higher vessel pressure ratings, pressure relief devices or containment dikes.

Active — Use of controls, safety interlocks and emergency shutdown systems to detect and control process deviations - for example, use a sensor, logic device and final control element aimed at preventing or mitigating a hazardous event.

Procedural — Use of standard operating procedures, safety rules, administrative controls, emergency response procedures and training to prevent an incident or intervene in the effects – for example, a confined space entry procedures.

SITE SECURITY IMPLICATIONS

ISP has received a lot of attention since the terrorist attacks in September 2001 as a strategy for reducing the vulnerability of chemical plants to sabotage or other attacks. While the concepts of ISP were developed with the idea of providing a robust safety strategy against accidental events, they clearly will also make it more difficult for somebody to intentionally cause a catastrophic event in a chemical plant. ISP does represent an important tool in the plant security toolbox, but it is not the only tool. As for process risk management in general, a good security program for a chemical plant is likely to require a mixture of inherent, passive, active, and procedural safety and security features. The appropriate combination of features will vary depending on the availability of alternative technologies, their state of development, the environment surrounding the plant, balancing of risks to different stakeholders and receptors, economic competitiveness of the plant and potential alternate technologies, and many other factors.

In 1984, a leak of methyl isocyanate (MIC) from a plant in Bhopal, India, killed over 2000 people and injured many thousands more. The leak was from a storage tank, which contained many tons of MIC. There is still controversy about whether the incident was accidental, or, as the company claims, the consequence of sabotage. With the increased concern about terrorist attacks against chemical plants in the past few years, all chemical plants are paying more attention to the potential for intentional hazardous material releases, whether by internal sabotage or outside assailants. But, for the people of Bhopal, it does not really matter whether the release of MIC was accidental or the result of a deliberate action. The large inventory of MIC represented a threat to the community, and the consequences of that threat could be realized either by accident or by an intentional act. The existence of a large inventory of this highly toxic material can be avoided completely, or greatly minimized, in several ways. Carboryl, the product manufactured at Bhopal can be made by alternate chemical synthesis routes which do not require the manufacture of highly toxic MIC as an intermediate. Note that these alternatives are not "inherently safe," since they still require the use of phosgene, which is very toxic but not as toxic as MIC. These routes, like all ISP alternatives, are better described as "inherently safer." Also, if a chemical synthesis route using MIC intermediate is used, there are alternative process designs which eliminate the need to store large quantities of MIC. An alternate commercial process generates MIC "as needed" as an in-process intermediate which is immediately consumed in the next reaction step. The total inventory of MIC in the process is only a few kilograms [2].

A common chemical hazard in communities throughout the United States is chlorine gas used to disinfect effluent from sewage treatment plants or drinking water. Many cities, including Washington, DC [9] and Philadelphia [10] have replaced elemental chlorine with sodium hypochlorite as a safer way of chlorinating water in some or all of their water treatment facilities.

Phosgene is a toxic gas which is useful as an intermediate for manufacture of many agricultural and pharmaceutical chemicals. A continuous process to manufacture phosgene "on demand" for batch phosgene consuming processes has been commercialized, eliminating the need to store large quantities of phosgene [11, 12].

ISP options clearly decrease the potential consequence of an accidental or intentional release of the material, as illustrated for two examples in Figures 1 and 2. ISP does represent an important tool for managing security risks in a chemical plant. However, we must recognize that it may not always be feasible or effective. In some cases, reducing inventory at a plant may increase other risks somewhere else. If a plant receives large shipments of a hazardous material by barge, it will need a large storage tank to hold the shipment. If the large tank is eliminated, and the shipments are by truck, the inventory of material at the plant will be reduced, perhaps by a factor of 10 or more. However, now there will be many more trucks carrying hazardous material on the highways. They will be subject to the ordinary risks of highway transport, and also to a security risk from potential attack by terrorists or hijacking. The plant has transferred the risk to the highway transportation system, and that risk might actually be more than the risk at the plant site. Designers must consider these conflicts and tradeoffs in making decisions.

INCORPORATING ISP INTO PSM

Many if not all companies in the Chemical Process Industry (CPI) have endeavored to incorporate the concept of inherently safer designs into their Process Safety Management (PSM) programs. For many it is an endeavor to make ISP a way of life and a way of thinking.

One way this can be done is to train process hazards review (PHR) teams in the concepts of ISP. One method that is commonly used in PHRs is the HAZOP or hazard and operability method. First introduced by Imperial Chemical Industries (ICI), this method uses an interdisciplinary team to identify and evaluate the significance of deviations to the normal design intention of a process. [13] During this review, process hazards that are inherent to the design can be identified. Ways to minimize these hazards are also identified and the process can be modified to make it inherently safer. The beauty of the HAZOP is in the interdisciplinary team that stimulates creativity and new ideas. It is then the responsibility of management to receive these new ideas and to unleash the resources necessary to pursue innovation in a way that is financially viable.

Another beauty of the HAZOP method and other rigorous process hazards analysis (PHA) methods is documentation. This is also true in the efforts around management of change as required by OSHA 1910.119. In the example of Bhopal, critical safety equipment was shutdown without proper documentation or thought. Stopping to think and document, requires recognition of inherent safety. What are the inherent hazards associated with the process/chemistry? Is the safety equipment in question passive or active? Are there other ways to moderate this hazard? Proper application of the requirements of management of change will go a long way to making chemical processes safer.

Another PHA tool is the Chemical Interaction Matrix. This is a simple tool for identifying interactions among specific chemicals or parameters. It is therefore necessary to understand the inherent hazards of the chemicals involved in a process. Unknown reactive chemical interactions can also be identified. Like the HAZOP, the quality of this type of review is highly dependent upon the make-up of the interdisciplinary team conducting the review. Training and experience are paramount even though there are efforts to make the identification of reactive chemistry more rigorous. The chemical interaction matrix can be used to identify areas where substitution of less hazardous chemicals is possible.

Because the application of the chemical interaction matrix yields greater results the earlier it is used, training of research and development personnel is vital. If the concepts of ISP can become second nature to the bench chemist, less hazardous solvents could be chosen first, instead of as an after thought.

Likewise the extension to training in academia can revolutionize an entire generation. Since the events of September 11, 2001, everyone no doubt sees the importance of minimizing risk. Current and future students at the high school and college level will welcome instruction on ways to make their work and home environments safer. Any effort that educates in this vane is an application of inherent safety.

PUBLIC ATTENTION AND REGULATION

ISP has received a great deal of public attention in recent months as a strategy for reducing the vulnerability of chemical plants to terrorist attack. ISP has been discussed on television (PBS, NPR, CNN, television stations in Boston, Philadelphia, and other major markets) and in newspapers (*Washington Post, Philadelphia Inquirer*, and others). Legislation requiring evaluation of ISP options has been proposed in the United States Congress [14]. The proposed "Chemical Security Act of 2001" was reported out of committee but did not pass, and the bill has been reintroduced in the current Congress. State legislation has been suggested in some states, including New Jersey. In Contra Costa County, California, the local industrial safety ordinance requires consideration of inherently safer technologies [15]. Moore [16] discusses experience with the Contra Costa County regulations.

The authors believe that regulation of ISP is problematical. ISP applies at many levels of technology, from the fundamental chemistry and process technology to detailed equipment design. There are often conflicts among different hazards and design alternatives, and there may also be conflicts with other areas of concern such as environmental impact, operability and reliability, and economic viability. To fully understand the ISP characteristics of process alternatives requires an intimate knowledge of the process technology, something which is impossible for any regulatory agency which must deal with hundreds or thousands of different facilities and technologies. We believe that ISP, as a design philosophy, is best implemented by individual engineers and chemists, and design teams, in the course of their design activities. As stated above, the key to future development of ISP is education of all engineers and chemists in this philosophy, so that they can implement ISP principles in their work.

CONCLUSIONS

ISP has been recognized as an important component of process safety management by the chemical engineering community over the past 25 years. With the recent increased concern about terrorism and security in the industry, ISP is also an important tool in reducing the vulnerability of the industry to attack or sabotage. However, we must remember that ISP is only one of many tools for both process safety management and chemical plant security. An effective overall process safety and security management program will incorporate a variety of strategies and approaches, including inherent, passive, active, and procedural risk management components.

Additional work is required in several areas to further advance the incorporation of inherently safer technologies in future generations of chemical plants. Some areas include:

- Ongoing development of improved chemical processing technology and understanding of the fundamental mechanisms of chemical processes. This knowledge will allow engineers to optimize processing operations, minimizing equipment size and conducting operations under less severe operating conditions. Fundamental research in diverse areas such as process intensification, catalysis, novel energy sources, biochemical engineering, and nanotechnology offer opportunities for future development of inherently safer manufacturing processes for many materials.
- Continued research on "green chemistry." Alternate chemical synthesis routes which use smaller quantities of materials, operate under less severe conditions, eliminate or minimize the use of hazardous solvents, and use reaction pathways which avoid hazardous raw materials and chemical intermediates will often result in inherently safer processes. While it is not universally true that a material or process which is more environmentally friendly will also be inherently safer, a "green" process often will also be inherently safer.
- Better measurement tools for inherent safety. Much of the current evaluation

of the relative inherent safety of different process alternatives is qualitative or semi-quantitative. Measurement tools that can easily and quickly be applied early in process research and development, when the inherently safer design philosophy can have the greatest impact on the ultimate plant design, are needed. These tools must be capable of measuring different, potentially conflicting, inherent safety characteristics in a clear and transparent fashion.

- Continuing education of chemical engineers and chemists on the philosophy and implementation of inherently safer design. This includes engineers and chemists currently working in industry and research, as well as university teachers, and particularly their students who will be the next generation of technologists in the chemical industry.

Figure 1. Downwind dispersion to ERPG-2 of a toxic cloud for 3 sizes of phosgene pipe 5 meters above ground level.

Meteorological conditions:
Wind speed – 5 m/sec, atmospheric stability Class D

Light Gray = ½ inch phosgene piping
Gray = 1 inch phosgene piping
Black = 2 inch phosgene piping

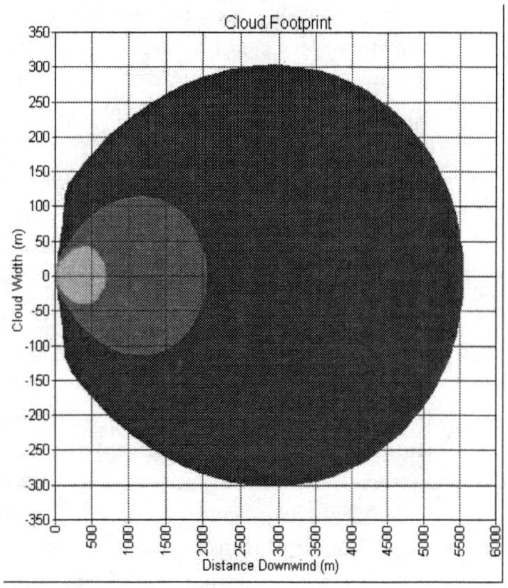

Figure 2. Downwind atmospheric concentration of ammonia resulting from rupture of a 2 inch pipe, for anhydrous ammonia and 28% aqueous ammonia.

Meteorological conditions:
Wind speed – 5 m/sec, atmospheric stability Class D

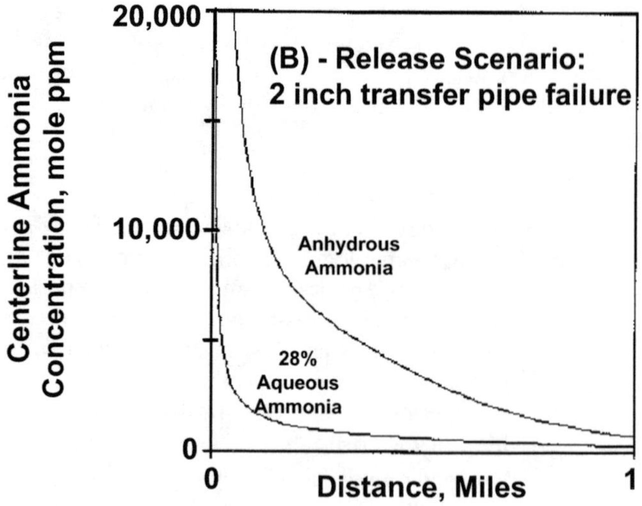

ACKNOWLEDGEMENTS

The authors would like to acknowledge the efforts and cooperation of Dave Clark of the DuPont Engineering Process Safety and Fire Protection group, David Cummings of the DuPont Safety Health and Environment Excellence Center, Art Burke, retired from DuPont, and Greg Keeports and Art Dowell of the Process Hazard Assessment Department of Rohm and Haas Company. We also thank DuPont and Rohm & Haas Company, for their ongoing support of our efforts to understand ISP, and their willingness to share those lessons with industry.

Finally, the authors would like to thank all the reviewers and contributors to this manuscript.

REFERENCES

1. Kletz, Trevor A. *Cheaper, Safer Plants or Health and Safety at Work.* Rugby, Warwickshire, England: The Institution of Chemical Engineers. 1984.
2. Kletz, Trevor A. *Process Plants: A Handbook for Inherently Safer Design.* Philadelphia, PA: Taylor and Francis. 1998.
3. Center for Chemical Process Safety. *Inherently Safer Chemical Process - A Life Cycle Approach.* New York: American Institute of Chemical Engineers. 1996.
4. Rolt, L. T. C. *The Railway Revolution: George and Robert Stevenson.* New York: St. Martin's Press. 1960.
5. Bain, D. H. *Empire Express: Building the First Trans-continental Railroad.* Viking: New York. 1999.
6. Kletz, Trevor A. *What Went Wrong? Case Histories of Process Plant Disasters.* Houston, TX: Gulf Publishing Company. 1998.
7. The Institution of Chemical Engineers (IChemE), and The International Process Safety Group (IPSG). *Inherently Safer Process Design.* Rugby, England: The Institution of Chemical Engineers. 1995.
8. Center for Chemical Process Safety. *Guidelines for Chemical Process Quantitative Risk Analysis.* New York: American Institute of Chemical Engineers. 1989.
9. Leonnig, C. D., and S. S. Hsu. *Fearing Attack, Blue Plains Ceases Toxic Chemical Use.* Washington Post. Washington, DC. November 10, 2001.
10. Fifield, A. *How to Reduce Risks of Toxic Disaster.* Philadelphia Inquirer. Philadelphia, PA. April 21, 2003.
11. Osterwalder, U. *Continuous Process to Fit Batch Operation: Safe Phosgene Production on Demand.* Symp. Pap. - Inst. Chem. Eng., North West. Branch, 6.1-6.6. Rugby, Warwickshire, England: IChemE. 1996.
12. Delseth, R. *Production Industrielle Avec Le Phosgène.* Chemia **52**, 12 (December), 698-701. 1998.
13. Center for Chemical Process Safety. *Guidelines for Hazard Evaluation Procedures.* New York: American Institute of Chemical Engineers. 1992.
14. United States Senate Bill S.1062, *Chemical Security Act of 2001.* October 31, 2001; reintroduced as S.157, *Chemical Security Act of 2003.* January 14, 2003.
15. Ordinance No. 98-48. *Industrial Safety Ordinance.* Contra Costa County Board of Supervisors, Martinez, California, December 15, 1998.
16. Moore, D. A. *Regulating Inherent Safety.* 17th Annual International Conference and Workshop: Risk, Reliability, and Security, Jacksonville, FL. October 8-11, 2002.

Proceedings of IMECE'03
2003 ASME International Mechanical Engineering Congress
Washington, D.C., November 15–21, 2003

IMECE2003-55044

A DESIGN APPROACH TO RISK ENGINEERING

Kristin L. Luthringer
Texas A&M University

T. R. Lalk
Texas A&M University

ABSTRACT

This article discusses the characteristics of the risk engineering process that would need to be generated in order to properly bring a designed system to market. First, the current status of risk engineering will be discussed from a design perspective by defining and explaining what is meant by key words such as "risk" and "risk engineering process". The general goals of the risk engineering process for an example fuel cell-based distributed cogeneration system (FC-Based DCG) will be noted before finally discussing the specific objectives, functions, procedure, & expected results of the risk engineering process for the system.

KEYWORDS: Design, Risk, Risk Assessment, Risk Analysis, and Risk Management

INTRODUCTION

Defining "Risk" is difficult because the general meaning is vague. After all, everyone has his or her own interpretation of what is "risky"! A literature review of the field of risk assessment and risk analysis demonstrates the there is not much consensus even among the experts. Much discussion centers around defining what the relevant risks are for the different situations to be analyzed and what weight to put on the values that will ultimately evaluate those risks. There is also dissention on which is the best method to use in a certain situation. The methodology of an engineering design approach should be a very useful tool in combining the risk engineering knowledge into a cohesive process. Since most risk personnel agree upon the principles of risk engineering but disagree on the details, the systematic design approach can give a framework of issues to address in risk engineering while leaving open the choice of tools to use when implementing the risk assessment, risk analysis, or risk management.

One key idea to take from the risk literature is to define upfront the risks at which one is looking and the values of the potential consequences. An important fact to keep in mind is that different kinds of risk exist, and it is the perception of those risks that is reality. The actual risks pertaining to a design can be over- or under-stated by how the design is viewed. Many of the aspects surrounding a technology, such as market pressure from consumers or product/company safety records, can change how the public views the risk associated with a particular design. Figure 1 graphically demonstrates the external pressures that affect how a risk is perceived by the public.

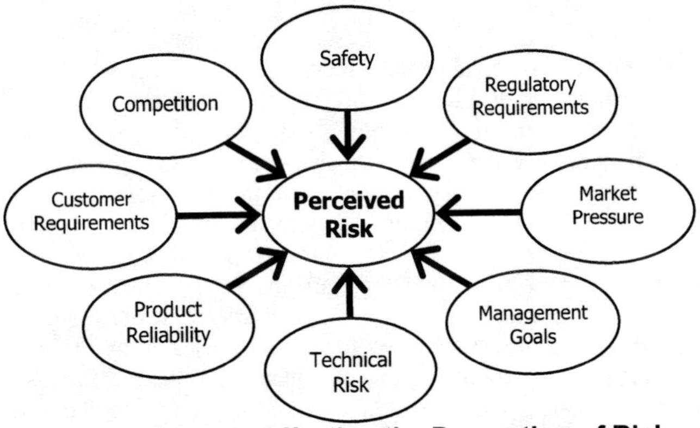

Figure 1: Factors Affecting the Perception of Risk [Wilson]

The difficulty in defining risk is compounded further because there exist several different types of risk. There is technical risk – associated with the equipment, technology, and science of systems; programmatic risk – including operational hazards and safety of designs; and supportability risk – associated with marketability and viability concerns. Cost risk – connected with financial impact in all aspects of the system; schedule risk – including the time impact of the design lifecycle; and political risk – associated with public opinion and governmental regulations; all can also be factored into the relative riskiness of a project. Although a general engineering division would be primarily focused on technical risk and to some extent programmatic risk, engineers must be aware of the other types of risk as they may also affect the total perceived risk of a project much like the aforementioned pressures on public risk perception.

NOMENCLATURE

This article advocates certain specific definitions of key vocabulary in order to minimize confusion among risk engineers, who are the practitioners of risk assessment, risk analysis or risk management in an engineering design company. Risk is defined to be the probability of occurrence and the consequences of an undesirable event, part of which can be reduced by design, part of which can be reduced by transfer, and the remaining residual risk that must be less than some acceptable level. [Wilson] The undesirable event is defined to be the occurrence of a failure, a partially unsuccessful occurrence, a problem, an accident, a loss, damage, or danger. The probability is defined to be the likelihood of occurrence, and consequences are defined to be the effects of the occurrence. The harmful effects of the occurrence of an undesirable event are usually of concern to engineers trying to prevent failure of their designs, whereas the non-harmful effects are not given a priority. This identifies an important point commonly misunderstood by the public – if an undesirable event is not harmful, then it is not a true risk. Likewise a hazard, or unsafe condition, is only a risk if someone or something is exposed to it. [Kolluru et. al.] Risk engineering is defined to be all the paraphernalia, procedures, tools, representations, and actions that involve the application of a broad spectrum of sciences in order to solve problems that must accomplish multiple, contradictory goals. [Gheorghe]

THE RISK ENGINEERING PROCESS

The three major functions of the risk engineering process have been identified as reviewing risk in the design, creating documentation of the risk, and managing (optimizing) the risk. Reviewing risk in the system entails both risk assessment and risk analysis. Creating documentation entails first understanding what is useful information, then obtaining, organizing, and presenting that information in a constructive manner. This middle function is necessary to ensure the data and ideas from the risk review function are useable by the risk management function and that the ideas to optimize the risk from risk management have enough information to be checked by another risk review. If the other two top functions are done properly this documentation can fall right out of the work without much effort; however if the documentation is improperly done, this can become a liability. Risk management entails evaluation of the risk data generated by the risk review, control of risk through prevention or mitigation, and monitoring both the non-design elements of risk reduction and the residual (i.e. "leftover") risk. Since risk reduction by decreasing the occurrence of the effects is weighed against cost an alternate way of looking at risk management is to think of it as implementing risk optimization. Recommendations for design improvements should be given from the risk review; however the management function must be carried out by the company (or other implementer of the evaluated technology) according to their values, practices, and policies. One of the major reasons for separation of the management function from the review function is to prevent bias in the risk review. This could happen if by knowing that the company can only spend a certain amount of money in fixing a problem (or is only capable of so much cleaning for a toxic chemical), a reviewer stopped assessing the risk once a certain number of problems (or toxic chemicals) were found. Further decomposition of the risk assessment and risk analysis functions under the risk review will be discussed further in this article, as these are the areas where the procedure is being developed. Figure 2 is a function structure displaying the top functions and their relationships to one another.

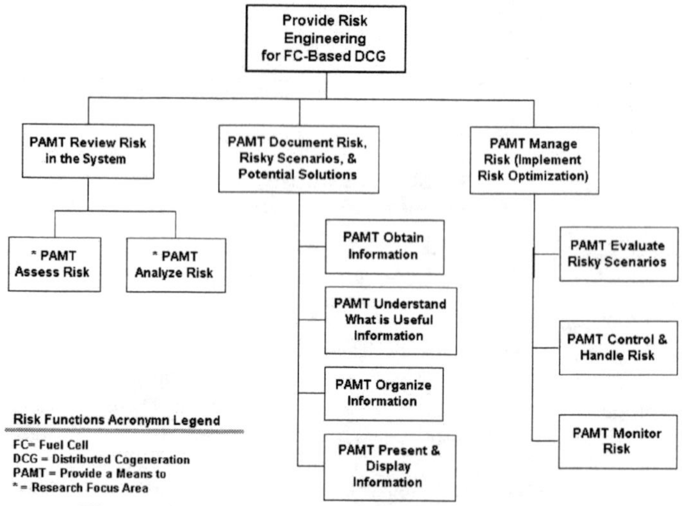

Figure 2: Top-Level Functions of the Risk Engineering Process

The function structure is a useful device in which the relationships between the necessary functions can be easily observed. It is also constructive to determine what needs to be done before deciding on how to proceed and this format potentially allows for the use of many risk-related tools. The function structure also displays the chronological process of risk engineering when read left to right, and allows for design iterations (or "looping") between risk management and risk review through the risk documentation.

Many industries use risk engineering as a tool in their daily and planning operations because of its many benefits. [Decisioneering] Risk engineering is a desirable tool because it may reduce waste of product, raw materials, and human resources (thereby cutting costs in the long term!). It can increase the chance of remaining in business and making a profit in the market. A good risk review can prevent failure through design, improve overall safety, increase reliability of equipment, and thereby decrease product liability. By identifying areas where the possibility of failure can be reduced, where the consequences of the failures can be controlled, or where the probability of occurrence can be modified, risk reviews feed information to risk management, which is then charged with ensuring the risk associated with the design is acceptable. When implemented correctly, the risk engineering process both improves public image and avoids possible legal problems.

Since FC-Based DCG is an emerging market in the energy industry, having a means to determine whether or not a particular system's configuration and implementation will likely be successful while still in the design stages would be invaluable as far as the costs and person-hours potentially saved from a "lessons-learned" experience. Table 1 lists several of the many industries that currently use some form of risk engineering in their decision making process.

Table 1: Industries that Use Risk Engineering

Aerospace	Construction
Consulting	Education
Energy, Oil & Gas	Environmental
Finance & Financial Planning	Facilities
Government	Insurance
Medical Devices & Healthcare	Military
Project & Cost Management	Pharmaceutical
Public Policy Research	Real Estate
Semiconductor & Microelectronics	Technology
Telecom	Utilities

It is useful to note some important differences in risk assessment and risk analysis as defined here and the words as defined by the prominent risk methodology prescribed by the United States Environmental Protection Agency (EPA). Here the focus is on a company's design risk as it would be addressed within a department and risk assessment describes the first, qualitative part of a risk review while risk analysis denotes the second, quantitative part. The EPA is primarily focused on human health risk and therefore the EPA risk assessment guidelines (RAGS) treat assessment and analysis as interchangeable words. The EPA focus is also more on the cancer and non-cancer (illness) effects of anthropogenic and natural chemicals in the environment, whereas the risk engineering process looks for all the potential end states of the design and tries to ensure a successful system. The four functions under an EPA-prescribed risk assessment are hazard identification, toxicity (or dose-response) assessment, exposure assessment, and risk characterization. [Kolluru et. al.] While this format works for a remedial situation where the damage is already done, it is not very useful for trying to troubleshoot a design in order to prevent damage.

GOALS OF RISK ENGINEERING FOR FC-BASED DCG

Electricity delivery in the United States of America presently occurs as allocation from a central plant through a grid of power lines, a power conditioning system, and transmission-boosting substations to the end user. Since deregulation of the energy market took effect the electric companies have argued over who will maintain this aging grid infrastructure and none are willing to invest in repairs or expansion. The centralized power plants have very high capital costs to install and have long lead-times to build and bring up to full capacity. So, as the demand for electricity keeps increasing, the centralized system is slow to respond. Americans have grown up used to having their electricity readily available and do not like being subjected to "brown-outs" or periods of no/low power due to a peak in electricity demand. One method proposed to boost electrical generation and provide better availability for the small, residential consumer is through the use of distributed generation – small-scale electricity generation for the individual residence or for a small group of neighbours. With the smaller transmission distance between the source and the end user it becomes cost-effective to use the waste heat from the distributed electricity generation processes at the residence. This dual use leads to the term distributed co-generation (DCG) system, alternatively called a combined heat and power system, and it can lead to a net energy savings. Several current technologies already exist for the power plant of a DCG system, but the research impetus has been for a fuel cell-based power plant in DCG systems due to the fact that ideal fuel cells do not need fossil fuels to operate. Ideally, hydrogen fuel gas would be derived from electrolyzed water, however it is currently reformed from natural gas or from methanol. Table 2 lists some of the barriers (disadvantages) for DCG systems with their respective power plant configuration.

Table 2: General & Specific Barriers for DCG [McKinley]

GENERAL DCG APPLICATIONS	FC-BASED DCG APPLICATIONS
Grid Interface	Technical Barriers
Ownership	Unproven Technology
Safety (Islanding)	Cost
Cogeneration Dilemma	(Use of Platinum, "Prototype" Mfg.)
Auxiliary Energy Storage	Sensitivity to Fuel Impurities
Regulatory Barriers	Maintenance

These issues are mostly self-explanatory for those involved with fuel cells. For the novice, the reference by Lariminie or the one by McKinley will provide further insight. The FC-Based DCG system is still in the design stages and therefore makes an ideal system to utilize the benefits of the risk engineering process.

There are several key benefits that a risk engineering process can specifically bring to DCG. The benefits include providing comparisons of alternative configurations in order to create a better design of the system, providing identification of alternative procedures, and providing documented and logical evaluation of trade-offs involved in the decision making process. These benefits would allow the risk management portion of the engineering process to provide the emerging DCG market with a design tool that will essentially troubleshoot the system design before creating the beta-test (preproduction) prototype for initial trials. A risk review should be able to identify which of the technical barriers can be overcome (failure prevention) and which are the most likely failure modes that should be mitigated. The risk management portion of risk engineering should present suggestions to make the system more reliable and safer.

There is a large issue to consider when creating and performing a risk review. This is summed up by the following quote:

> "Many risk-related decisions are driven by perceptions, not necessarily objective risk as defined by [calculations]. Perceptions of consequences tend to grow faster than

```
the consequences themselves – that is,
several small accidents are not
perceived as strongly as one large one,
even if fatalities are identical."
[NASA Systems Engineering Handbook]
```

Two major implications can be gleaned from this quote. First, it is necessary not only to create a risk review of the system, but to do it right! Sloppy work will lead to negative perceptions of the system, which is not good for either the design or the company producing the design. It will also increase the difficulty in improving the design if risk managers try to use information generated by a poor risk review. Second, risk engineers must use their hearts as well as their minds, especially when making judgments between alternatives. If we omit from consideration the public perception of our work, a grounded engineering decision may appear to be without base.

OBJECTIVES, FUNCTIONS, PROCEDURE, & EXPECTED RESULTS

The three major general objectives for the risk engineering process are to identify the risks involved in the system, to categorize those risks, and to effect changes, if necessary, thereby reducing the risks below background levels. Most experts in all fields of risk engineering will agree upon as much. Applying the design process at this point, the risk objectives are stated as dual aspects of the following need statement for the example system:

```
There exists a need to identify the
risks involved with a fuel cell-based
distributed co-generation system using
reformed natural gas and to categorize
those risks in order to use them as a
decision tool for selecting between
different design configurations.
```

Therefore, the necessary tasks to satisfy the objectives are: to identify the necessary elements of a risk engineering process, to develop a procedure for reviewing risk, to perform both a risk assessment and a risk analysis on at least one example system configuration to test the procedure, and to make recommendations according to the data generated from the analysis.

The identification of the necessary elements of the risk engineering process has already begun with the previous introduction (see Fig. 2) thus completing the first task! The first portion of the risk engineering process is the risk review. In the example system, the risk review must start by identifying and categorizing risks in order to meet the process objectives. So, in order to develop the procedure and complete the second task, the risk review is further broken into assessing risk and analyzing risk as what needs to be done in order to identify and categorize the risk associated with the available different system configurations for FC-Based DCG. The next two figures (3 & 4) show the functional decomposition of what needs to be addressed in order to perform the risk assessment and the risk analysis, respectively.

The risk engineering procedure for risk assessment, as shown in Fig. 3, centers on the idea that assessment is a qualitative measure of a system's risk. A "top-down" identification and characterization of risky scenarios will be conducted using a literature review of current FC-Based DCG technology system configurations and operational characteristics. Some typical methodologies employed for this stage of a risk review are: fault tree analysis, potential failure modes and effects analysis, expert interviews, risk templates, and lessons learned.

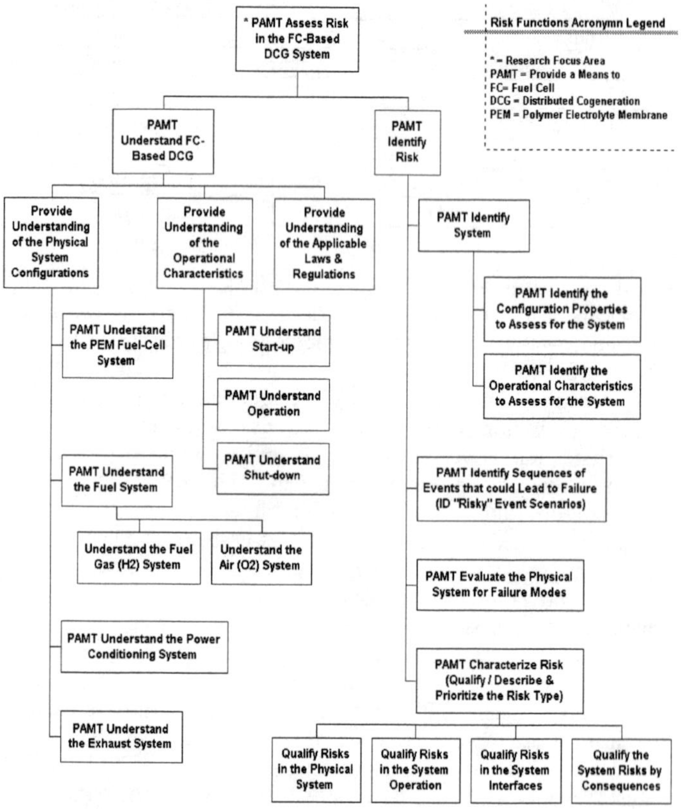

Figure 3: Risk Review Assessment Functions (Research Focus)

The expected results for the risk assessment portion are: a developed sequence of events, a set of potential modes of failure, and several scenarios for analysis. Since there are currently many different ways that could be used to configure and operate the FC-Based DCG, each scenario from the risk assessment should be of the form of a set of specific physical system configurations (PC), operating characteristics (OC), legal requirements (LR), failure events (FE), failure modes (FM), and failure consequences (FC). On way to represent this is using the set

$$\{PC, OC, LR, FE, FM, FC\} \quad (1)$$

where each capital letter is the set of characteristics (or failures) for the specific design for which the risk is to be determined. For instance, the physical system configuration set would be

$$\{PC = pc_1, pc_2, \ldots, pc_n\} \quad (2)$$

where pc_1 could be the type of fuel cell used (Solid Oxide or Polymer Exchange Membrane), pc_2 could be the type of fuel gas system (pure hydrogen or reformate), etcetera, for n such characteristics needed to define the design. At this point in the design of a risk review process, the function structure is not yet at its lowest level. Therefore, some work remains in order to complete this half of the second task.

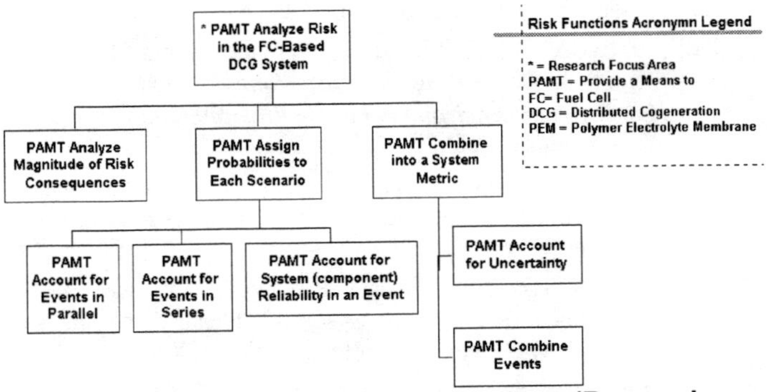

Figure 4: Risk Review Analysis Functions (Research Focus)

The risk engineering procedure for analysis, as shown in Fig. 4, centers on the idea that analysis is a quantitative measure of a system's risk. A "bottom-up" rating for each scenario previously identified by the assessment will be conducted at this point using standard risk industry practices. Some typical methodologies employed for this portion of the risk review are: event trees, failure rates, probabilistic risk analysis, scheduling network analysis, Monte Carlo simulation, reliability ranges for each component, Ishikawa "fishbone" diagram, hazard analysis, and trade-off studies.

Four example calculations follow (Eq. (3) – (6)) which demonstrate some of the probability principles that may be useful in finding an overall value of risk for the system. As all products degrade over time (t), the probabilities are shown as functions of time. First, the sum of the probability of failure ($P_f(t)$) & of success ($P_s(t)$) will always be unity.

$$P_f(t) + P_s(t) = 1 \qquad (3)$$

Second, as most component data will be of the form of the mean time between failure (MTBF) of tested parts, some way to convert this into a probability is needed. A Monte Carlo simulation can be quite useful in turning failure data into a MTBF. The generally accepted equation to convert from MTBF to a probability of success is as follows

$$P_s(t) = e^{(t\,/\,MTBF)} \qquad (4)$$

Third, many systems today are very complex, creating rather complex event structures leading to failures. If the sub-events are independent, i.e. no common causes, a means to account for the complexity of these sub-events is to multiply each sub-event probability. Alternately, if all sub-events have the same probability, raise the success probability for a single event to a power equal to the number of sub-events (n).

$$P_{cmplx}(t) = \Pi\ P_s(t)_i \text{ (with i varying from 1 to n)} \qquad (5a)$$

$$\text{or } P_{cmplx}(t) = [P_s(t)]^n \qquad (5b)$$

Last, to account for the probability of not detecting a failure, when using component reliability information (R) and inspection frequency (I) the probability can be estimated with the equation below. The "not detecting" means the system would be falsely tested as okay and sent to the customer destined to fail.

$$P_f(t) = (1 - R)^I \qquad (6)$$

The expected results for the risk analysis portion are: a set of end states for each scenario ranging from complete success to total failure, the probabilities associated with for each scenario, and the distribution of that probability (relating to the confidence in the system's risk rating). The probability of each scenario should allow the engineers to discover which are the most important risks to do something about. What to do about them is precisely what the risk management function will determine.

Recalling that when the risk review and the risk management functions are fulfilled properly the risk documentation naturally is completed, the next portion of the risk engineering process to discuss is the risk management functions. The three functions necessary in order to optimize risk are: evaluation of the risk review information, control of the calculated risks of concern, and monitoring both the changes and the residual risk in the system design. Risk evaluation connotes deciding what is acceptable risk. The identification and decision of acceptability limits for the system design must be made within societal norms and with respect to the background levels of risk currently accepted by potential users of the design. At this point in the process, potential trade-offs in the system are identified. Risk control is preventing or mitigating the risks of concern for the system. Design mitigation changes try to ameliorate the effects of the failure but do not change the failure mode, i.e. a safety factor for ensuring strength. On the other hand, design prevention changes modify the probability of the failure or even eliminate the failure mode itself. The identification of alternatives or other changes is now decided upon. Risk monitoring involves tracking the non-design risk reduction methods and the residual risk. During this portion of the risk engineering process, such things as inspections, maintenance schedules, tracking checklists, or other procedural changes to transfer risk would be implemented. It is important to identify the residual risk – the level of risk that is "leftover" once all design and non-design reduction methods have been implemented. It is this residual risk that must be at or below the acceptable level determined in the risk evaluation portion.

The expected result of risk management is a system design optimized for the minimal risk with respect to minimal cost. The trade-off criteria and the design changes to optimize risk will have been identified. Alternative configurations and/or components will have been proposed and the best ones chosen for implementation. The controls to optimize the systems risk, such as inspections, procedures, tracking, etcetera, will have been proposed.

At the present, only the first task has been fully completed towards realizing the risk objectives – the risk engineering process has been identified with the design procedure. While advocating some changes to the process used by the EPA, this risk engineering process holds true to the objectives and allows for creativity in finding the problems and solutions in a system.

The completion of two of the remaining tasks – completing the risk assessment function structure, and reviewing an example system to test the process and to verify the expected results of the risk review portion – will be the subject of the main author's Master's Thesis for Texas A&M. The final task to complete the risk engineering process objectives – the risk management portion – is one for an implementer of the system as only they are able to determine how much risk reduction is worth the cost to them.

REFERENCES

Decisioneering, Inc, 2000, Denver, CO. http://www.decisioneering.com

Gheorghe, Adrian V., and Mock, Ralf, 1999, *Risk Engineering: Bridging Risk Analysis with Stakeholder Values*, Kluwer Academic Publishers, Boston, MA.

Kolluru, Rao V, Bartell, Steven M., Pitblado, Robin M. and Stricoff, R. Scott, 1996, *Risk Assessment and Management Handbook for Environmental, Health, and Safety Professionals*, McGraw-Hill, St. Louis, MO.

Larminie, James and Dicks, Andrew, 2000, *Fuel Cell Systems Explained*, John Wiley & Sons, New York, NY, chapters 1, 4, 7, 8, & 9.

Luthringer, Kristin, 2003, "A Risk Review For Fuel Cell-Based Distributed Co-Generation Systems", Thesis Proposal, Texas A&M University, College Station, TX.

McKinley, Nathan, 2002, "A Summary of the 2001 Developmental and Market Status of Fuel Cell Based Distributed Power Generation", MEEN 685, Texas A&M University, College Station, TX.

NASA Special Publications, 1995, *NASA Systems Engineering Handbook*, NASA SP-610S.

Wilson, Anne, 2002, "Risk Analysis Methods for Medical Devices", Biomedical Engineering Seminar, Texas A&M University, College Station, TX.

Proceedings of IMECE'03
2003 ASME International Mechanical Engineering Congress
Washington, D.C., November 15–21, 2003

IMECE2003-41330

NIOSH AutoROPS 3rd Generation Static Testing and Human Interaction Element

McKenzie*, Jr., E.A., Etherton, J.R., Harris, J.R., Cantis, D.M., Lutz, T.J.

Abstract

To address the need for rollover protective structures (ROPS) on farm tractors that are easily adapted to low overhead clearance situations, the Division of Safety Research (DSR), National Institute for Occupational Safety and Health (NIOSH), developed an automatically deploying, telescoping ROPS (AutoROPS). The NIOSH AutoROPS at the present is in the third generation design and static testing phase, and the first phase of human subject (human operator) testing and manufacturing. The static testing is based on the SAE J2194 standard for testing ROPS for agricultural tractor use. The nature of the NIOSH AutoROPS is to be in a retracted position until an overturn is determined to be imminent. It is during the deployment time period that potential safety hazards exist that are not present in a traditional fixed ROPS and not addressed in the standards. Human interaction is a key ingredient in refining the design to be both functional and desirable while considering possible hazards. Feedback from farmers who have operated a tractor with the NIOSH AutoROPS installed and in the ready state will enhance the design and acceptability. NIOSH's goal is to reduce the number of fatal agricultural overturns by increasing the percentage of tractors with ROPS and seatbelts which operate in low clearance environments. This design has met laboratory static testing criteria of the SAE J2194 standard for ROPS on agricultural tractors. Field evaluation of the AutoROPS use by poultry farmers (N=32) in eastern West Virginia showed favorable results and a preference for wanting to purchase and use the NIOSH AutoROPS compared with a currently available manually foldable ROPS.

This paper discusses the overall performance of the NIOSH AutoROPS as subjected to the SAE J2194 standard and human interaction/feedback of operating an agricultural tractor with this added safety device.

Introduction

Since the inception of the NIOSH AutoROPS research program, the need for a passive protective device has not diminished. In 1993 the average fatal agricultural overturns in the United States was 132 (Myers and Snyder, 1993). Myers (2003) showed that tractor rollover fatalities remain high (between 150 and 200 deaths per year), while efforts to equip the 2.5 million tractors without ROPS are moving forward at an unacceptably slow pace. The protection of the farmer/worker from possible injury or death in tractor rollovers in low clearance work environments is still a significant concern. Low profile tractors working in low clearance environments are exempt from OSHA's ROPS regulations. Such work sites are typically inside farm buildings, poultry barns, greenhouses, in orchards or vineyards where overhead obstructions exist and where vertical clearance is not sufficient for safe operation with a fixed ROPS. All tractors sold after 1985 come with a ROPS installed in either a fixed or a foldable configuration. Low clearance tractors typically come with a foldable ROPS and are operated with ROPS in the down position while in the low clearance work environment. A pre-1985 model tractor may not have a ROPS at all. Low clearance tractors may still be operated in a manner that needs the constant protection of a ROPS. The Washington State FACE program investigated 3 tractor rollover related fatalities at work between 1998 and 2001. Two of the three incidents involved low profile orchard tractors that are not required to have ROPS because of low clearance work.

The current design of the NIOSH AutoROPS represents the third generation and is a dramatic change from its predecessors.

Third Generation NIOSH AutoROPS

The current design (Figure 1) is dramatically different in looks, material, and fabrication from its proof-of-concept predecessor shown in Figure 2.

Figure 1 NIOSH AutoROPS 3rd Generation Design

Figure 2 NIOSH AutoROPS 2nd Generation Design

The AutoROPS is constructed out of square and rectangular steel tubing and closely resembles commercially available fixed ROPS. The telescoping deployable section is constructed from (3.5 x 3.5 x 0.1875 inch) square mild steel tubing. The fixed section is constructed from (2 x 3 x 0.25 inch) rectangular steel tubing, making this similar in material and dimensions to the commercially available fixed ROPS. This design is an inverted design of the 2nd generation. As shown in Figure 1, the deployable section covers (receives) the fixed section when retracted. Figure 2 displays how the fixed section of the 2nd generation design covers (receives) the deployable section when retracted. The latching pins are now located inside the fixed section and push out: the 2nd generation had the latch pins on the outside pushing in. The latch and release mechanism (LRM) utilizes the same ball and groove design adapted to square tubing instead of the round tubing used in the second generation design. The retraction method used to set the 3rd generation AutoROPS is an inexpensive external web hoist (max load 1500 pounds) as compared to two internally mounted hydraulic cylinders. The overall benefits to this design are ease of manufacturing, lower manufacturing cost, reduced environmental effects (rust, mud, etc.) on the AutoROPS, ease of operation, and overall visual appeal. The changes can easily be seen in Figure 3 which shows a side-by-side comparison of a single post of the 2nd and 3rd generation AutoROPS as well as the LRMs. A detail of the inner workings of the 2nd generation NIOSH AutoROPS can be found in McKenzie and Etherton (2002).

Figure 3 Side-by-Side Comparison of 2nd and 3rd Generation AutoROPS and LRMs Respectively

SAE J2194 Testing

The main purpose of the SAE J2194 static laboratory testing is to simulate field upset in a controlled and repeatable environment (SAE J2194 1997). During the SAE J2194 static laboratory testing the loads are applied slowly over time with the ROPS displacement and applied force collected. From these measurements the energy absorbed by the ROPS can be calculated. The four static loading sequences (a 90° rear overturn, 90° side overturn and a 180°) simulate field all-wheels-up overturn. The rear and side overturn are governed by energy absorption and the 180° all wheels up is governed by applied force. The static loading sequence consists of four tests: (1) longitudinal loading; (2) 1st vertical loading; (3) transverse loading; and (4) 2nd vertical loading. It is important to note that the ROPS cannot be altered (bolts tightened , material repairs, etc.) during any of the four phases of static testing. A complete description of the test procedures can be found in the SAE J2194 standard and in Etherton et.al. (2002). The highest energy criterion (and the most challenging for design) occurs in the transverse test which is shown in Figure 4.

Figure 4 Transverse Loading of NIOSH AutoROPS

Results from Static Testing

The NIOSH AutoROPS has passed the static loading requirements and is ready for field upset testing at the NIOSH Pittsburgh Research Laboratory (PRL). The energy absorbed by the structure during the transverse test is shown in Figure 5 and is illustrative of typical testing results.

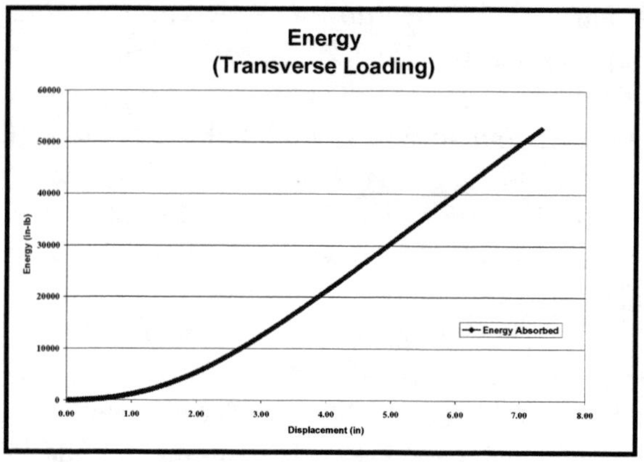

Figure 5 Energy Absorbed During Static Testing (Transverse Loading)

NIOSH AutoROPS End-User Field Evaluation Overview and Objectives

A controlled field evaluation compared a manually adjustable ROPS with the AutoROPS, including operation of an AutoROPS-equipped tractor on level ground, and obtaining responses to preference questions about the manually adjustable ROPS and the AutoROPS. The objective of this study was to determine the acceptability of the AutoROPS among a sample of potential users of the new technology. The study collected potential user responses to questions about how the AutoROPS functions, how an AutoROPS-equipped tractor operates in simple tasks, and preferences in how the AutoROPS was designed.

The study was conducted at large indoor arenas in Grant County and Berkely County, West Virginia. The test subjects were tractor operators who normally work in low clearance environments. Their low clearance work included orchard spraying and harvesting as well as cleaning poultry houses, dairy cattle stalls, and beef cattle feedways. Each of the 32 participants was shown a manual ROPS and asked to operate it (Figure 6). They were then shown an AutoROPS deployment while standing beside the tractor and asked to operate it (Figure 7). Finally they were asked to drive the tractor over a short course and back the tractor up as they would to hitch an implement (Figure 8).

Figure 6 Lowering Manually Adjustable ROPS

Figure 7 Lowering the NIOSH AutoROPS

Figure 8 Driving the NIOSH AutoROPS Equipped Tractor

Ratings were made by the test subjects at the completion of each task. The ratings helped answer the following questions:

- After having observed AutoROPS deployments, does the device appear to provide a more effective way than manually adjusted ROPS to prevent fatalities in tractor rollover events?

- After having performed the reset (relatching) of the AutoROPS, is this task perceived as one that is more acceptable than manually adjusted ROPS in normal use of the device?

- After having initiated a deployment manually, is this task perceived as one that is more acceptable than manually adjusted ROPS in normal use of the device?

- Is required use of a seatbelt perceived as more acceptable in normal use of the AutoROPS than manually adjusted ROPS?

- After having operated the AutoROPS-equipped tractor on level ground, is normal use of the device perceived as more acceptable than manually adjusted ROPS?

- Is there a perception that operating an AutoROPS-equipped tractor would be safer than operating the manually adjustable ROPS-equipped tractor?

- If a new tractor were being purchased, would an AutoROPS-equipped tractor be of more positive interest than a manually adjusted ROPS-equipped tractor?

Results from the Field Evaluation

The farmer group was of the opinion that the AutoROPS deployment is more effective than the manual ROPS alternative ($p<.0001$) and that the protection effectiveness provided by AutoROPS will be superior to the protection provided by manual ROPS ($p<.01$). Of great prevention importance was the clear increase in interest in purchasing a tractor with an AutoROPS compared to purchasing a tractor with manual ROPS ($p<.0001$). This result is a strong indicator that this new technology can successfully achieve wide use on the farm especially among those with low clearance needs (Figure 9 and 10). Farmer opinions also indicate the need for further design work to improve seating restraint and method for lowering the structure.

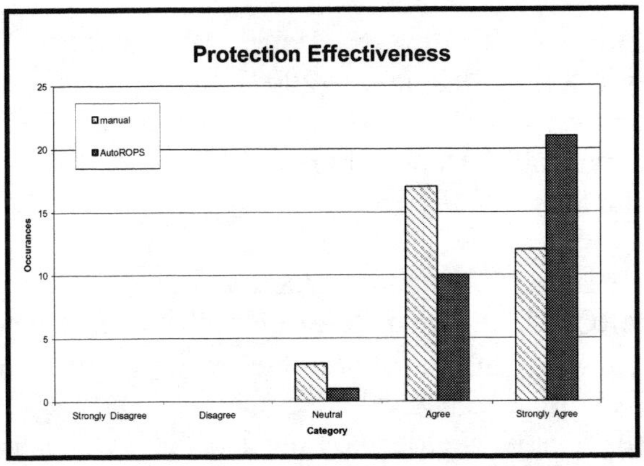

Figure 9 Protection Effectiveness

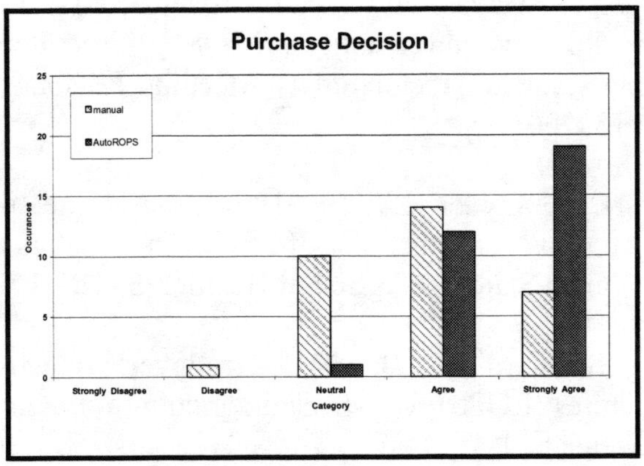

Figure 10 Purchase Decision

Conclusions

The third generation NIOSH AutoROPS has shown both structural integrity by passing the laboratory testing portion of the SAE J2194 standard and end user acceptance based on the field evaluation results.

Future Work

The refinement of this design to ease manufacturing and operational challenges is continuing at the time of this writing. The main focus is to have the NIOSH AutoROPS as an Original Equipment

Manufacturer (OEM) option for the end user in the near future.

References

Etherton JE, Cutlip JR, Harris JR, Ronaghi M, Means KH, Gillispie A [2002] Static Load Test Performance of a Telescoping Structure for an Automatically Deployable ROPS. J Agric Saf Health 8(1) 119-126.

McKenzie, EA and Etherton JR, "NIOSH AutoROPS Latch and Release Mechanism: Second Generation". In the Proceedings of 2002 American Society of Mechanical Engineers, Congress and Exposition. New Orleans, Louisiana, November 17-22, 2002.

Myers, J., "Update on NIOSH Statistics on Tractor-related Injury and Fatality", NIOSH Agricultural Safety Centers Tractor Safety Meeting, Pittsburgh, Feb,13, 2003

Myers JR, Snyder KA [1993]. Roll-over protective structure use and the cost of retrofitting tractors in the United States. J Agric Saf Health *1*(3):185-197.

SAE Standard [1997]. J2194, rollover protective structures (ROPS) for wheeled agricultural tractors. Warrendale, PA: SAE.

.

Proceedings of IMECE'03
2003 ASME International Mechanical Engineering Congress
Washington, D.C., November 15–21, 2003

IMECE2003-42563

WHAT KIND OF PROPULSION REDUNDANCY IS NEEDED FOR TANKER

Mitja Kožuh
University of Ljubljana
Faculty of Chemistry and Chemical Technology
Aškerčeva 5, Ljubljana, Slovenia

Stojan Petelin
University of Ljubljana
Faculty of Maritime Studies and Transportation
Pot pomorščakov 4, Portorož, Slovenia

ABSTRACT

In the last decade a lot of emphasis was given to the propulsion redundancy of tankers. Since there are several levels of redundancy, it not always clear to what extent these redundancies can help in difficult situations and at last what is the objective of these redundancies.

When we look at the costs optimization can be made minimizing the operating costs or it can be done in such a way that probability of spill and the size of its consequences are minimal. With this approach we are minimizing environmental risk.

With environmental risk in mind only the higher level of redundancy is acceptable and lower levels are not enough to first prevent and second to mitigate the consequences of possible spills.

INTRODUCTION

Every ship needs some sort of safety propulsion for the case in which it would be lost without propulsion or it would cause damage to life and most of all to the environment. The last experience with tanker Prestige that polluted Spain coast showed that environment is very sensitive to such treats and that the cost of such pollution goes much above the insurance than can be paid by companies.

When we talk about redundancy we have to compare risks of different options that compete for the same final result and for this result we have to decide what our goal is. Most of the time everything is fine but once in its lifetime ship can experience event for which its abilities are surpassed. Either propulsion is lost and the ship cannot control its behavior during the storm and the final results are lost lives or the ship is lost with a polluted environment.

One cannot predict the size of the storm and the circumstances that will find the ship on the sea but we can predict the consequences of the cargo to the environment. It is through that most of the time ship lines are on the open sea where the pollution is not the issue but certain part of the time they travel along the cost of inhabited areas where pollution cannot be measured by the money lost since the price is so high and consequences nearly irrecoverable. Therefore new thinking has to be adopted for this kind of risk calculations. The world cannot afford 2% of all accidents to end like Exxon Valdez or Prestige or some other tanker. These kind of accidents should not happen and propulsion redundancy has to contribute to this goal.

PROBLEM DESCRIPTION

Recent Redundancy Rules for Classification Societies from Det Norske Veritas (DNV) in 1996, Registro Italiano Navale (RINA) in 1996, American Bureau of Shipping (ABS) in 1997, Lloyd´s Register of Shipping (LR) in 1999, and Germanischer Lloyd (GL) in 2000 changed a lot in the approach to propulsion redundancy which no longer is a hazy notion but is well defined and has its definition. There are several classes which differ according to the approach and to the requirements. Based on the class, insurance level is defined. Mostly the damage which is insured cannot cover the damage to the environment which in certain cases is much higher than all the other costs. In our work we will concentrate only on those cases where environment is endangered.

In this case all other costs are more or less negligible and we have to prevent consequences due to spillage of the crude oil near the coast which then pollute the coastal part of land.

The costs from the three spills are following [8]:

Accident	Date	Size of spill [tonnes]	Clean-up costs [ECU]
Amoco Cadiz spill, France	16th March 1978	220.000 of crude oil	199 Million
Exxon Valdez spill, Alaska	24th March 1989	40.000 of crude oil	2,4 Billion
Haven incident, Italy	11th April 1991	10.000 of crude oil	61 Million

HAZARD FROM TANKERS TO THE ENVIRONMENT

In our work we will consider Aframax tanker (Average Freight Rate Assessment) with its dead-weight between 80.000 and 100.000 tons. From [1] we took that the average one way distance is 6.500 Nm [Nautical miles] and that the division of operational time is following:

In loaded condition 46%
In ballast condition 26%
Time in port 28%

Additional assumption is that at least 30 % of time tanker is near the coast and there is potential threat to the environment. Time used for this journey is around 450 hours.

THREE LEVELS OF PROPULSION REDUNDANCY

If the ship is to be recognized as to have redundant propulsion it must fulfilled the following conditions:

The are three levels of propulsion redundancy as defined by Germanischer Lloyd are:

A ship is provided with independent Take home auxiliary system
A twin screw ship is provided with independent propulsion plants
A Power Take In (PTI) operational mode is enabled by the capability of the same shaft generator to work as an electric motor too.

For the Take home auxiliary system there are following requirements set by Germanischer Lloyd:

Germanischer Lloyd requirement No.1:

The ship must be able to maneuver into position of less resistance to the weather and must be able to keep this position under wind speed of up to and including 11m/s (8 on Beaufort scale) and significant wave height of 5.4 m with an average wave period of 8.3 seconds.

Germanischer Lloyd requirement No.2.:

The ship must be able to attain a speed of 7 knots under normal weather conditions, regarding as corresponding to wind speed of up to and including 11 m/sec (5 on Beaufort scale) and significant wave height of 2,8m and average wave period of 6.7 seconds.

According to explanation of these rules [2] Beaufort 5 covers 80% of the North Atlantic weather and Beaufort 8 covers 98% of the North Atlantic weather conditions.

RELIABILITY OF THE PROPULSION PLANTS ON TANKERS

Data for the reliability of propulsion plants on tankers are taken from different data bases [7] and show that propulsion plants are not so reliable as one would expect.

JSRIC (Japan Ship Reliability Investigation Committee) has collected data from 1982 to 1997 for 224 ships that have collected 7 billion propeller hours. Research has showed that there have been 6800 single failures on propulsion plant or on generators from which approximately 25 % were critical due to fact that most of the ships were not equipped with redundant parallel propulsion system.

Also other collected data [7] show that a ship experience on the average at least one problem with propulsion plant per year. All of these facts make us believe that loosing propulsion is not a rare event and therefore it has to be treated with great care. This in conjunction with the fact that 25 % are critical show the real image of what is going on. When we read comment to the data we can conclude that along with the reported problems there is also a vast number of unreported events that remain hidden in the memory of the maritime companies.

WHAT IS ACCEPTABLE RISK IN THE CASE OF TANKERS

There are several possible views on the above subject. The most common is that any redundancy is cost which burdens the operating company through the years and that this kind of approach is not to follow. The argument is advocated mostly by poorer companies which in turn have the oldest ships and that would be ruined if propulsion redundancy is being made mandatory. On the other hand rich companies operating much younger fleet which is already safer then the previous one can afford some kind of redundancy but the question is against which threats it is designed.

We can say that in most cases companies are largely afraid of Business losses, on the second place it is Safety (loss of lives) and on the last it is Environmental pollution. Of course the order is normally changed with Safety on first and Environment on the second place but the facts tend to show different picture.

When redundant propulsion plant is designed it is mostly designed against the possibility that the ship will loose propulsion and there will be a vast cost to bring the ship to the nearest harbor for the repair. By optimizing the ratio between the cost of redundant system and the saved costs for the time lost during the repair we certainly can save a lot of money on equipment but on the other hand we have still not covered the small but not negligible risk for environmental pollution when the ship gets stranded and the oil pollutes the coast.

Environmental pollution is very expensive for the company, for the Classification Society and for the society at large. The problem is that the money cannot restore the things back to normal in one day but it takes time and money to heal the wounds to the environment.

Even the Safety costs are not really defined since the compensation for the lost lives is negotiable and it also depends on the law skills of the lawyers. Therefore only the Business costs can be defined with relative correctness. In [1] it was assumed that the costs to the Environment for Aframax tankers are somewhere around 35 million ECU which is in comparison with the above table rather on the modest size of the costs.

The key question is whether the world can afford such accidents and to what extent we can tolerate accidents to happen. We think that after the Prestige accident there is very little tolerance left.

CHARACTERIZATION OF THE MARINE SYSTEM

As Formal Safety Assessment showed that an intensified maintenance is the most cost effect measure to eliminate or mitigate risks caused by main engine failure. With the regard to that, studies show that 80 % of marine accidents are caused by human error and it is normally that more emphasis is given to hardware since it is there, than to human resources on which nobody is very keen to depend on. [4]

In [2] it is written that marine systems have error inducing character which lies in the social organization of the personnel aboard the ship, the economic operating pressures, the structure of the industry and insurance and the difficulties of the national and international regulation.

The problem with this system is, that when people from the industry answered the questionnaire 99,6 % of them responded that they had sailed on the ship that they personally knew was unseaworthy. Based on that we can deduct that few accidents that had happened are just the top of an iceberg of unknown size. If we deduce the size of the problem from experience from safety issues where the accident ratio is as follows: on 1 serious injury there are 10 minor injuries and 30 property damages and 600 near misses with no reported damage. This shows that the picture is quite different [5].

APPROXIMATE EVALUATION OF THE PROBABILITY FOR ENVIRONMENTAL CONSEQUENCES FROM TANKER SPILL

We have made for the moment simple probabilistic model from which we can deduct approximate size of the problem in form of probability of tanker to pollute environment. In the first stage of our work we have identified the following factors having impact on the probability for environmental pollution:

1. Exposure to hazard
2. Probability of bad weather
3. Probability of failure of propulsion plant
4. Probability of not starting redundant propulsion
5. Probability of redundant propulsion not to fulfil the mission

From the above facts we can see that the exposure is approximately 0,5, since 50 % of the time tanker is loaded and on high sea. Probability for bad weather is assumed to be 25 % of the time. Probability propulsion plant failure is assumed to be 14 % [7]. Probability for redundant propulsion not be able to start is to be 3% and not to run for 72 hours is 9 % [9].

From the above figures we have calculated that the probability of spill from tanker to be 2.10^{-3} per voyage which in our opinion is not at all small figure. From this estimation we think that the redundant propulsion should be designed for 99 % and not for 98 %, which is the case for PTI, of the weather conditions since we have to optimize it for the risk of environmental pollution. In other cases we are optimizing on Business losses which are far less demanding in the consequence size.

We are aware that our ad hoc analysis was conservatively done but we think that it opens new view on the topic giving different priorities for optimization of the size of redundant propulsion plant. Data taken for the analysis were taken from generic data bases and some from specific tanker data from register companies therefore they are not that far away from the real life.

CONCLUSION

From articles that deal with redundant propulsion plants for tankers most of them are optimized with respect to the Business loss and not taking into account the Environmental pollution costs. Most of the judged costs are more on the modest size. Problems with propulsion plants are not so rare and the known data are reduced by companies that do not report every single event when the ship lost propulsion plant. From the informal evidence we can assume more cases than reported, therefore the model was made to incorporate the above assumptions. From the model we got the probability for spill (assuming that every ship wreck results in a spill which is conservative assumption) from tanker which is not at all small. As the result from this study we can suggest that the optimization of the redundant propulsion plant should be made on the Environmental costs and not only on Business have losses which gives us rather small redundant propulsion planted which cannot fulfil the mission in extremely bad weather conditions.

REFERENCES

1. Erik Moller: Redundancy of propulsion plants: Can you afford not to?, The Motor Ship 2000 pp.28-67,
2. Peter Securius: How to Conform to Redundancy Rules, The Motor Ship 2000, pp.69-78,
3. Horst Michael Becker: Additional Redundant Propulsion Systems on Chemical Tanker Newbuildings, The Motor Ship 2000, pp. 79-87,
4. Charles Perrow: The Normal Accidents, Basic Books 1984, Chapter 6, Marine Accidents, pp. 170-192
5. Lloyd's Register: Guidelines on Risk Issues, The Engineering Council, 1993, pp.21.
6. James Reason: Managing the Risks of Organizational Accidents, Ashgate, Aldershot 1997
7. Marko Perkovič: Propulsion Model of Ship's Propulsion for Risk Reduction, Master Thesis, Portorož 2001, pp.104-113
8. Europe's Environment, The Dobříš Assessment, European Environment Agency Copenhagen 1995, ISBN 92-826-5409-5, pp.537
9. Roberts H.R.,Vesely W.E., Haasl D.F., Goldberg F.F.: Fault Tree Handbook, U.S. Nuclear Regulatory Commission, NUREG-0492, 1981, pp. XI-9

Proceedings of IMECE'03
2003 ASME International Mechanical Engineering Congress
Washington, D.C., November 15–21, 2003

IMECE2003-42597

ON THE SAFETY OF INFEEDING COMMERCIAL TREE CHIPPERS

Dennis B. Brickman
Triodyne, Inc.
5950 West Touhy Avenue
Niles, IL 60714
(847) 677-4730 (voice)
(847) 647-2047 (facsimile)
brickman@triodyne.com

ABSTRACT

This paper addresses a commercial mobile tree chipper with a mechanical infeed system that is manually fed. OSHA records associated with commercial tree chipper accidents indicate tree workers have alleged that their arms have been pulled into the chipper feed wheels by branches during the feeding process. An infeed extension pan has been proposed to prevent a tree worker's arm from being pulled into the chipper feed wheels if his gloved hand gets caught in the branches being fed into the infeed hopper. Experiments performed with human test subjects and an anthropomorphic dummy are presented under arm pull-in scenarios to study the safety implications of the proposed infeed extension pan.

INTRODUCTION

The author's previous research has studied whether an operator's arm can be pulled into the infeed hopper of a consumer chipper/shredder [1]. Further research has analyzed whether a tree worker's leg can be pulled into the feed wheels of a commercial tree chipper equipped with an infeed extension pan if the worker places his foot on the extension pan or kicks branches entering the infeed hopper [2]. This paper addresses the question of whether an infeed extension pan on a manually fed commercial tree chipper will prevent a tree worker from being pulled into the feed wheels if his gloved hand gets caught in the branches being fed into the tree chipper infeed hopper. A typical commercial tree disc chipper infeed hopper with a mechanical infeed system is shown in Fig. 1. During the feeding mode, the operator will manually feed the tree branches into the infeed hopper to the feed wheels which in turn bring the branches to the cutting disc. According to the requirements of ANSI Z133.1-2000 [3], brush chippers equipped with a mechanical infeed system shall have a quick stop and reversing device on the infeed system. The

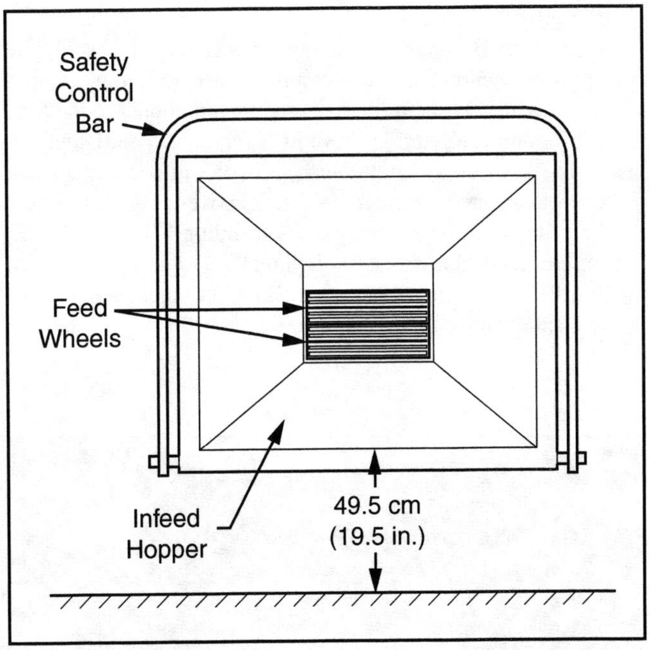

Figure 1 - Commercial Tree Disc Chipper Infeed System

activating mechanism for the quick stop and reversing device shall be located across the top, along each side, and close to the feed end of the infeed hopper within easy reach of the worker. The safety control bar illustrated in Fig. 1 satisfies the requirements of ANSI Z133.1-2000 by quickly stopping and reversing the feed wheels when the bar is pushed toward the feed wheels. A review of the Occupational Safety and Health Administration (OSHA) records associated with commercial tree chipper accidents indicates tree workers have alleged that their arms have been pulled into the feed wheels by tree branches during the feeding process [4]. Safety specialists have hypothesized that the presence of an infeed extension pan attached to the chipper infeed hopper will prevent a tree worker's snagged gloved hand from being pulled into the feed wheels with the branches. The purpose of this paper is to test this hypothesis through a series of five experiments performed using an anthropomorphic dummy and a commercial tree disc chipper infeed hopper equipped with a proposed infeed extension pan. In addition, further tests with eleven human subjects were conducted to determine their strength resistance when their hand was being pulled into the infeed hopper both with and without an infeed extension pan. Finally, a critical review of the OSHA records associated with commercial tree chipper accidents was performed to analyze whether these accidents have occurred on chippers equipped with an infeed extension pan.

DUMMY ARM PULL-IN TESTING WITH INFEED EXTENSION PAN

Test Setup

Figure 2A shows the test setup that was utilized to study the arm pull-in phenomenon in a commercial mobile tree chipper equipped with a 91.4 cm (36 in.) long infeed hopper to the feed wheels inrunning nip and an 83.8 cm (33 in.) long infeed extension pan. The bottom horizontal surface of the infeed hopper test fixture was located 49.5 cm (19.5 in.) above the ground at the leading edge of the infeed hopper. A standing 175.3 cm (69 in.) tall anthropomorphic dummy weighing 95.3 kg (210 lb) is posed behind the infeed extension pan with both of its feet resting on the ground and its torso facing the infeed hopper. The pulling force of the tree chipper infeed rolls is applied to the dummy's left wrist by a winch and pulley system. A cable coming from the winch is attached to a rope which is wrapped around the left wrist of the dummy.

Testing

Five arm pull-in tests were conducted with the anthropomorphic dummy situated at the left, left center, center, right center, and right of the infeed extension pan. Figure 2 is a photographic sequence with the dummy originating at the left of the infeed extension pan. Photograph 2A shows the position of the dummy before the force is applied to its left wrist. In the next photograph, 2B, the dummy's left hand and head have entered the infeed hopper and its right leg has been pulled off the ground. The final photograph, 2C, depicts the dummy's left hand passing through the back of the infeed hopper which simulates the left hand entering the feed wheels. In addition, both of the dummy's legs have been pulled onto the infeed extension pan and the dummy's head and torso have entered the infeed hopper. Similar results were obtained when the dummy originated at the left center, center, right center, and right positions of the infeed extension pan during the testing. Figure 3 displays a photographic sequence with the dummy starting at the center position of the infeed extension pan and Fig. 4 shows a photographic sequence with the dummy starting at the right position of the infeed extension pan. Results of these five tests indicate that the tree worker's arm can be pulled into the tree chipper feed wheels even with an infeed extension pan given the conditions set forth in this test program.

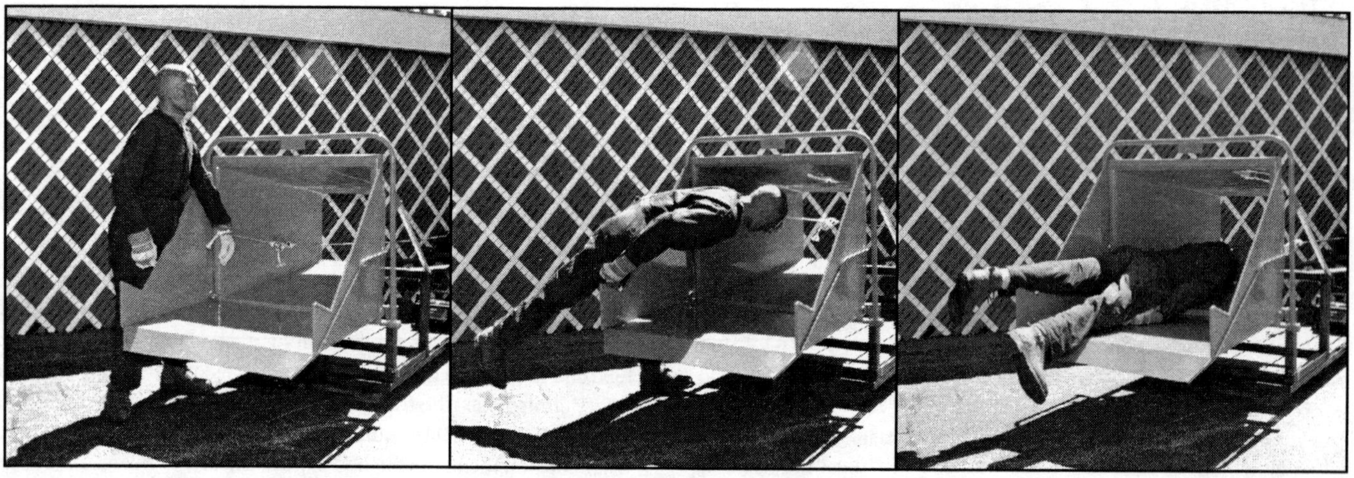

A) Unloaded Arm B) Arm Pulled Into Infeed Hopper B) Final Position

Figure 2 - Arm Pull-In Test with Infeed Extension Pan (Dummy at Left)

A) Unloaded Arm B) Arm Pulled Into Infeed Hopper B) Final Position

Figure 3 - Arm Pull-In Test with Infeed Extension Pan (Dummy at Center)

HUMAN ARM PULL-IN TESTING

Test Setup

Tests were performed with 11 human subjects whose collective average was 182.9 cm (72 in.) in height and 94.8 kg (209 lb) in weight as shown in Table 1. Each test subject participated in two tests where the infeed hopper was equipped with an infeed extension pan and without. The infeed test fixture was equipped with a safety control bar located close to the feed end of the infeed hopper within easy reach of the worker as specified in ANSI Z133.1-2000 [3]. The initial position of the human subjects had both feet on the ground with their torso facing the infeed hopper. Similar to the arm pull-in testing with the anthropomorphic dummy, a cable coming from the winch is attached in series to a load cell and rope which is wrapped around the human subject's gloved left wrist as shown in Fig. 5A. The load cell is connected to a digital display which records the peak tension force exerted by the human subject. Each human subject's left wrist is pulled into the infeed hopper, after which the human subject attempted to exert maximum resistive effort as depicted in Fig. 5B. Peak resistance forces were recorded with and without the infeed extension pan. If the human resistance force exceeds the connection resistance between the snagged worker's gloved

A) Unloaded Arm B) Arm Pulled Into Infeed Hopper B) Final Position

Figure 4 - Arm Pull-In Test with Infeed Extension Pan (Dummy at Right)

A) Infeed Hopper - Unloaded Arm

B) Infeed Hopper - Loaded Arm

C) Infeed Extension Pan - Unloaded Arm

D) Infeed Extension Pan - Loaded Arm

Figure 5 - Arm Pull-In Testing with Human Subject No. 2

Table I - Human Test Subject Resistance

Test Subject	Height - cm (in.)	Weight - kg (lb)	Resistance With Infeed Chute - N (lb)	Resistance With Infeed Extension Pan - N (lb)
1	180.3 (71)	89.8 (198)	1673 (376)	1388 (312)
2	172.7 (68)	83.9 (185)	761 (171)	520 (117)
3	167.6 (66)	83.9 (185)	752 (169)	614 (138)
4	200.7 (79)	127.0 (280)	1508 (339)	863 (194)
5	188.0 (74)	90.7 (200)	1099 (247)	930 (209)
6	182.9 (72)	90.7 (200)	974 (219)	836 (188)
7	182.9 (72)	86.2 (190)	1041 (234)	854 (192)
8	185.4 (73)	88.5 (195)	1415 (318)	1339 (301)
9	182.9 (72)	92.5 (204)	1441 (324)	899 (202)
10	188.0 (74)	104.3 (230)	1201 (270)	1179 (265)
11	185.4 (73)	106.1 (234)	1094 (246)	1045 (235)
Average	182.9 (72)	94.8 (209)	1179 (265)	952 (214)

hand and the branch, the worker can withstand the pull of the chipper feed wheels.

Testing

The maximum resistance forces for the 11 human subjects for the infeed hopper only and for the infeed extension pan are displayed in Table 1. All 11 human subjects recorded higher resistance forces without the infeed extension pan with an average differential of 51 pounds (265 pounds with infeed hopper and 214 pounds with infeed extension pan). As shown in Fig. 6B without the infeed extension pan, human subject number 10 is able to pull with both hands while simultaneously bracing his right shoulder and right leg against the leading edge of the infeed hopper. In contrast, Fig. 6D depicts human subject number 10 with the infeed extension pan where he is off-balance with his left foot off the ground and his right shoulder unsupported. In Fig. 5B showing the infeed hopper condition, human subject number 2 is able to brace his right foot against the bottom leading edge of the infeed hopper and brace his right hand against the safety control bar for maximum resistance effort. In contrast, Fig. 5D depicts human subject number 2 where the infeed extension pan causes the subject to lean forward with both feet on the ground and his right hand against the top leading edge of the infeed hopper in an off-balance condition. Clearly, the presence of the infeed extension pan presents a downside with respect to strength resistance and positioning of the tree worker.

COMMERCIAL TREE CHIPPER ACCIDENT STATISTICS

The Occupational Safety and Health Administration (OSHA) has conducted accidents associated with commercial tree chippers [4]. Selected OSHA accident abstracts reported for commercial tree chippers equipped with a mechanical infeed system and an infeed hopper extension pan are presented in this section. With the exception of the deletion of the indentity of the commercial tree chipper manufacturers, these abstracts are quoted as they appear in the OSHA accident investigations for commercial tree chippers. The text has not been edited; misspellings and incomplete sentences are unaltered.

On 4/23/97 at approximately 5:00 pm, employee was feeding pruned tree limbs into a Wood Chipper. Employee made contact

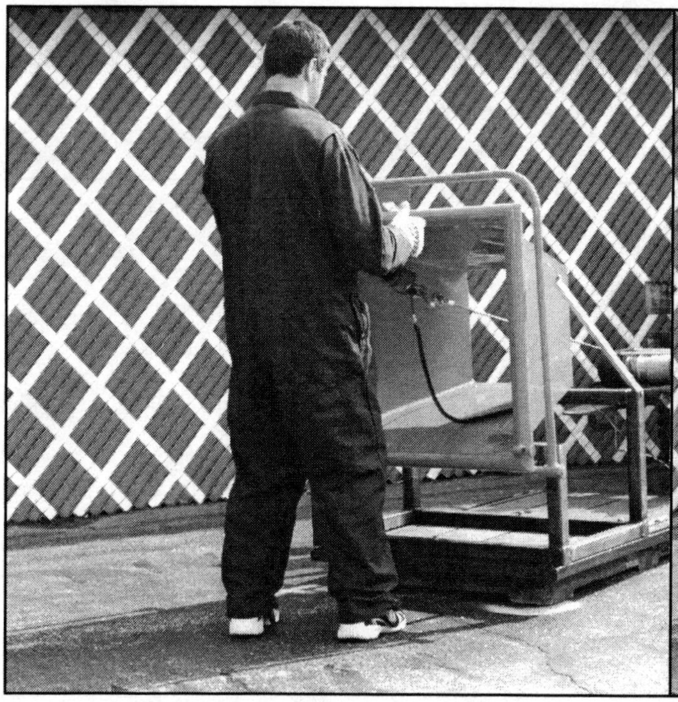

A) Infeed Hopper - Unloaded Arm

B) Infeed Hopper - Loaded Arm

C) Infeed Extension Pan - Unloaded Arm

D) Infeed Extension Pan - Loaded Arm

Figure 6 - Arm Pull-In Testing with Human Subject No. 10

with the feeder drum, located 63 inches from where he was standing on the ground at the infeed tray and was grabbed by the feed wheel. The feeder wheel fed employee to the chipper drum located inside of the machine where he was chopped to death by the chipper blades. Employee's remains were then deposited into the wood chip receiver truck through the discharge spout of the chipper. there is a reversing safety bar for the feeder drum. It is located at the mouth of the infeed and is bail shaped and goes down both sides and across the top of the feeder.

On May 27, 1998, a 28-year-old groundsman died after he was caught and pulled into a wood chipper. The victim and two coworkers, a foreman and a climber, were cleaning up limbs after pruning a tulip poplar behind a townhouse. The coworkers were behind the building gathering loose branches while the victim was in front operating the chipper. The coworkers began dragging brush to the chipper. As the climber approached the chipper, he saw the victim's legs sticking out of the infeed hopper. Investigators concluded that the victim either fell or reached into the infeed hopper while feeding short branches. His hands were caught by the feed mechanism, forcing his head and upper torso into the chipper knives.

On 4/22/99, two employees (one of which is crew leader) take a wood chipper into the lemon orchard to trim and mulch the branches. Crew leader trims the branches and brings them to the victim to put in the chipper. Crew leader turns around to go back to trimming and hears a different hum comming from the chipper. Crew leader turns back to see what the problem is and noticed that the victim was gone. Victim had been sent through the chipper. Fatal.

On 11/19/99, employee was pulled into a chipper and suffered major injurries to his head and arms.

ON 15 JULY 2000 AT APPROXIMATELY 1330 HOURS, EMPLOYEE NO. 1 SUFFERED A GRUESOME FATAL WORK INJURY AS RESULT OF A CONTACT WITH THE BRUSH CHIPPER'S BLADES. THE EMPLOYER WAS ENGAGED IN TREE TRIMMING OPERATION IN THE PAST 20 YEARS. A CREW OF 8 WERE ASSIGNED TO DO A HALF A DAY TRIMMING OF EUCALYPTUS AND FINE TREES AT THE SIDE AND FRONT OF AN OFFICE COMPLEX. THE TRIMMING CREW WERE STACKING TREE BRANCHES BY THE TREES, AND THE VICTIM WAS GOING AROUND CHIPPING THE PILES OF BRANCHES. THE BRUSH CHIPPER STAYED HITCHED INTO AN ENCLOSED DUMP-TRUCK AND DEPOSITS THE CHIPS ONTO THE DUMP-TRUCK. BY 1:30 P.M., THE VICTIM WAS CHIPPING A PILE OF MEDIUM AND SHORT FINE TREE LIMBS BY THE FRONT ROAD. HE WAS WORKING ALONE WHILE FEEDING A PILE OF SHORT LIMBS, WITH HIS GLOVES, INTO THE TWO FEED ROLLERS WHEN HE EITHER LOST HIS BALANCE DURING A SNAG AND WAS PULLED INTO THE SHREDDING BLADES OR THE VICTIM WAS LEANING FORWARD ON THE SHORT FEED TABLE WHILE PUSHING, WITH HIS GLOVES, A PILE OF SHORT LIMBS THROUGH THE FEED ROLLERS WHEN HIS GLOVES WERE CAUGHT BY THE FEED ROLLERS THEN PULLED INTO THE SHREDDING BLADES. THERE WERE NO ACTUAL WITNESSES TO THE ACCIDENT. THE FEED ROLLERS FEED TABLE WAS ABOUT 1.5 FEET HIGH, 6.5 FEET WIDE AND 5 FEET LONG. THE 5'8" TALL EMPLOYEE COULD EASILY REACHED THE FEED ROLLER'S BLADES. THE TWO FEED ROLLERS WERE DESIGNED TO OPEN BY SPRINGS SO AS TO EASILY GRAB AND PUSH A LARGE 1-1/2 FEET DIAMETER LOG INTO THE SHREDDING BLADES. THE VICTIM'S ENTIRE BODY WAS SLOWLY CHOPPED INTO SMALL PIECES AT ABOUT 4 TO 6 INCHES PER SECOND. THE EMPLOYEE WAS NOT USING A PUSH- STICK OR ANY LONG TREE BRANCH TO PUSH THE SHORT PIECES OF LIMBS THROUGH THE INTAKE BLADES IN ORDER TO PREVENT THE OPERATOR'S HAND FROM ENTERING THE DANGER ZONE. THE EMPLOYEE WAS OPERATING A HEAVY DUTY BRUSH CHIPPER. THE MACHINE WAS IMPOUNDED BY THE COUNTY CORONER'S OFFICE UNTIL THEY WERE ABLE TO CONCLUDE THAT MOST OF THE VICTIM'S BODY, ABOUT 80, WAS ACCOUNTED FOR. THE, 34 YEARS OLD & UNMARRIED, EMPLOYEE WAS HIRED IN 29 APRIL 1993 AND HAS 7 YEARS EXPERIENCE IN OPERATING BRUSH CHIPPERS. THE CAUSAL FACTOR WHICH CONTRIBUTED TO THE ACCIDENT INCLUDED THE EMPLOYER'S FAILURE IN PREVENTING THE EMPLOYEE FROM PUSHING A PILE OF SHORT LIMBS, WITH HIS HANDS, INTO THE FEED ROLLER'S BLADES. THE EMPLOYER WAS FOUND IN VIOLATIONS OF T8CCR 342(a) AND 3428(a). CITATIONS WERE ISSUED WITH PENALTIES.

On 5/24/02, an employee of tree service while operating a wood chipper which had no safety control bar was pulled into the feeding wheel up to his left arm elbow before an employee was able to reverse the feeding wheel.

The following observations can be made regarding the commercial tree chipper accidents presented:

1. Accidents have occurred on commercial tree chippers equipped with a mechanical infeed system and an infeed hopper extension pan.

2. The commercial tree chipper accident statistics corroborate the test results presented in this paper.

3. Tree workers who were pulled into the feed wheels were unable to successfully use the safety control bar to prevent themselves from being pulled in.

4. Investigators concluded that tree workers reaching into the infeed hopper while feeding short branches was a contributing factor in some accidents.

CONCLUSIONS

1. Experimental testing using an anthropomorphic dummy under arm pull-in conditions has disproved the hypothesis that an infeed extension pan will prevent a snagged tree worker's gloved hand from being pulled into the chipper feed wheels given the conditions set forth in this test program.

2. Testing using human subjects revealed that a greater strength resistance to arm pull-in can be achieved without the presence of the infeed extension pan. The strength resistance results obtained in the human subject arm pull-in test program exceed the static connection resistance test results for gloves nailed to branches reported in a prior study [1].

3. OSHA accident investigations associated with commercial tree chippers equipped with a mechanical infeed system and an infeed hopper extension pan corroborate the test results presented in this paper.

4. Manufacturers of commercial tree chippers have admonished workers not to place their body parts into the infeed hopper. Similar admonitions are contained in the commercial tree chipper safety literature [5-7]. In addition, ANSI Z133.1-2000 [3] requires that "arborists, workers and mechanics shall not reach into the infeed hopper when the cutter disc or rotary drum or feed rollers are moving." Following warnings and instructions of this type will contribute to reducing tree worker arm injuries.

REFERENCES

1. Brickman, D.B. and Barnett, R.L., "On the Safety of Infeeding Vertical Garden Shredders," IMECE 2001/ SERA-2400, Proceedings of 2001 ASME International Mechanical Engineering Congress and Exposition, American Society of Mechanical Engineers, New York, Nov. 11-16, 2001, pp. 1-6.

2. Brickman, D.B., "Commercial Tree Chippper: The Leg Pull-In Hypothesis," IMECE2002-32459, Proceedings of IMECE 2002 ASME International Mechanical Engineering Congress & Exposition, American Society of Mechanical Engineers, New York, Nov. 17-22, 2002, pp. 1-5.

3. "American National Standard for Arboricultural Operations–Pruning, Repairing, Maintaining, and Removing Trees, and Cutting Brush – Safety Requirements." ANSI Z133.1-2000, American National Standards Institute, New York, Oct. 19, 2000.

4. "Chipper Accident Search Detail," Occupational Safety & Health Administration, U.S. Department of Labor, Washington, D.C., 1984 to 2001.

5. "Chipper Use and Safety," The Treeworker, No. 125, Nov. 1992, p. 2.

6. Guyette, J.E., "Chipper/Shredder Safety Precautions," Landscape Management, Vol. 32, No. 5, May 1993, pp. 24 & 26.

7 "Brush Chippers Deserve Respect," The Treeworker, No. 140, Feb. 1994, pp. 1-2.

ACKNOWLEDGEMENT

Bill Brown and Matt Ulmenstine are mechanical laboratory technicians at Triodyne Inc. The author would like to acknowledge the technical contributions of Mr. Brown and Mr. Ulmenstine on this project.

Proceedings of IMECE'03
2003 ASME International Mechanical Engineering Congress
Washington, D.C., November 15–21, 2003

IMECE2003-42633

ON THE SAFETY OF CONSUMER DEEP FRYERS

Dennis B. Brickman
Triodyne, Inc.
5950 West Touhy Avenue
Niles, IL 60714
(847) 677-4730 (voice)
(847) 647-2047 (facsimile)
brickman@triodyne.com

ABSTRACT

A 27 month old boy sustained burn injuries from a consumer electric cooker which was pulled off a kitchen countertop. Approaches utilized in the safety analysis include accident reconstruction, critical accident statistics analysis, UL 1083 standard research, and an evaluation of design alternatives. This paper investigates the safety implications of the new UL required break-away power cord for electric cookers/deep fryers.

INTRODUCTION

This paper addresses a household accident in which a 27 month old boy sustained burn injuries from a consumer electric cooker/deep fryer which was pulled off a kitchen counter. Shortly before the accident, the injured boy's grandfather placed the pot filled with grease on the kitchen counter and suspended the cord over the edge of the counter to plug the power cord into the electrical outlet on the adjacent wall as depicted in Fig. 1. While the oil in the pot was heating up, the injured boy's grandfather went outside the home to rake leaves, leaving his grandson unattended. Even though the injured boy's grandmother, mother, and aunt were sitting at the kitchen table in the parlor adjacent to the kitchen pantry where the accident occurred, they claim not to have witnessed the accident. After the accident, the pot and grease were found on the kitchen pantry floor and the two year old boy sustained second and third degree burns to his face, scalp, arms, and back. It was hypothesized that the two year old boy pulled the pot off the counter by pulling on the power cord.

Figure 1 - Cooker Accident Setup

WARNINGS AND INSTRUCTIONS

Figure 2 depicts the warning tag that was permanently attached to the originally supplied power cord within 51 mm (2 in.) of the plug in accordance with the requirements contained in UL 1083 (3/1/94) Safety Standard for Household Electric Skillets and Frying-Type Appliances [1]. It should be noted that at the time of the accident, a substitute power cord, obtained from a different electric kitchen appliance, was being utilized because the knob broke off the original temperature control on the cord. In addition to the warning tag attached to the original power supply cord, the manual for the subject cooker contained the following instructions:

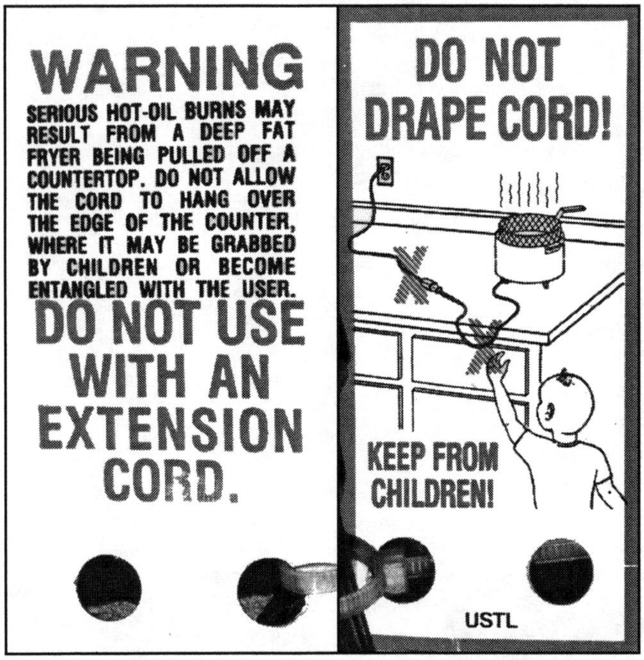

Figure 2 - Power Supply Cord Warning Tag

1. Close supervision is necessary when any appliance is used by or near children.

2. Do not let cord hang over edge of table or counter or touch hot surfaces.

Furthermore, the label for the shortening used with the cooker contains the warning: "DO NOT leave unattended while heating."

ACCIDENT RECONSTRUCTION

An accident reconstruction analysis was conducted using an exemplar cooker and a 96.5 cm (38 in.), 15.0 kg (33.1 lb), 28 month old boy who was similar to the injured boy in age, height, and weight. The exemplar cooker was filled with water and placed on a laminate countertop located 90.9 cm (35.8 in.) above the floor in a similar fashion as depicted in Fig. 1. The exemplar cooker power cord was not plugged into an electrical outlet. The accident reconstruction was photographed and videotaped. Results of the accident reconstruction are as follows:

1. While standing on the floor, the 28 month old boy was able to grab the power cord and pull the pot off the edge of the counter.

2. While standing on the floor, the 28 month old boy was able to grab the foot of the pot as shown in Fig. 3 and pull the pot off the edge of the counter.

Figure 3 - 2 Year Old Boy Pulling Pot Foot

3. While standing on the floor, the 28 month old boy was able to grab the temperature control as shown in Fig. 4 and pull the pot off the edge of the counter.

4. Based on the above, there are three potential explanations for how the injured boy pulled the cooker off the counter at the time of his accident. It is inconclusive that the 102.5 cm (40.4 in.), 15.0 kg (33.1 lb), 27 month old injured boy pulled the cooker off the counter by pulling on the power cord at the time of his accident.

Figure 4 - 2 Year Old Boy Pulling Temperature Control

ACCIDENT STATISTICS

On July 4, 2001, a news release from the U.S. Consumer Product Safety Commission (CPSC) was issued indicating that the CPSC has reports of 7 deaths and 64 serious burn injuries to children under 10 from deep fryers since 1980 [2]. According to this CPSC news release, these deaths and injuries resulted from young children pulling on the power cord, tipping over the deep fryer and its hot contents. In order to determine if all the reported deaths and serious burn injuries to children under 10 were associated with children pulling on the power cord, a critical analysis was performed on the CPSC spreadsheets which contain narratives as generated by the National Electronic Injury Surveillance System (NEISS) [3]. All 71 cases from the CPSC spreadsheets are presented in the Appendix breaking the cases out into the three categories utilized by the CPSC as follows:

1. Cases that definitely involved a child pulling on the cord (22 cases).

2. Cases that definitely did not involve a child pulling on the cord (9 cases).

3. Cases which probably/possibly involved a child pulling on the cord (40 cases).

The narratives are quoted as they appear in the CPSC spreadsheets. The date of the accident and the age of the injured child precede each narrative. The text has not been edited; misspellings and incomplete sentences are unaltered. It should be noted that there were approximately 20 million deep fryers in U.S. households in 1998 and approximately 2,625,000 new deep fryers sold in 1999 [4].

UNDERWRITERS LABORATORIES UL 1083

According to the July 4, 2001 CPSC news release [2], "The U.S. Consumer Product Safety Commission (CPSC) worked with Underwriters Laboratories (UL) to develop a new safety standard that calls for 'break-away' power cords for electric deep fryers. This new power cord will break away if a child pulls on it, preventing the deep fryer from tipping over and spilling scalding hot cooking oil on the child. The new requirements also apply to electric fondues and electric multi-purpose pots that use cooking oil. The new 'break-away' power cord is held to the deep fryer by a magnet." The revised requirements of the fourth edition of UL 1083 [5] which call for a detachable power supply cord employing a magnetic appliance coupler became effective on May 30, 2001. The fourth edition of UL 1083 [5] revised requirements for the detachable power-supply cord contained in section 36A.4 specify that "the force required to separate the detachable power supply cord from the appliance, as determined in 36A.1, shall be at least 5 percent less than the force required to overcome the static friction of the appliance, as determined in 36A.2." [5].

DETACHABLE POWER SUPPLY CORD

Testing

A new cooker similar in construction to the subject accident cooker equipped with a magnetic detachable power supply cord and rubber foot pads was obtained for the test program. The magnetic detachable power cord on the new test cooker meets the revised UL 1083 requirements effective on May 30, 2001 and is covered by United States Patent No. US 6,250,931 B1 issued on June 26, 2001 [6]. The test cooker was placed on a laminate counter and the magnetic detachable power cord was pulled by a force gauge and by the same 28 month old boy who participated in the accident reconstruction program. Figure 5 depicts three frames from a video sequence where a straight 10.2 N (2.3 lb) force is horizontally applied to the end of the magnetic detachable power cord to pull the test cooker filled with water off the counter. It should be noted that the magnetic power cord detached from the pot after it fell off the edge of the counter as shown in Fig. 5C. Figure 6 depicts three frames from a video sequence where the 28 month old boy pulled on the magnetic power cord while standing on the floor, causing the test cooker filled with water to fall off the counter. It should be noted that the power cord did not detach from the pot even after the pot fell off the edge of the counter as shown in Fig. 6B. These test results indicate that the presence of the magnetic detachable power cord as required by the fourth edition of UL 1083 effective on May 30, 2001 does not absolutely

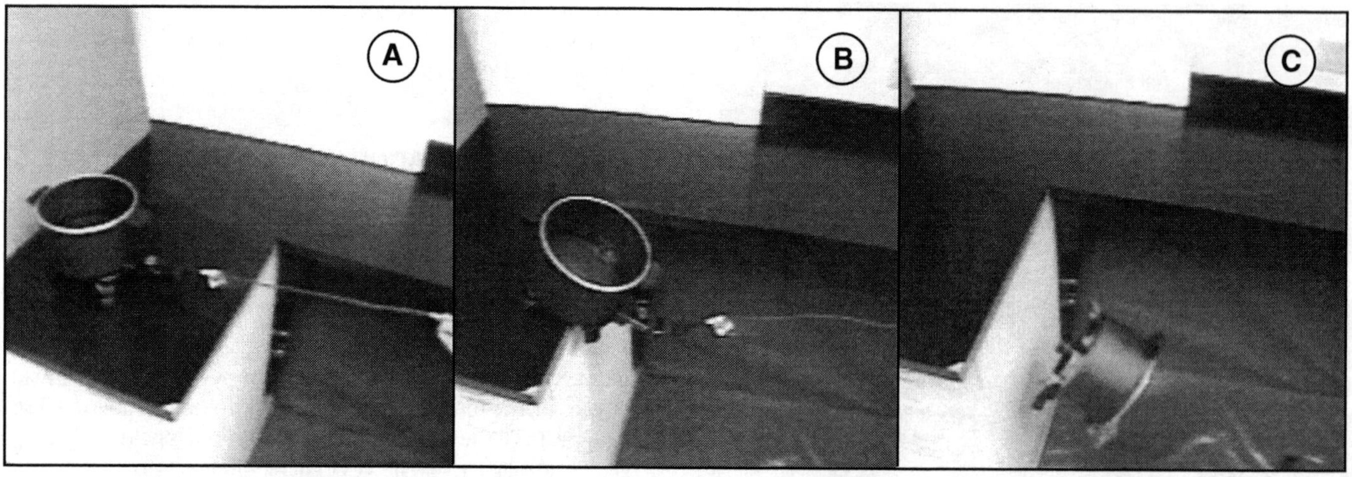

Figure 5 - Test Sequence - Cooker with Detachable Cord Pulled Off Counter with 10.2 N (2.3 lb)

prevent a child from sustaining a burn injury by pulling a cooker filled with hot oil off the counter. This conclusion is agreed with in both comments from the CPSC and responses to these comments by UL [7].

Safety Analysis

In addition to not absolutely preventing a child from pulling a cooker off the counter, the presence of the magnetic detachable power cord creates new dangers as described in the instruction manual for the new test cooker. Some of the potential safety downsides associated with the magnetic detachable power cord on an electric cooker/deep fryer are as follows:

1. Presents an electric shock hazard as a child places the detached power cord into his mouth while the cord is plugged into an electrical outlet.
2. Presents a burn hazard as the magnetic power cord is attached to a hot pot while the cord is plugged into an electrical outlet.
3. Presents an explosion hazard as a spark creates an ignition source for combustible gases.
4. Presents a fire hazard.
5. Presents a trip and fall hazard on the detached power cord.
6. Produces a false sense of security leading to decreased supervision of children around cookers.

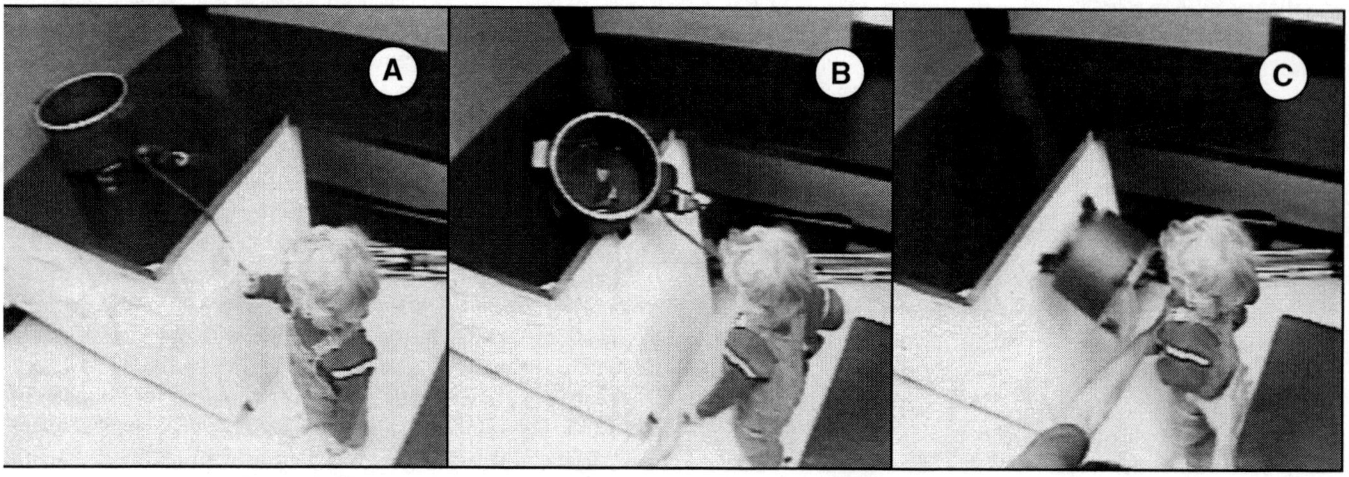

Figure 6 - Test Sequence - Cooker with Detachable Cord Pulled Off Counter by 2 Year Old Boy

7. Leads to misuse as "Off" control by detaching magnetic power cord from pot instead of disconnecting power cord plug from electrical outlet or turning temperature control to the off position.

In addition to the potential safety downsides listed above, some potential functional downsides associated with the magnetic detachable power cord on an electric cooker/deep fryer are as follows:

1. Nuisance detachments of the magnetic power cord from the pot.

2. Temperature control fractures due to impact after detachment of the magnetic power cord from the pot.

3. Lack of interchangeability with other electric kitchen appliance power cords. It should be noted that a substitute power cord was being utilized at the time of the subject accident.

4. Loss of attachment capability between magnetic power cord and pot due to food residue or cleaning agents.

DESIGN ALTERNATIVES

Latching/Locking Lid

One of the design alternatives that was considered, but rejected, for inclusion in the revised UL 1083 standard effective on May 30, 2001 was a latching/locking lid for a consumer electric cooker/deep fryer for minimizing the release of or retaining the hot oil in the vessel if it is pulled off the counter. In considering the efficacy of the locking lid proposal, the CPSC conducted some testing of deep fryers equipped with latching lids to see what would happen when the units were tipped over a counter onto the floor [8]. Preliminary testing revealed that the latching lid opened in some tests and allowed the escape of oil during the descent [8]. Furthermore, the CPSC accident data on deep fryers did not support the call for latching because having a latching lid might not have changed the situation in 5 of the 9 incidents the CPSC summarized [8]. In addition, the manufacturers mentioned several reasons why requiring a latching lid may not be effective such as [8]:

1. Many of the units in the incident reports were being emptied or were in cool-down, when consumers do not close the lid.

2. A large open latching lid could make the unit more able to be knocked about.

3. The unit would have to grow in size to comply with stability tests.

4. Making the units more expensive by adding the new lids may force some consumers to fry more foods on the range top and could increase the number of grease fires.

5. With added lids, consumers would have more interaction with the appliance, thereby increasing the possibility of tip-over or spillage.

Rubber Feet

Another design alternative that was considered, but rejected, for inclusion in the UL 1083 standard was rubber feet to increase the pot's resistance to sliding on a counter. Underwriters Laboratories provided some reasons why they do not believe the ability of a deep fryer to slide across a countertop should be criticized as follows [9]:

1. The ability of a deep fryer to slide on a countertop allows the deep fryer to be easily repositioned on the countertop without the need for lifting the appliance. If the coefficient of friction between the countertop and the deep fryer is increased it will be necessary to lift the deep fryer to reposition it out of the way when use is completed. The probability of spilling hot oil is greater if it is necessary to lift the deep fryer to reposition it as opposed to being able to slide it across a countertop.

2. If a deep fryer is impacted, it may either be placed into translation or caused to tip over depending on the point of impact with respect to the center of gravity and the coefficient of friction between the legs of the deep fryer and the supporting surface. Increasing the coefficient of friction will increase the likelihood that an impact will cause the deep fryer to tip over and spill hot oil rather than being put into translation.

A new cooker equipped with rubber foot pads is being tipped over in Fig. 7. Here a 28 month old boy is able to tip the cooker over by pulling on the temperature control. This can be compared to Fig. 4 where a similar pot without rubber foot pads was put into translation across the countertop when the same 28 month old boy pulled on the temperature control.

CONCLUSIONS

An accident reconstruction utilizing an exemplar cooker, countertop, and a 28 month old boy indicates that there are multiple potential explanations for how the subject cooker was pulled off the kitchen counter at the time of the subject accident. Since there were no eyewitnesses, it is inconclusive that the injured boy pulled the cooker's power cord to cause his accident. Indeed, the majority of the burn injuries and deaths to children under 10 associated with deep fryers reported by the CPSC fall

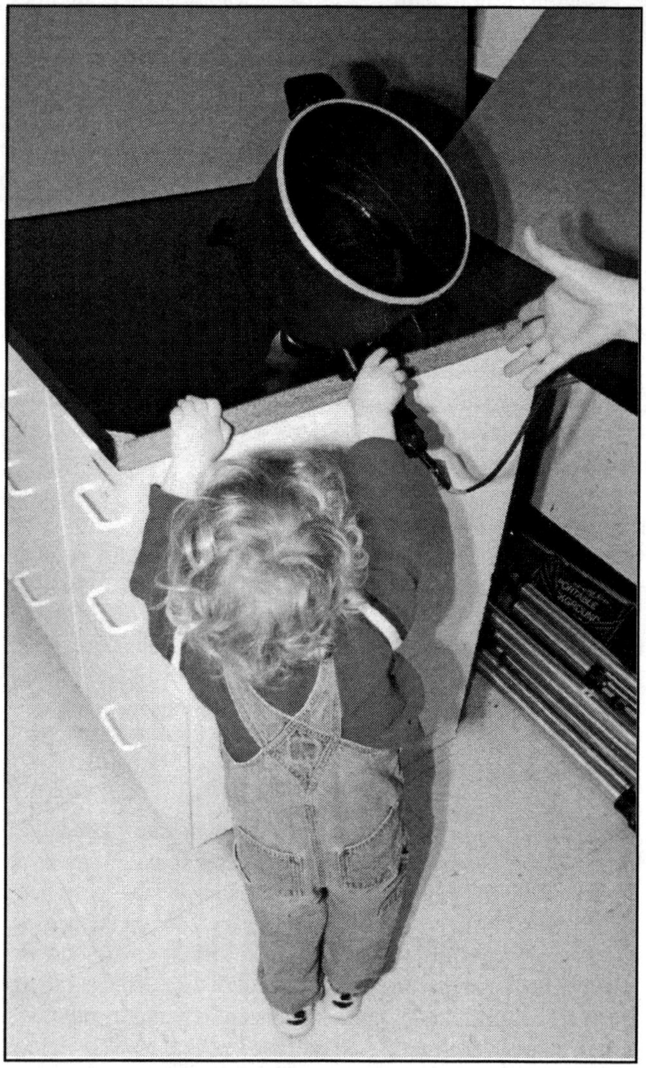

Figure 7 - Tip Over of cooker with Rubber Feet

into the category where it is not definitive that the child pulled on the cord (40 out of 71 cases). Based on these 71 reported cases of burn injuries to children under 10 over approximately a 20 year period, Underwriters Laboratories revised its standard 1083 effective on May 30, 2001 to require a detachable power supply cord. Testing of a new cooker equipped with this new magnetic detachable power cord reveals that a child can still pull the pot off the counter and sustain burn injuries. Because of the small sample size of CPSC reported child burn injuries associated with deep fryers and the novelty of the detachable power cord on cookers/ deep fryers, it is far too early to tell whether the revised UL Standard 1083 requirements will have a significant effect on these injuries. On the other hand, the real safety downsides associated with the detachable power cord identified in the new cooker/deep fryer instruction manuals must be monitored. According to the Dependency Hypothesis [10-11], the introduction of the detachable power cord on consumer electric cookers/deep fryers gives rise to a statistically significant pattern of user dependence. In this case that might mean decreased supervision by caregivers of children around the cookers and their hot contents. The question should be asked: Will more caregivers now leave their children unattended with the new style cooker in use as the injured boy's parent and grandparent did at the time of the subject accident?

REFERENCES

1. "Standard for Household Electric Skillets and Frying-Type Appliances," UL 1083, Third Edition, Underwriters Laboratories, Northbrook, IL, Mar. 1, 1994.

2. "CPSC Announces New Standard for 'Break-Away' Power Cords on Electric Deep Fryers," Release #01-184, U.S. Consumer Product Safety Commission, Office of Information and Public Affairs, Washington, DC, Jul. 4, 2001.

3. "National Injury Information Clearinghouse - Deep Fryers," U.S. Consumer Product Safety Commission, Washington, D.C., 1980-2000.

4. "22nd Portrait of the U.S. Appliance Industry," Appliance Magazine, Sept. 1999, pp. 76-81.

5. "Standard for Household Electric Skillets and Frying-Type Appliances," UL 1083, Northbrook, IL, May 30, 2001.

6. Mendelson, L.A., "Detachable Power Supply Apparatus," U.S. Patent 6,250,931 B1, Jun. 26, 2001.

7. Trotta, A.M., "Comments on Proposed Requirements for Automatic Toasters and Deep-Fat Fryers from October 29, 1999 Underwriters Laboratories Bulletin, Subject 1026," U.S. Consumer Product Safety Commission, Washington, D.C., Dec. 16, 1999.

8. Morris, W., "Summary of Deep Fryer Meeting with CPSC," Association of Home Appliance Manufacturers, Washington, D.C., Jul. 29, 1998.

9. Jungblut, J.L., et al., "UL Review of the Electrically Heated Deep-Fat Fryer Hazard Assessment Artech Report J8522.01-FR," Underwriters Laboratories, Northbrook, IL, Jun. 24, 1985.

10. Barnett, R.L., et al., "The Dependency Hypothesis – Misuse," ASAE Paper No. 85-1624, American Society of Agricultural Engineers, St. Joseph, MI, Dec. 17 – 20, 1985, pp. 1-15.

11. Barnett, R.L., et al., "The Dependency Hypothesis – Expected Use," ASAE Paper No. 86-5021, American Society of Agricultural Engineers, St. Joseph, MI, Jun. 29 – Jul. 2, 1986, pp. 1-13.

APPENDIX

Table 1 – Cases That Definitely Involved Child Pulling On Cord

3/1/81 – 2 years. WHILE WALKING IN THE KITCHEN, THE VICTIM'S FOREHEAD CAME INTO CONTACT WITH THE CORD OF THE ELECTRIC DEEP FRYER WHICH WAS DRAPED OVER THE END OF THE COUNTER. THE VICTIM'S CONTACT AND CONTINUED MOVEMENT CAUSED THE FRYER TO SLIDE ACROSS THE COUNTER. ALTHOUGH THE VICTIM'S FATHER WAS ABLE TO STRIKE AT THE FRYER TO KEEP IT FROM HITTING THE VICTIM, THE HOT OIL IN THE FRYER SPILLED OVER THE VICTIM.

12/11/83 – 16 months. PULLED CORD ,DEEP FRYER FELL OFF TABLE,HIT ARM

9/7/85 – 23 months. 38% TOTAL BODY SURFACE THERMAL INJURY – ALLEGEDLY GRABBEDELECTRIC CORD PULLING DEEP FAT FRYER ON HIMSELF – AUTOPSY YES

4/20/87 – 7 months. A 7-MONTH BOY WAS HOSPITALIZED AFTER HE PULLED ON A CORD ATTACHEDTO A DEEP FRYER SPILLING HOT GREASE ON HIM.

7/17/89 – 12 months. A 1-YEAR OLD GIRL WAS HOSPITALIZED AFTER SHE PULLED CORD OF A DEEP FRYER SPILLING HOT GREASE ON HER.

3/28/91 – 7 months. PULLED DEEP FRYER BY THE CORD ONTO HIMSELF SUSTAINING LOWER TRUNK BURNS.

10/15/92 – 13 months. PT GRABBED ELECTRIC CORD FROM DEEP FRYER WITH HOT GREASE PULLED IT ONTOFACE, 1ST DEGREE BURNS.

10/20/92 – 10 months. ONONLY DIAPER BURNS OF ALL EXTREMITIES CHEST ABD FACE. DX. 2 DEGREE BURNS

12/28/92 – 8 months. THE 8 MONTH OLD FEMALE VICTIM DIED FROM 3RD DEGREE BURNS TO 27% OFOF HER BODY SKIN AREA WHEN SHE GRABBED THE CORD OF A FRYER FILLEDWITH HOT GREASE, THAT WAS SITTING ON TOP OF A KITCHEN COUNTER TOPIN HER HOME, AND PULLED FRYER ON TOP OF HER. HOT GREASE SPILLED ONHER HEAD, CHEST, AND ARMS. THE VICTIM WAS TAKEN TO THE EMERGENCYROOM WHERE SHE DIED 3 DAYS LATER.

5/27/94 – 8 months. AN 8 YEAR OLD MALE RECEIVED SEVERE SCALD BURNS FROM HOT OIL IN A DEEPFAT FRYER. THE CHILD WAS IN HIS BABY WALKER WHEN HE GRABBED THEFRYER'S POWER CORD AND PULLED THE FRYER OFF THE COUNTER SPLASHINGHIMSELF WITH THE HOT OIL. SINCE THE INCIDENT THE CHILD HAS UNDERGONE EXTENSIVE MEDICAL TREATMENT FOR HIS BURNS.

11/21/94 – 9 months. A 9 MONTH OLD MALE WAS IN A WALKER AND SOMEHOW TRAVELED TO AREA NEARKITCHEN COUNTER. HE WAS ABLE TO GRAB THE POWER CORD OF AN ELECTRICDEEP FRYER AND PULL THE DEEP FRYER OFF OF THE COUNTER. THE HOT OILINSIDE POURED ONTOTHE VICTIM. HE WAS TAKEN TO BURN UNIT OF A LOCALHOSPITAL, WHERE HE DIED 3 WEEKS LATER OF HIS INJURIES.

2/3/95 – 6 months. A 6 MOM SUFFERED A SERIOUS UPPER BODY BURNS FROM HOT OIL IN A DEEP FRYER THAT SPILLED ON HIM AFTER HE PULLED ON THE CORD OF THE FRYER.

9/30/95 – 12 months. A 12MOM WAS PAINFULLY BURNED WHEN HE TUGGED ON THE CORD OF A DEEPFRYER SPILLING THE CONTENTS ON HIMSELF. HIS MOTHER WAS ALSO BURNEDLESS SERIOUSLY.

3/11/96 – 6 months. 6 MTH OLD MALE SITTING IN A HIGHCHAIR, PULLED ON CORD OF A Deep fryerAND SUSTAINED A LEFT UPPER LEG BURN WHEN HOT GREASE SPLASHED.

5/27/96 – 3 years. 5% BSA 1ST DEGREE BURNS RT NECK/SHOULDER-PULLED FRY DADDY FULL OFHOT GREASE ON HIMSELF AND BROTHER.

3/29/97 – 7 months. A 7 MONTH OLD MALE SUFFERED 2ND AND 3RD DEGREE BURNS TO OVER 45% OF HIS BODY WHEN HOT OIL FROM A DEEP FAT FRYER SPILLED ON HIM. THECHILD WAS REPORTED TO HAVE PULLED ON THE FRYER'S POWER CORD, WHICHPULLED THE FRYER OFF A COUNTER, SPILLING ITS CONTENTS ONTO THECHILD. THE INCIDENT OCCURRED IN THE KITCHEN OF THE CHILD'S HOME.

5/2/97– 2 years. SECOND DEGREE BURNS TO BACK – PLAYING WITH ANOTHER CHILD, CAUGHTCORD OF FRYER WITH HOT GREASE, SPILLED ON BACK

5/2/97 – 2 years. SMALL 1ST DEGREE BURN LOW BACK – PLAYING WITH ANOTHER CHILD, CAUGHT CORD OF FRYER, HOT GREASE SPILLED ON BACK

7/26/97 – 11 months. A DEEP FRYER WAS KNOCKED OFF & FELL ON PT DX-BURN TO UPPER EXTREMITIES (Investigation indicated that the child pulled the cord).

1/17/99 – 12 months. 15% TBSA 2ND DEGREE BURNS TO FACE AND TRUNK-TRIPPED ON CHAIR & PULLED CORD OF Deep Fryer HOT GREASE SPLASHED CHILD-AT HOME

10/18/00 – 2 years. AS A 2-YEAR-OLD FEMALE WALKED INTO THE KITCHEN, HER ARM CAUGHT THEHANGING CORD OF A DEEP FRYER. THE FRYER TIPPED OVER ON TOP OF HER.SHE SUFFERED 3RD DEGREE BURNS OVER 60% OF HER BODY. SHE ISRECOVERING IN A BURN CENTER HOSPITAL.

Unknown – 0 years. DEEP FAT FRYER DOESN'T HAVE A TOP FOR USE DURING OPERATION AND CORD WAS LONG ENOUGH TO ALLOW CHILD TO PULL APPLIANCE OVER ON HERSELF.

Table 2 - Cases That Definitely Did Not Involve Child Pulling On Cord

3/17/81 – 6 years. Deep fryer SITTING ON FLOOR-PT WALKED INTO IN, SPILL GREASE

9/22/82 – 12 months. ASPIRATION AND CARDIOPULMONARY ARREST; 2ND AND 3RD DEGREEBURNS OVER 30% BODY SURFACE AREA – FELL INTO DEEP FAT FRYER WITH WATER— AUTOPSY YES.

4/20/84 – 3 years. A THREE YEAR OLD MALE SUFFERED 2ND DEGREE BURNS ON HIS CHEST AND BACKWHEN A DEEP FAT FRYER TIPPED OVER ON HIM IN THE KITCHEN OF HISPARENT'S HOME. THE FRYER WAS UNPLUGGED AND CORD HAD BEEN REMOVED.THE FRYER WAS RESTING ON A WOODEN PULL-OUT BREAD BOARD ABOUT 34" ABOUT 34" ABOVE THE FLOOR. THERE WAS NO EYE WITNESS AND IT ISUNCERTAIN WHAT THE VICTIM WAS DOING WHEN THE ACCIDENT OCCURRED. THEVICTIM WAS HOSPITALIZED FOR MORE THAN 60 DAYS, AND HAS HAD SEVERALSKIN GRAFT OPERATIONS.

4/18/87 – 6 years. SPILLED A DEEP FAT FRYER WHTEN OPENING ANTOHER

1/1/88 – 4 years. A 4-YEAR OLD BOY AND HIS MOTHER WERE HOSPITALIZED AFTER THE BOY BUMPED A DEEP FAT FRYER DUMPING HOT GREASE.

10/17/92 – 3 years. PT CLIMBED UP ON KITCHEN COUNTER AND ACCD. KNOCKED OVER FRYER WITH HOTGREASE ONTO HAND/ FOREARM/ FACE, 2 DEGREE BURNS

1/7/96 – 12 months. PT. PULLED THE CORD OF A PORTABLE FRYER UNIT AND IT SPILLED CONTENTS SUSTAINED GREASE BURNS TO UPPER TRUNK, FACE AND NECK. E924.0 (Investigation indicated that this was not a cord case).

4/20/97 – 3 years. BURNS OVER BODY – KNOCKED GREASE FRYER OVER AND HOT GREASE SPILLED

Unknown – 5 years. AFTER COOKING IN DEEP FAT FRYER CHILD REACHED UP TO COUNTER – TIPPEDIT OVER – BURNED BY HOT GREASE - LAWSUIT

Table 3 - Cases Which Probably/Possibly Involved Child Pulling On Cord

7/4/80– 6 years. CHILD BURNED – GREASE FROM DEEP FRYER SPILLED ON HER

3/1/81 – 2 years. APPLIANCE FELL OFF COUNTER AND HOT GREASE BURNED VICTIM.

9/4/81 – 2 years. PULLED COOKER OFF TABLE FIRST AND SC DEGREE BURNS 15 PERCENT TBSA

4/29/82 – 16 months. PULLED DEEP FAT FRYER OFF COUNTER ONTO SELF

12/15/82 – 12 years. BURNED WITH HOT GREASE FROM FYER AT HOME

1/11/83 – 12 years. TIPPED OVER deep fryer DEEP FRYER,GREASE SPILLED

1/18/83 – 0 years. INDIVIDUAL WAS BURNED FROM HOT COOKING OIL WHEN CHILD PULLED FRYEROFF COUNTER.

4/4/83 – 15 months. BOY INJURED BY ACCIDENT INVOLVING HOT GREASE & DEEP FAT FRYER.

8/9/83 – 11 months. THE CHILD PULLED A PAN OF HOT GREASE ALL OVER HIM.

1/26/84 – 8 months. BOY UPSET DEEP FRYER FILLED WITH HOT OIL. HE WAS HOSPITALIZED FOR BURNS.

3/17/85 – 3 years. GIRL WAS BURNED WHEN SHE PULLED A DEEP FRYER OVER SPILLING THE HOT OIL ON HER.

4/10/85 – 7 years. BOY PULLED DEEP FRYER FILLED WITH HOT GREASE OVER ON HIMSELF.

7/1/85– 13 months. BOY, 13 MONTHS, WAS BURNED WHEN DEEP FAT FRYER SLID OFF EDGE OF COUNTER AND POURED HOT OIL ON HIM.

1/1/86– 0 years. A MALE CHILD WAS SEVERELY BURNED WHEN HE PULLED A DEEP FRYER OF A TABLE TOP SPILLING THE CONTENTS ONTO HIM.

6/5/86 – 12 months. CHILD PULLED DEEP FRYER AND SPILLED HOT OIL ON HIMSELF.

6/14/86 – 10 months. PULLED SMALL HOT FRYER ON HIMSELF

8/1/86 – 12 months. BURN ABDOMIN - CHILD PULLED deep fryer OVER ON SELF – GREASE BURN

11/11/87— 2 years. PULLED ELECTRIC DEEP FRYER ONTO HEAD. BURNS TO FACE, EARS, ARMS.

1/1/88– 15 months. A 15-MONTH OLD BOY DIED DUE TI BURNS HE RECEIVED AFTER HE PULLED A MINI-FRYER FULL OF HOT GREASE OVER ON TOP OF HIM.

2/3/90– 0 years. A FEMALE INFANT WAS SERIOUSLY BURNED WHEN AN ELECTRIC DEEP FRYER FELLON HER SPILLING HOT OIL.

1/19/91 – 12 months. COUNTER ONTO SELF 2ND DEGREE BURNS 60 PER CENT BODY/BURN CENTER BY AIR

2/22/91 – 22 months. PT WAS AT HOME WHEN HOT GREASE FROM deep fryer SPILLED OVER-GREASE BURNS 1ST AND 2ND DEG. SHOULDERS & ARMS 10% BODY AREA

6/14/91 – 4 years. A 4 YEAR OLD MALE WAS BURNED AFTER UPSETTING A DEEP FAT FRYER OF HOTGREASE IN HIS HOME

5/24/92 – 2 years. A 2 YEAR OLD FEMALE WAS SEVERELY BURNED WHEN A DEEP FRYER SPILLED OVER ON HER.

10/4/93 – 20 months. A 20 MONTH OLD FEMALE WAS SERIOUSLY BURNED AFTER PULLING A DEEP FRYERFULL OF HOT GREASE OFF A COUNTER ON HER.

1/20/94 – 10 months. GREASE BURN FROM A HOT FRYER AT HOME.DX: SECOND DEGREE BURN TO RT. LEG (CALF AREA).

6/28/95 – 2 years. PATIENT PULLED LOT-GREASE FROM FRYER OFF OF COUNTER TOPONTO HIS BACK – MULTISYSTEM ORGAN FAILURE TO INCLUDE RENAL, RESPIRATORY & LIVER FAILURE; ACUTE RESPIRATORY DISTRESS FAILURE/SYNDROME – SEPTIC SHOCK SYNDROME – AUTOPSY NO

1/5/96– 4 years. A DEEP FRYER, FILLED WITH HOT COOKING OIL, WAS SET ON A KITCHENCOUNTER. THE PRODUCT HAD BEEN USED TO PREPARE FRENCH FRIES. A FOURYEAR OLD FEMALE WAS SEVERELY BURNED WHEN THE DEEP FRYER TOPPLEDOVER, SPILLING OIL ON THE VICTIM. THE PRODUCT'S PLASTIC LID WAS NOTON AT THE TIME.

4/5/96 – 6 years. THIS INVESTIGATION WAS INITIATED THROUGH A NEWSPAPER ACCOUNT OF A SIXYEAR OLD FEMALE WHO WAS BURNED OVER 12% OF HER BODY WHEN THE DEEPFAT FRYER HER MOTHER WAS GOING TO USE TO COOK FRENCH FRIES FOR THEEVENING MEAL FELL OVER ONTO THE FLOOR. THE HOT OIL IN THE FRYERSPILLED ONTO HER CHEST, FACE, ARMS, AND FEET, SEVERELY BURNING HER.THE FRYER, WHICH WAS APPROXIMIATELY FOUR YEARS OLD AT THE TIME OF THEINCIDENT DID NOT HAVE A COVER AND WAS DISPOSED OF BY THE PARENTS.THE FRYER WAS BEING PREHEATED, USING A REPLACEMENT CORD PURCHASED BY THE MOTHER.

4/22/96 – 2 years. SECOND DEGREE BURNS TO BACK AND BOTH FEET, ELECTRIC DEEP FRYER TURNEDOVER AND HOT GREASE SPILLED ON PT

5/27/96 – 2 years. 10% BSA 1ST AND 2ND DEGREE BURNS LT TORSO-PULLED FRY DADDY FULL OF HOT GREASE ON HIMSELF AND BROTHER

10/7/96 – 3 years. 3 YR FEMALE SUSTAINED BURNED FEET WHEN TIPPED OVER DEEEP FRYER.

10/18/96– 13 months. BURNED FOOT WHEN PULLED FRYER SPILLING GREASE

11/13/98 – 2 years. 2 YOM PULLED DEEP FRYER OVER ON HIMSELF. BURNS TO CHEST.

3/18/99 – 11 months. PT. AT HM WHEN HE PULLED A Deep Fryer OIL COOKER FILLED W/HOT OILIN IT ONTO SELF, PT SUST 35% 2ND/3RD DEG BURNS TO BODY.E924.0TIME12:30

7/23/99 – 8 months. BURNS TO RIGHT SIDE OF BODY BY HOT GREASE FROM DEEP FRYER.MAJOR BURN/ ?VSD

7/31/99 – 12 months. PT'S FATHER FOUND HIM ON KITCHEN FLOOR NEXT TO DEEP FRYER, WHICH HAD BEEN PLUGGED IN (PER CHART) DX: 2ND DEGREE BURNS TO 15% OF BODY

12/3/99 – 6 years. PULLED DEEP FRYER ON HIMSELF/ PARTIAL THICKNESS BURNS NECK ANDABDOMEN

7/22/00 – 4 years. FAMILY WAS HAVING A FISH FRY AND PT TRIPPED OVER AN EXTENSION CORD& KNOCKED OVER THE POT HOT GREASE & FISH. D:2ND DEGREE BURN R FOOT.

Unknown – 0 years. CHILD WAS BURNED WHEN DEEP FRYER TIPPED OVER.

Proceedings of IMECE'03
2003 ASME International Mechanical Engineering Congress
Washington, D.C., November 15–21, 2003

IMECE2003-42078

OPTICAL (CAMERA-BASED) TECHNOLOGY FOR SEISMIC RISK ASSESMENT

Thomas Wischgoll
Electrical Engineering & Computer Science
University of California, Irvine
Irvine, CA 92697-2625
Email: twischgo@uci.edu

Tara C. Hutchinson
Civil & Environmental Engineering
University of California, Irvine
Irvine, CA 92697-2175
Email: thutchin@uci.edu

Falko Kuester
Electrical Engineering & Computer Science
University of California, Irvine
Irvine, CA 92697-2625
Email: fkuester@uci.edu

ABSTRACT

Due to an increasing bandwidth for data transport in recent computers, optical (camera-based) sensors with high frame rates and reasonable resolutions can nowadays be used with off-the-shelf computers. By monitoring buildings, bridges and other infrastructure with such sensors, they can greatly assist in risk assessment. In this paper, several field application examples are described, largely encompassing the areas of civil infrastructure monitoring. Optical (camera-based) systems can be particularly powerful for monitoring both local and global movements within a scene or environment. Therefore, natural hazards induced by such movements, for example during earthquake events, are an important application area for these new technologies. First, a clear definition of seismic risk assessment is provided. Subsequently, two important fields where cameras may be useful in seismic risk assessment are described. Specifically, the context of (i) early warning systems and (ii) post-earthquake assessment are addressed. An example of using the optical record from a series of large shake table tests is provided and comparison with other methods discussed.

INTRODUCTION

The overall performance of buildings, bridges, and other infrastructure subjected to earthquake motions has dramatically improved over the years, largely due to an increased understanding of the behavior of primary structural systems. Although the primary structural system may perform fairly well, other (nonstructural) hazards due to imposed earthquake motions may result in building closure and/or lead to other dangerous hazards. Specific examples include the breakage of pipes or elements within the pipe network, which may result in water or other fluid leakage throughout the building, rendering the structure temporarily unusable. The application of a network of sensors that is able to detect such a situation has the potential for reducing any subsequent damage by activating shut-off valves in the fluid supply. Sensors that detect hazardous situations, such as gas leaks or dangerous chemicals that have spilled can also be used to warn people from entering a hazardous area.

A particularly promising device for use in these situations is an optical (camera-based) sensor. Optical sensors can be of great potential when monitoring movement of all different types of systems. In this paper, the term optical sensor is used to denote any type of image-based sensor [e.g. video cameras, charged-couple-device (CCD) cameras, infrared or other cameras]. Due to the readily available high-speed computational platforms today, a huge amount of data can be processed in near-real time. These computational advantages, combined with an increased availability of high quality optical sensors have promoted their usage in many practical field applications. Common field examples might be broadly categorized as either observation or surveillance. The most pervasive example of the former is transportation or traffic observation while the most common example of the later is surveillance in public buildings. However, optical sensors are advantageous beyond inspection or surveillance applications. A promising application is field monitoring. For example, their non-intrusive nature may be beneficial for monitoring certain systems critical to the infrastructure, such as pipe(s) or networks of pipes, electrical systems, or other service system equipment. Perhaps the simplest example is an optical sensor that has detected a broken pipe. This information may be used to close an upstream valve to prevent further fluid leakage and subsequent damage to the area. An important life-threatening example is an area containing hazardous substances. These areas can be monitored if, for instance, there is potential that dangerous chemicals may spill. After identifying such a hazardous situation, the affected area may then be isolated in order to prevent humans from entering this hazardous section.

Since optical sensors capable of capture rates of more than 30 frames per second are available at reasonable prices, the proposed monitoring system can be installed permanently in several rooms of a building at relatively low cost, compared to the overall cost of the structure. With such an infrastructure, an additional safety system can be provided that is capable of reducing or eliminating the types of hazards described previously.

In this paper, concepts are presented for using optical sensors to reduce earthquake hazards (particularly secondary hazards) by increasing our assessment ability. However, the methods described are not restricted to this particular application area. The concept holds for any type of application where movement is imposed upon structures. Initially, an overview of optical sensors used in various field applications is provided. Subsequently, selection of optical sensors that reasonably meet the requirements for monitoring structural (and nonstructural) movements and techniques for evaluating optical data are discussed. A precise definition of risk assessment is provided and examples from past earthquakes are described where deployment of these systems would have been extremely beneficial. Finally, challenges related to implementing optical sensors and computer vision algorithms in this application area are discussed.

FIELD APPLICATION EXAMPLES

The increased speed and resolution, combined with dramatically decreasing cost of optical sensors in recent years has resulted in an increased usage of these systems in practical field applications. Examples include observing public areas for security reasons, monitoring traffic or bridges, inspecting sensitive systems, or detecting changes in the environment over time. This section describes specific field applications in these areas.

As an example, optical sensors may be used to enhance security in public buildings. Surveillance cameras can be installed to monitor the pass through of people. These systems may also be interfaced with face recognition software capable of identifying runaway teenagers or suspects in criminal activities [1].

Another application is monitoring traffic on a surface road or highway. This can aid in detecting unauthorized use of a roadway or in recognizing special traffic situations, such as excessive congestion. Several publications are available describing vehicle tracking in the field (e.g. [2] [3] [4]). The RoadWatch project described by Beymer et al. [5] uses one camera to track multiple cars at once. Vision algorithms allow for tracking only parts of a vehicle to avoid occlusion problems. This enables the system to count the number of vehicles on a highway and detect volume and location of traffic congestion. Figure 1 shows an example of several cars being tracked by this system.

Similarly, optical sensors have been used to monitor bridge structures [6]. In this example project, streamed digital video images provide a mechanism to monitor the traffic moving over

Figure 1. Cars tracked by RoadWatch (Image courtesy of David Beymer [5]).

critical areas of the bridge. The optical sensors are combined with other types of measurement devices such as accelerometers, mechanical displacement and force transducers, each strategically located on the structure. With the optical sensors it is easy to identify heavy trucks that are traveling over the structure and measure their velocities. The bottom half of Figure 2 illustrates the synchronization of images and vibration data used to identify the response of a heavy truck crossing the bridge.

Optical sensors may also be used in areas such as inspection of various train and railway related systems [7]. Applications exist that inspect the rail profile [8], [9], rail gap [10], contact wire positions [11], or wear [12]. Additionally, the train itself may be inspected or monitored. Applications include the inspection of the wheels or the thickness of the brake pads. During

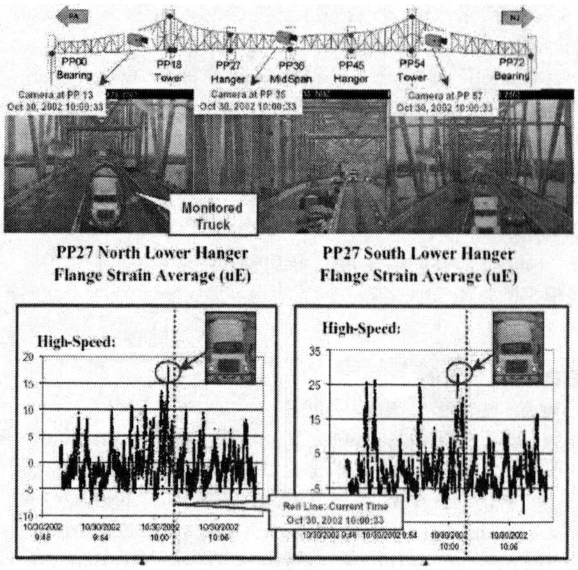

Figure 2. Bridge monitoring using cameras and conventional sensors (Commodore Barry Bridge, monitoring system by Aktan et al. [6]) (Schematic courtesy of E. Aktan).

Figure 3. Growing ice at the shore highlighted by an edge detection algorithm (Image courtesy of U.S. Geological Survey [16]).

operation, optical sensors can detect locking wheels or overheating brake systems. To observe the maximum occupancy inside a train, optical sensors have also been used to count the number of passengers (e.g. [13] [14] [15]).

For environmental monitoring changes in the shorelines and costal bluffs can be observed, based on images collected from continuously streaming cameras monitoring, for example, the water-beach interface [16]. In this example, a system was developed that is able to monitor breaking waves, alongshore currents, rip currents, and beach-face profiling, especially in stormy weather. Figure 3 illustrates an example of image data captured during a cold winter season, where the line indicated by an arrow outlines the ice-shoreline interface.

OPTICAL SYSTEMS FOR SEISMIC APPLICATIONS

Based on the dominant range of frequencies of movement generated during earthquakes and satisfying optimal sampling rates (i.e. to ensure aliasing of images does not occur), data acquisition rates of at least 30 frames per second are necessary. Ideally, higher frequencies (greater than 60 frames per second) are desirable to allow for differentiation of acquired movement data to obtain velocities and accelerations. In addition, the optical sensors and associated processing system should be capable of capturing image data at high resolutions. Given only these specifications, there are hundreds of optical sensors available. However, given other desirable attributes, such as low cost, small size, high shock rating, readily available and compatible interfaces (such as IEEE firewire or USB-2 interfaces), the selection greatly reduces. Table 1 lists a few camera models that would fit the needs for an application described in this paper. All of these cameras use a low resolution to facilitate higher frame rates at a reasonable price.

For the overall capturing system, limiting factors include interface transfer hardware (e.g. PC-specific hardware such as the PCI bus, hard drives, interface cards, etc.). These elements may have the potential for limiting the system bandwidth and subsequently the capture rate. For example, capturing with one of the cameras listed in Table 1 results in data rates of up to 54 MB/s. The PCI bus in standard computers is able to transfer at

Table 1. Comparison between different available camera models.

Manufacturer	Basler	Pulnix	Cooke
Model	A301fc	TM-6700CL	PixelFly VGA
Resolution	658×494	640×494	640×480
Frame rate	80 fps	60 fps	40 fps
Color depth	8 bit	24 bit	24/36 bit
Data rate	25 MB/s	54 MB/s	35/47 MB/s
Price (U.S.)	$ 1,895	$ 1,795	$ 6,090

speeds of up to 80 MB/s in typical applications. To store the images captured by the camera on the hard drive, the data needs to be transfered to main memory first before it can be saved on the hard drive. Consequently, it has to cross the PCI bus twice. Therefore, a computer with two PCI buses may be necessary depending on the types and number of cameras used. In addition, available hard drives have very fast transfer rates for reading data but are comparably slow when it comes to writing data. Depending on the data rate produced by the camera, a redundant array of independent/inexpensive disks (RAID) consisting of several hard drives may be required, such that only part of the data needs to be written to one specific hard drive resulting in an increased transfer rate.

TECHNIQUES FOR INTERPRETING OPTICAL DATA

Optical sensor networks within a building can greatly aid in monitoring sensitive systems. At any given time, a single sensor may be used to observe multiple systems or system components simultaneously. This may provide a pre-event inventory of a particular structure or room, including both occupancy statistics as well as details regarding the interior infrastructure.

According to the Federal Emergency Management Agency [17] assessing the condition of nonstructural components within a building after an earthquake event is performed by visually inspecting one to three samples of a particular component. This depends on the availability of detailed drawings of the component and its attachment. However, this does not take into account the pre-earthquake condition of the component. Using optical sensors the performance of the component or any other building equipment can be monitored throughout the duration of an earthquake event and any time before or after that event.

Figure 4 shows a series of sequential images captured by an optical sensor. In this example, a retail store is reconstructed on a shake table and the contents of the scene observed. During simulated earthquake motions, a heavy box falls from one of the shelves and rolls onto the floor. These changes in the environment can be detected and highlighted. In this case, rectangles show object movement over time, annotated with a vector indicating the primary direction of movement. Computer vision

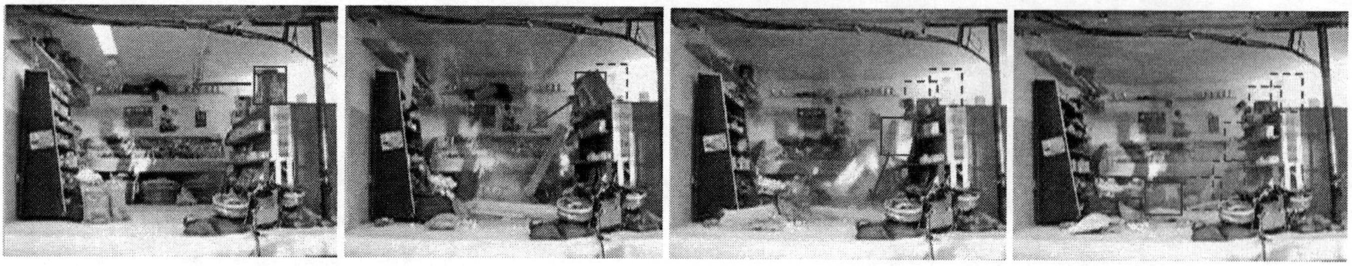

Figure 4. Object tracking and detection from digital streamline images taken during shake table experiments. Current position shown solid, tracked (previous) position shown dashed. (Digital movie courtesy of K. Mosalam, UC Berkeley).

techniques allow tracking of various objects using a set of image sequences. This may include large objects undergoing significant positional changes, as well as other types of nonstructural elements such as networks of pipes or suspended lights. Other elements falling from furniture inside the monitored room can be tracked as well. This facilitates the detection of hazardous situations if it is known what types of elements are present in the environment. For increased efficiency and accuracy, a pre-calibration of the environment may be required.

Figure 5 shows another example environment with a smaller field of view. The optical sensor is directed towards a portion of a shelf holding several glass containers that may contain potentially hazardous chemicals. During shaking, one of the containers topples and spills some of its contents. The monitored container is circled in the images shown in Figure 5(a). The upper images show single shots of the captured video stream. To track the container an edge detection algorithm has been used, resulting in the processed images shown in Figure 5(b). For this example, the algorithm introduced by Canny [18] provides reasonable results. The figure shows that the edges of the monitored container can be extracted in order to track the target object and identify it toppling.

If at least two cameras are present and located in the same area, a three-dimensional model can be computed from the images if no occlusion occurs. This, of course, assumes that the optical sensors are synchronized and every sensor takes an image at exactly the same time or that precise time stamps are available, allowing for image synchronization. Using edge detection algorithms, objects in the scene can be identified and tracked when they move. Furthermore, it is possible to measure the absolute movement of objects in the scene while tracking. Therefore, a reference distance in the scene is necessary for each of the three dimensions. Using optical sensors with higher frame rates, differentiation and signal processing allow the calculation of velocity and acceleration of objects in the scene.

RISK ASSESSMENT

Many definitions are available to quantify the term risk assessment. Generally, risk is defined as the uncertainty of a specific occurrence times the consequence resulting from this occurrence. Uncertainty must be determined for each risk individually. This can be done using mathematical models based on probability estimates. Consequence is most often defined as the financial measure of the effect of the occurrence. One must distinguish between risks that affect people and risks that affect only material (objects). Financial consequence affecting material (objects) may be fairly challenging to estimate, whereas risk associated with people, such as fatalities, may be defined with a greater level of certainty. When people are involved, one can precisely define risk as the hazard severity level times the likelihood of occurrence times the number of persons exposed to the risk [19].

SEISMIC RISK ASSESSMENT

In the context of seismic risk assessment, risk is often thought of as the identification of the risk of damage incurred during a particular earthquake event. An alternative measure may be the number of people injured or fatalities caused by the specific event. To assess this type of risk, the likelihood of occurrence and the hazard severity level must be determined in the same fashion as conventional risk assessment. In the context of earthquake hazards, not only is consideration of direct impacts required (such as damage to primary structures), but post-earthquake hazards (such as chemical spills or broken pipes) must also be considered. These hazards may often result in substantial damage, which may be avoided using the appropriate assessment tools.

POTENTIAL FOR OPTICAL SENSORS TO REDUCE RISK OF EARTHQUAKE RELATED HAZARDS

The purpose of this paper is to identify specific issues and areas where cameras and camera-based technologies and associated image processing algorithms can assist in assessing the risk due to either natural or man-made disasters. The focus of this paper is on earthquake hazards, specifically secondary (post-earthquake) hazards created within the interior of building structures. In this case, one can differentiate between damage that is associated with elements directly attached to the structure (such as connected piping networks or attached equipment) and elements not attached to the structure (such as building contents or unattached

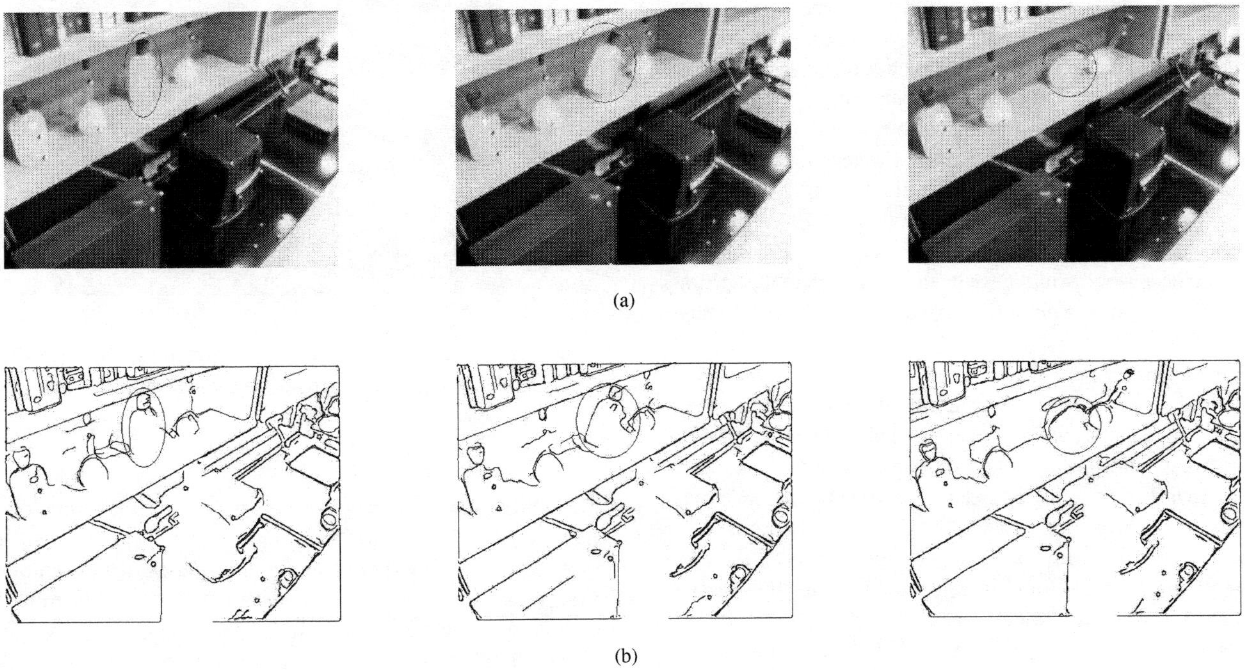

Figure 5. Chemicals observed by an optical sensor during a shake table experiment: (a) original images and (b) images with only edges identified. The tracked object is circled. Both, single shots of the video stream and detected edges are shown in subsequent order.

equipment). The following discussion provides specific examples from past earthquakes with reference to both attached and unattached elements.

After analyzing the damage resulting from the 1971 San Fernando earthquake, California enacted the California Hospital Act. The objective of this act was to provide a level of design and construction that improved the resistance of hospitals in California compared to structures built before that time. The 1994 Northridge earthquake illustrated that structural systems in hospitals built after 1971, considering the new California Hospital Act, performed very well. However, the piping and air handling systems in many hospitals suffered fractures along individual pipes or at joints, resulting in temporarily closure of many hospital buildings [20]. Figure 6(a) and 6(b) show examples of sprinkler pipe leakages. In Figure 6(a), a sprinkler inside the Olive View hospital at the ceiling level is shown, while Figure 6(b) shows leakage on the outside of a building near Olive View Hospital resulting from a sprinkler system damaged during an aftershock. To ensure that the fire prevention system of a building is still functional after an earthquake, guidelines provided by the Federal Emergency Management Agency [17] points to improved standards and the need for flexibly mounted piping. However, older structures may not conform to current understanding of earthquake demands resulting in failure of sprinkler systems. Newer structures being built according to modern design standards, such as the Uniform Building Code [21] or International Building Code [22], meet these requirements.

Figure 6. (a) Broken sprinkler in Olive View hospital after Northridge earthquake in 1994.(b) Sprinkler pipe leakage after an aftershock of the 1994 Northridge earthquake near Olive View Hospital. (Courtesy of National Information Service for Earthquake Engineering (NISEE), University of California, Berkeley [23])

During the 1994 Northridge earthquake, 2500 water heaters

were damaged and subsequently introduced wide spread natural gas leaks. By optically monitoring and correlating measured movements to tolerable limits, the potential for post-earthquake fires in this situation would be greatly reduced.

The 2001 Nisqually earthquake also resulted in widespread nonstructural damage. In one example, a water pipe ruptured in the mechanical room on the roof of a hotel causing 3000 liters of water in a storage tank to flood several floors [24]. Figure 7 shows the ruptured pipe and the storage tank. The proposed system of optical sensors would have been able to detect the spilling water and shut down the piping system to prevent further damage to the building.

Optical sensors may also reveal the occupancy in public areas and subsequently aid in determining if large equipment has toppled or slid and buried occupants. This would be particularly helpful in guiding earthquake reconnaissance teams to areas of distress while removing the need to survey unoccupied areas. Figure 8 shows an example of a failed ceiling element that may have buried several occupants. This photograph was taken at the Instituto Politecnico National in Mexico City after the Mexico earthquake in 1957. An optical sensor may be able to tell if people were present in the area now covered by the ceiling element.

Additionally, using optical sensors, a blocked escape route can at least be recognized and occupants can be advised to use another escape route. In order to keep escape routes clear the Federal Emergency Management Agency [17], for instance, recommends to not use certain materials such as hollow clay or unreinforced masonry around stairs, elevators, and corridors. However, it is well known that these materials are used in such areas nonetheless.

Telecommunication is another important area where optical sensors may assist in evaluating and minimizing seismic risk, particularly due to the strong linkage with today's economy. A loss of communication services can be extremely costly on both a local and national scale. Consequently, backup systems are generally provided for added redundancy. Depending on the type of services provided, the backup system is either (i) activated after failure of the main system is detected or (ii) operated in parallel. Optical sensor networks can also assist in identifying failures within a telecommunication system. This information may then be provided to early warning systems to prevent the entire system from failing. For example, if one monitors a communications rack with hundreds of cables connected to it, the movement of this rack can be tracked. If the movement exceeds a predetermined limit, a warning can be issued to maintenance staff and wires and cabling surveyed for potential damage.

Figure 7. Ruptured supply line pipe for a 3000 liter storage tank after the 2001 Nisqually earthquake (Image courtesy of Filiatrault et. al. [24]).

Figure 8. Ceiling of the eastern 4-story unit of the Instito Politecnico National – 1957 Mexico City earthquake (Courtesy of NISEE, University of California, Berkeley).

Another important example is related to potentially hazardous areas, such as biological or chemical science laboratories. These laboratories often contain hazardous chemicals, unrestrained and mounted on shelves or bench tops. Spillage of these chemicals may result in subsequent hazardous areas within the structure. To further exasperate this, fire may also result where these chemicals are exposed to each other or combined with leaking gas or other fluids. While other devices may be more usable for such tasks, optical systems can serve as a redundant or backup device in many instances. They can assist in detecting situations mentioned above and thus reduce hazards by turning of gas or other piping systems or warning people to prevent them from entering such hazardous areas. This facilitates a rapid method to inform the emergency personnel to be aware of such types of critical situations and be more attentive to primary hazard zones.

USING OPTICAL RECORDS FOR RISK ANALYSIS

An example of applying the optical record to risk analysis is shown in Figure 9. Part (a) of this figure shows one frame of a video stream captured during a shake table experiment. It highlights a glass container that originally resided on the shelf but falls down during shaking. Upon hitting the bench top, the container breaks as can be seen in the image. Different image processing techniques may be utilized to extract further details from this image. By subtracting a reference image depicting the original arrangement from the final image it can be determined that there is a glass container missing in the upper shelf. One may also discover that there is additional broken glass on the bench top. Of course, a robust thresholding is necessary to avoid false alarms.

(a)

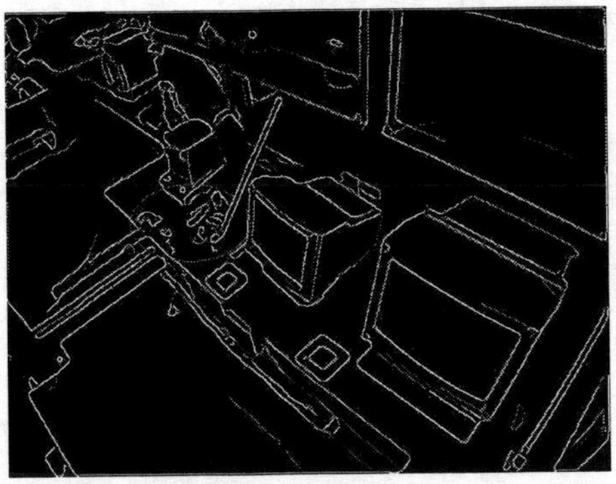

(b)

Figure 9. Glass container hitting the bench top during one of our shake table experiments; (a) shows the original image while (b) detects the breaking glassware using an edge detection algorithm.

Continuous monitoring using edge detection algorithms can aid in revealing this situation, as shown on in Figure 9(b). The path of the container can be tracked and its changing physical shape observed. In this case, even the individual parts of the broken glass container can be tracked using the video stream. Consequently, it is possible to determine that this container is broken. In case of any dangerous material being stored in this container, a hazard warning can be issued by the system automatically pinpointing that a potential risk exists in this area.

COMPARING OPTICAL TECHNIQUES WITH OTHER METHODS

Techniques available for monitoring the interior of building structures during earthquake motions, thus providing data for risk analysis, include primarily discretely located analog sensors (wired or more recently wireless, force, displacement, acceleration, etc.). It may be difficult with such sensors (primarily due to the mass of the sensor itself) to mount and detect the damage to the glassware described in the previous section. However, if such an incident results in secondary hazards (such as a fire), smoke detectors will reveal the situation. With conventional sensors it is not possible to determine the reason for this hazard nor it is possible to estimate the risk because it is not known why the fire occurred. The continuous monitoring approach proposed in this paper can provide detailed information about what exactly caused the hazard. This allows the system attached to the sensors to provide a more detailed risk analysis.

IMPENDING CHALLENGES

Several challenges for using optical sensors during earthquake events exist, such as the shaking of the sensor itself, the large amount of data, or the lighting condition. Since the sensor, as part of the scene itself, is also shaking during an earthquake, all captured images need to be aligned before making any assertions about object movement within the scene. Therefore, some type of reference point in the scene is necessary. Besides movement of objects, a building structure after an earthquake experiment can be investigated. It is very important to understand the vulnerability of the interior in such a structure. Several cameras can be applied to monitor the experiment and analyze the behavior of both structural and non-structural elements inside that building.

The use of optical sensors at high frame rates to monitor nonstructural elements or other building equipment result in a huge amount of data, even when low-resolution images are acquired. Ideally, this information should be processed in real-time. In other words, the system must be fast enough to estimate the whole scene captured by the sensor at almost the same time the images are taken. In addition, most optical sensors require robust lighting conditions. During nighttime or when building lights are not available, acquiring images with an optical sensor may be difficult. Night vision devices may resolve this problem. Finally, architects and designers need to be convinced that such systems are useful in emergency situations and that the benefits outweigh the additional costs.

SUMMARY

In this paper, potential areas in which optical (camera-based) technologies can assist in risk assessment were identified. Several field application examples are described, largely encompassing the areas of civil infrastructure monitoring. Since optical (camera-based) systems can be particularly powerful for monitoring both local and global movements within a scene or environment, natural hazards such as earthquakes are an important application area for these new technologies. Two important areas where optical sensors may be useful in seismic risk assessment are (i) early warning systems and (ii) post-earthquake assessment. Specific examples include: automatically shutting certain systems down to reduce the hazard, avoiding hazardous situations altogether, or warning people or rescue personnel of such impending hazards.

ACKNOWLEDGMENTS

We gratefully acknowledge the support of the Consortium of Universities for Research in Earthquake Engineering and Kajima corporation (CUREE-Kajima phase V). In addition, the helpful comments of project manager Dr. Katsuhisa Kanda are appreciated. Support for the shake table experiments shown in Figure 5 and 9 was provided in part by the Earthquake Engineering Research Centers Program of the National Science Foundation, under Award Number EEC-9701568 through the Pacific Earthquake Engineering Research Center (PEER).

REFERENCES

[1] D. Comaniciu and V. Ramesh. Robust detection and tracking of human faces with an active camera. In *Third IEEE Workshop on Visual Surveillance, Dublin, Ireland*, pages 11–18, 2000.

[2] W Bell, P. F. Felzenszwalb, and D. Huttenlocher. Detection and long term tracking of moving objects in aerial video. *http://www.cs.cornell.edu/vision/wbell/identtracker/default.html*, 1999.

[3] M Irani and P. Anandan. A unified approach to moving object detection in 2d and 3d scenes. *IEEE Transactions on Pattern Analysis and Machine Intelligence*, 20(6):577–589, 1998.

[4] D. Koller, J. Weber, and J. Malik. Robust multiple car tracking with occlusion reasoning. In *Third European Conference on Computer Vision*, LNCS 800, pages 189–196. Springer-Verlag, 1994.

[5] D. Beymer, P. McLauchlan, B. Coifman, and J. Malik. A real-time computer vision system for measuring traffic parameters. In *Computer Vision and Pattern Recognition*, pages 495–501, 1997.

[6] E. Aktan, F. N. Catbas, K. Grimmeslamn, M. Pervispour, J. Curtis, K. Shen, , and X. Qin. Sensing communication, computing and information systems for infrastructure health, performance and security monitoring. pages 1–16, 2002. Di3 position paper.

[7] C. Mair and S. Fararooy. Practice and potential of computer vision for railways. In *IEE Seminar Condition Monitoring for Rail Transport Systems (Ref. No.1998/501)*, pages 10/1–3, 1998.

[8] D. L. Magnus. Non-contact technology for track speed rail measurement (orian). In *SPIE Proceedings*, volume 2458, pages 45–51, 1995.

[9] G. S. Bachinsky. The electronic bar gauge (a customized optical rail profile measurement system for rail grinding applications). In *SPIE Proceedings*, volume 2458, pages 52–63, 1995.

[10] H. Sasama, M. Ukai, and Y. Okimura. The development of rail-gap inspection system. *Quarterly Report of RTRI*, 32(1):21–29, 1991.

[11] M. Ostermeyer. Berührungsloses Messen der Fahrdrahtruhelage mit hoher Geschwindigkeit. *Elektrische Bahnen*, 81(11):343–348, 1983.

[12] J. M. Van Gigch, C. Smorenburg, and A. W. Benschop. The contact wire thickness-measuring system (aton) of the netherlands railways. In *Rail International*, pages 20–31, 1991.

[13] X. Zhang and G. Sexton. Automatic pedestrian counting using image processing techniques. *Electronic Letters*, 31(11):863–865, 1995.

[14] L. Khoudaour, J. P. Depairs, and L. Duvieubourg. Linear image sequence analysis for passengers counting in public transport. In *International Conference on Public Transport Electric Systems*, volume 425, pages 100–104. IEE Conference Publication, 1996.

[15] C. Ottonello, M. Peri, C. Regazzoni, and A. Tesei. Integration of multisensor data for overcrowding estimation. In *Proceedings of the IEEE International Conference on Systems, Man, and Cybernetics*, pages 791–796, 1992.

[16] U.S. Geological Survey. remote video monitoring project. http://coastal.er.usgs.gov/rvm/, 1991.

[17] FEMA. *Prestandard and commentary for the seismic rehabilitation of buildings – FEMA 356*. Federal Emergency Management Agency, prepared by ASCE, Reston, Virginia, 2000.

[18] J. Canny. A computational approach to edge detection. *IEEE Trans. On Pattern Analysis and Machine Intelligence*, 8(6):679–698, 1986.

[19] H. Campbell. Risk assessment: subjective or objective? *Engineering Science and Education Journal*, pages 57–63, 1998.

[20] D. Todd, N. Carino, R. M. Chung, H. S. Lew, A. W. Taylor, W. D. Walton, D. J. Cooper, and R. Nimis. 1994 Northridge earthquake – performance of structures, lifelines, and fire protection systems. NIST publication 862, National Institute of Standards and Technology, 1994.

[21] International Conference of Building Officials. *Uniform Building Code*. Whittier, CA, 1997.

[22] International Code Council. *International Building Code 2000*. Falls Church, VA, 2000.

[23] National Informtaion Service for Earthquake Engineering. National Information Service for Earthquake Engineering, Berkeley (NISEE). http://nisee.berkeley.edu.

[24] A. Filiatrault, C. Christopoulos, and C. Stearns. Guidelines, specifications, and seismic performance characterization of nonstructural building components and equipment. Structural Systems Report Series Report no. SSRP-2001/13, University of California, San Diego, 2001.

Proceedings of IMECE'03
2003 ASME International Mechanical Engineering Congress
Washington, D.C., November 15–21, 2003

IMECE2003-42567

Using Risk Analysis to Identify Cost Effective Improvements to a Semiconductor Fabrication Chemical Distribution Facility

Fred L. Leverenz, Jr. / Pacific Northwest National Lab

ABSTRACT

In today's semiconductor industry, the business loss associated with one full day of lost production can have a significant impact on revenues. Key factors affecting overall profitability are unplanned fabrication downtimes and contamination resulting in defective products. It is critical, therefore, to implement effective risk management strategies for the fabrication process to eliminate yield-limiting defects and to improve the overall production.

This paper will discuss a study completed for a semiconductor chemical supplier. Battelle applied a methodology based on the Process Event Tree/Hazard and Operability Study (PET-HAZOP) methodology, identifying numerous ways to avoid unnecessary outages and lost production at a specific client's semiconductor fabrication facility. This methodology uses a systematic approach to identifying risks in a process, and risks are quantified in terms of expected dollars per year. Within the analysis, the importance of individual equipment items in the process is measured with respect to risk. Two risk measures are used in the analysis: Risk Reduction Worth (RRW) and Risk Achievement Worth (RAW). The RRW identifies opportunities for areas of risk reduction and the RAW identifies critical equipment that must be maintained to prevent the overall risk of the system from increasing significantly.

Results were provided in a cost-benefit form for the top items on the RRW and RAW lists. For example, replacing existing chemical distribution pumps with completely non-metallic pumps (cost of about $60,000) would result in savings, due to avoidance of possible chemical contamination, of $430,000 per year. This, and other findings, proved that a systematic approach to evaluating risks for the semiconductor front-end process can identify valuable risk reduction alternatives resulting in significant cost savings.

INTRODUCTION

An Operability Risk Assessment was conducted on the chemical management operations at a semiconductor manufacturing facility. The purpose of this assessment was to determine the impact of failures of various components of the chemical management operations, such that contaminated chemicals were introduced into the semiconductor manufacturing process. In this study, the focus was on the bulk sulfuric acid system.

The analysis was conducted using combination of event-trees and the HAZOP study method, termed the Process Event Tree based Hazard and Operability Study, or PET-HAZOP (Leverenz, 1996). The analysis also used Importance Measures to help identify the relative importance of equipment as well as operational activities.

The analysis addressed the bulk sulfuric acid hardware, including storage, distribution and control systems. It also addressed key procedures, such as receipt of sulfuric acid and key maintenance activities.

The study focus was to determine the cost-benefit of postulated improvements in hardware and operational changes. In some cases, these changes increased the risk of contamination, but resulted in a cost savings.

SCOPE OF ANALYSIS

The operational risk analysis was conducted on the bulk sulfuric acid system of the semiconductor fabrication facility. This chemical was chosen due to its high usage and impact on fabrication. The system analyzed in the study included the following:

- Bulk sulfuric acid tank and recirculation loop
- Model FSI-1000 Chemical Distribution Unit (CDU)
- Flow from FSI-100 through recirculation loop back to the bulk sulfuric acid tank

- Drum Filling
- Model FSI-4000 Chemical Distribution Unit (CDU)
- Booster station Chemical Distribution Unit (CDU)
- Distribution valve boxes.

In addition to the hardware, several operating procedures were evaluated as part of the analysis. The procedure-based portion of the analysis addressed all procedural actions that could result in process deviations not already addressed in the equipment-based portion of the study. Procedures were selected for evaluation based on the potential for human error during the procedure causing contamination of the process or impacts to worker safety. The following procedures were evaluated:

- Bulk unload procedure for sulfuric acid
- PLC 1000 filter replacement
- Pump replacement procedure
- Chemical Dispense Unit valve replacement procedure.

ANALYSIS ASSUMPTIONS

Several assumptions were necessary for consistency during the study. These assumptions were:

- The business loss associated with one full day of lost production in the fabrication unit was assumed to be one million dollars. Downtimes of less that one full day were estimated at incremental values of this amount (i.e., 12 hours of lost production was valued at $500,000.)
- Failure modes of tank sampling included: analytical results are incorrect (the results indicate chemical is good when it is contaminated), samples get mixed up during analysis, and the sample is overlooked (no sample taken).
- When evaluating the failure of valve internals such that the sulfuric acid is contaminated, it was assumed that there was one chance in 5000 that a valve fails and will contain enough contaminates to affect the process.
- In the event of a filter failure, it was assumed that the fabrication monitors particulate on each lot and will be aware of high particulate content after one lot has been processed.
- When determining cause frequencies in the CDUs, the frequencies were doubled to account for the presence of two CDUs in the process.

STUDY TECHNIQUE AND APPROACH

The analysis methodology used in this study was the Process Event Tree Based Hazard and Operability (PET-HAZOP) methodology. The PET-HAZOP study combines the concepts of the HAZOP study with an Event Tree Analysis (AIChE, 1992). The brainstorming/guide word approach of the HAZOP study is used to determine initiating events, or causes, to investigate. Given a particular cause, the concepts of the Event Tree are used to determine the relevant set of scenarios. This combination allows the analysis to consider both the risk and the potential for risk reduction actions. The resultant scenarios are recorded in the standard HAZOP format (Causes, Consequences, Safeguards, Actions).

For each scenario, the risk associated with that scenario was estimated on an order-of-magnitude basis. The risk estimate was formed by assigning order-of-magnitude scores to the initiating events, consequences and safeguards. Table 1 provides the Cause scores and their meanings.

Score	Description
1	Expected to occur periodically or frequently (10 times/yr)
0	Expected to occur on an infrequent basis (once/yr)
-1	May occur a couple of times during facility life (1/10yr)
-2	Unlikely to occur during facility life (1/100 yr)
-3	Very unlikely to occur during facility life (1/1,000 yr)
-4	Extremely unlikely to occur during facility life (1/10,000 yr)

Table 1: Scenario Cause Score Definition

Engineering and administrative controls applicable to the process hazards were documented as Safeguards. These controls are intended to prevent or minimize the consequences once the cause has occurred. For each scenario, the level of protection for these controls is estimated using the scores in Table 2.

Score	Description
0	No safeguards
1	Safeguards require single operator response
2	Safeguards utilize a single train of hardware
3	Safeguards utilize a single train with operator backup, or a passive emergency relief device.
4	Safeguards utilize at least two redundant, independent hardware trains.

Table 2: Scenario Safeguard Score Definition

For each scenario, the severity of the consequence is estimated and included in the scenario. Two consequence estimates were assigned to each scenario, one for worker safety and a second for business loss. The consequence estimates were determined for each scenario by the team, considering the following:

- The number of workers expected in the area when the accident occurs
- The expected activities at the time of the accident (i.e., maintenance)
- The proximity of the failure to other equipment
- The speed with which the accident is expected to develop.

Each scenario was assigned a worker safety consequence score based on the categories listed in Table 3.

Score	Severity Descriptor	Unlikely, but Possible Effect on One Person Onsite (About 10% chance)	Effect Reasonably Expected to Occur to 1 or 2 On-site Personnel	Effect Reasonable Expected to occur to 5 to 20 On-Site Personnel
6	Extremely High	⇨	⇨	Fatalities
5	Very High	⇨	Fatality	Immediate impairment; permanent health effects
4	High	Fatality	Immediate impairment; permanent health effects	Severe injury requiring hospitalization; significant lost time.
3	Intermediate	Immediate impairment; permanent health effects	Severe injury requiring hospitalization; significant lost time.	Injury requiring medical treatment
2	Low	Severe injury requiring hospitalization; significant lost time.	Injury requiring medical treatment	Minor injury; first-aid case
1	Very Low	Injury requiring medical treatment	Minor injury; first-aid case	Probably none
0	Negligible	Minor injury; first-aid case	Probably none	None

Table 3: Scenario Consequence Score Definition for Workers

The business impact from each scenario considered the following factors:

- The value of the process materials
- The expected level of equipment damage
- The expected downtime for ordering replacement equipment and/or repair
- The expected amount of time before contamination is identified.

The scoring system for business loss used a logarithmic scale. Fractional scores were used where necessary to more accurately reflect the economic consequences associated with a particular scenario. Table 4 shows the economic consequence scores.

Score	Descriptor	Economic Impact: Equipment Damage or Production Loss
6	Extremely High	$1,000,000
5	Very High	$100,000
4	High	$10,000
3	Intermediate	$1,000
2	Low	$100
1	Very Low	$10
0	Negligible	None

Table 4: Scenario Consequence Score Definition for Economic Loss

The total consequence score was estimated as the largest score of either the worker consequence or the economic loss. In all cases the economic loss dominated the consequence. The overall scenario risk (termed the Risk Assessment Code or RAC) was calculated by adding the scenario Cause score and the Consequence score, then subtracting the safeguard score.

Since the scores are the logarithm of the values, the RAC can be interpreted as actual risk in dollars per year. For example, an RAC of 2 represents a risk of $100 per year. Further, the definition of risk used in this analysis is:

Risk = (Cause Frequency) x (Protection Failure Probability) x (Consequence)

In addition to evaluating the risk of each scenario and the total risk, Importance Measures were utilized. Importance Measures (Lambert, 1975 and Vesley, 1983) were developed as a way to add understanding to the results from risk and reliability studies. While they are based on numerical calculations, the numerical values associated with the results have no direct meaning. There are many Importance Measures that could be used; however, two were believed to be the most useful for this study. These two are Risk Reduction Worth and Risk Achievement Worth.

Risk Reduction Worth. Risk Reduction Worth (RRW) is intended to provide the decision-maker with information on how to most effectively reduce risk. When the causes and safeguards associated with accident scenarios in a facility are ranked by using the RRW, the decision-maker has insight as to which causes should be made less likely, and which safeguards should be made more reliable, given a need to reduce risk. The relative rankings provided by the RRW analysis should be considered as approximate in that there may be little difference in items close to each other on the list.

However, there is a greater assurance in the impact of taking action with respect to items that are widely separated on the list. The RRW may be expressed in equation form as:

RRW = [Risk of the system as-is] / [System risk given an item of interest can be made perfect]

Risk Achievement Worth. Risk Achievement Worth (RAW) provides the decision-maker with information on how to most effectively keep the risk from increasing. In this case, the decision-maker would learn which causes and/or safeguards should receive the most attention in test and maintenance to prevent degradation in the risk from the current level. The RAW may be expresses in equation form as:

RAW = [System risk given an item of interest is failed] / [Risk of the system as-is]

ANALYSIS RESULTS

The analysis was completed in two basic steps. First a baseline risk was completed from the scenarios. Next, cost-benefit analyses were completed for key situations identified by the team.

The RAW and RRW were calculated for all pieces of equipment included in the PET-HAZOP. When these results were obtained, the team reconvened to determine which equipment items warranted further study. The team chose seven changes for cost-benefit analysis the RAW and RRW as well as specific operational considerations. The seven postulated changes investigated were:

- Backup drums of sulfuric acid are no longer readily available, and cannot be considered as a safeguard in the event of a failure of the bulk sulfuric acid system.
- Drums of sulfuric acid are stored on-site, thereby improving the reliability of the backup drums as protection
- A system is in place that automatically notified the operators when a CDU drum level was low for more than one hour, indicating that there is a problem in automatic drum filling
- Pumps are bench tested after rebuilds prior to installation.
- The Backup ChemFill computer is no longer available, and cannot be considered a protection in the event of a failure of the Master ChemFill computer.
- Non-metallic pumps are installed that do not contaminate the process when an internal pump leak occurs. (However, it was determined that contamination may still occur (although it would be less likely) if the leak continued and leaked through to the valve shuttle.)
- An in-line contamination detector is installed at the outlet of the CDU booster station to detect contamination before contaminated materials reached the fabrication facility. (No such detector was known to be available; however, its value was of interest so an arbitrary reliability of 50% was assumed.)

Table 5 provides the values associated with the seven postulated situations.

Scenario	Estimated Risk $/yr	Cost Savings (Increase), $/yr
BASELINE RISK	$2,087,220	-
Backup drums stored on-site	$2,057,784	$29,435
Automatic operator notification of low drum level	$1,952,724	$134,496
Pump bench testing procedure after rebuilds	$2,069,868	$17,351
Installation of non-metallic pumps	$1,656,544	$430,675
Installation of an inline contamination detector	$1,818,513	$268,706
Backup drums no longer available	$5,322,613	($3,235,393)
No backup ChemFill computer	$4,062,529	($1,975,310)

Table 5: Risk Estimates for Situations of Interest

Once the cost savings (or in some cases increases) were calculated, an attempt was made to come up with a rough estimate of the costs associated with the implementation of these situations. The comparisons are shown in Table 6. With the exception of on-site storage of backup sulfuric acid drums, the cost savings for all of the situations evaluated outweighed the estimated costs. No cost was estimated for the installation of an in-line contamination detector, as the availability of such a device was unknown.

Scenario	Estimated Cost	Cost Savings (Increase), $/yr
Backup drums stored on-site	$100,000	$29,435
Automatic operator notification of low drum level	$10,000 - $15,000	$134,496
Pump bench testing procedure after rebuilds	$5,000	$17,351
Installation of non-metallic pumps	$30,000 - $60,000	$430,675
Installation of an inline contamination detector	???	$268,706
Backup drums no longer available	?¢/lb (small)	($3,235,393)
No backup ChemFill computer	$5,000	($1,975,310)

Table 6: Implementation Costs versus Savings

ANALYSIS CONCLUSIONS

The following conclusions were provided to the facility management, which resulted in related actions.

- Both the backup drums and the backup ChemFill computer are very important to avoid increasing the risk beyond any savings of eliminating these safeguards.
- The greatest reduction in risk can be achieved by the installation of non-metallic pumps.
- When the addition of an inline contamination detector was analyzed, it appeared at the top of the RAW and RRW lists. This indicates that if an inline detector were available, it would be very important to maintain its reliability. Additionally, at only 50% reliability, increasing the reliability of such a device would greatly reduce the risk. If such a device could be found that is 99% reliable, the savings would be over $500,000/ year.

While this study resulted in direct payback to the facility operator in terms of actions to reduce operational risk, it also demonstrates the value of using what is generally considered a qualitative study to make good quantitative decisions. In this case, order-of-magnitude estimates of risk were used to discriminate between a number of situations at the facility, with clear results. At a lower level, the study allow an estimation of the value of an non-existing safeguard simply by inserting this safeguard into the scenarios that would be impacted if it did exit.

ACKNOWLEDGMENTS

As with any team approach to a hazard analysis, the key to achieving a high quality result is having an experienced team leader and a knowledgeable team. Hence, the author acknowledges the entire team involved in this study. Besides the team leader and scribe from Battelle, Ashland Chemical Company's Chemical Management Services Division staff working at their client's fabrication facility made up this team, and facilitated the valuable results. In addition it was Ashland's foresight that initiated the study to look for ways to improve their services to their clients.

REFERENCES

Leverenz, Fred. L. and G. Bradley Chadwell, 1996. "Process Event Tree Based HAZOP Studies," Process Plant Reliability Conference, Organized by Gulf Publishing Company and HYDROCARBON PROCESSING, November 11-13, 1996.

AIChE, 1992. *Guidelines for Hazard Evaluation Procedures, Second Edition with Worked Examples.*

Lambert, H.E., 1975. "Measures of Importance of Events," in *Reliability and Fault Tree Analysis*, ed. R. E. Barlow, J.B. Fussell, and N.D. Singpurwalla, SIAM Press, Philadelphia, pp 77-100.

Vesley, W.E. et. Al, 1983. "Measures of Risk Importance and Their Application," NUREG/CR 3243, Nuclear Regulatory Commission.

Proceedings of IMECE'03
2003 ASME International Mechanical Engineering Congress
Washington, D.C., November 15–21, 2003

IMECE2003-42761

COMMON COMPONENTS OF HUMAN ERROR IN DESIGN, MAINTENANCE, OR OPERATIONS

Joseph P. Balkey
1330 Wilkie Drive
Charleston, WV
304/345-1699
JPBASB@aol.com

ABSTRACT

Often, public reports of accidents only identify the last, obvious failure or immediate cause of the accident. If human error is the immediate cause or final failure, further assessment of accident contributors may stop, and an enhanced training program is often determined to be the primary solution for preventing further accidents of this type. However, in many cases, the accident is the final result of many inputs, decisions, actions and inactions.

To demonstrate this characteristic of accidents, the 20 stories in a publication titled "Set Phasers on Stun" have been categorized into action errors and planning errors that involve designers, mechanics, or operators. For each story, the hazard and the number of simultaneous failures are listed. Then two of the 20 stories are assessed in detail; one story involves an action error and the other one involves a planning error. In each of these two stories, the system is first described as it should operate and then its risk is quantitatively assessed to identify findings, lessons learned, recommendations, analogies to the other 18 stories, and applications.

This paper has three immediate goals. One, to recognize the difference between an action error and a planning error. Two, to recognize that most accidents involve 2 to 4 simultaneous failures. Three, to appreciate that quantifying the failure frequency serves two benefits. Because it is usually difficult to find out exactly what happened after an accident, the calculated frequency can help confirm what actually happened. When various alternatives are recommended, it can also help to select the most economic ones.

This paper has two long term goals. One, consider assessing the failure rates of near misses. By reducing near misses, larger accidents will be reduced. Two, consider assessing the failure rates of personal near misses because you know what actually happened.
Keywords: human error, accident investigation, risk analysis.

INTRODUCTION

Recently I was invited to substitute for a professor who taught a human factors course at Marshall University. As part of his course format, his students qualitatively discussed a few stories in the book "Set Phasers on Stun" (1). When I taught the class, my students quantitatively assessed the stories. To my surprise, the quantitative analysis revealed far more lessons than I expected. Because many copies of this book have been sold since it was first published in 1993 and because many organizations are reluctant to state their failures publicly for analysis, I want to show how to assess such stories in this paper.

TERMINOLOGY

This section defines the terminology that is used throughout the paper. It begins with the term "risk" and then breaks it into components which are then broken into smaller ones.

Risk

In this paper, risk consists of two components: a hazard that has a consequence and a scenario that leads to the hazard at some frequency (2). Although risk is defined many different ways in the literature, to me, the results are the same.

Hazard consequence

This is an unwanted consequence.

Scenario frequency

A scenario (what can go wrong) involves mitigating and initiating events. To calculate the frequency (how likely), the mitigating events are probabilities but the initiating event is a frequency. The number of simultaneous mitigation and initiating events in these stories range from 2 to 4.

Latent failures. Happened some time ago and remain undetected. This is usually the design or maintenance person. The sum of these probabilities are less than or equal to unity.

Initiating event. Appears in the news because it was the last event that initiated the accident and is easily detected. This usually involves the operating person. The sum of these frequencies could be greater than unity.

A scenario usually involves combinations of instrument failures, structural failures, and people failures. Latent within the instrument and structural failures are human errors by designers and/or maintenance technicians. This paper will only address human errors.

Human Error Description

The two broad categories of human error are action errors and planning errors.

Action errors. People work toward a correct goal but fail along the way. These errors are further divided into failures of input, diagnosis, or response.

- Input and response errors are a function of the number of available choices (4, 6) or attention and memory failures (5).
- Diagnostic errors are a function of the available time to diagnose and the person's experience level (3).

Awareness. Knowing that a hazard can be life-threatening usually decreases the chance of an action error. A person is more cautious to avoid failure.

Recovery. When we error, feedback and/or another person can help us to correct our error before it can lead to a hazard. Types of feedback that help us correct mistakes are checklists, tests, and compelling. When we depend on one or more other people to correct our mistakes, they may not be as good as we think they are. The very next person may correct our mistake but the person after that is much less likely to correct our mistake.

Adjusters. For a given error probability, training may reduce it by half and stress may double it. In most of these stories, training and stress are unimportant.

Planning errors. People decide to head toward the wrong goal. This error is further divided into unintentional and intentional planning errors.

Unintentional planning error. This error is a mistake in solving a problem.

Awareness. Similar to an action error, awareness of a hazard may decrease the chance of an unintentional planning error.

Recovery. Feedback will decrease the chance of an unintentional planning error.

Intentional planning error. This error is either without prior intent or with prior intent.

Awareness. Unlike an action error, awareness of a hazard may NOT decreases the chance of an intentional planning error. The person acts without being more cautious.

Recovery. Feedback may NOT decrease the chance of an intentional planning error.

CATEGORIZE HUMAN ERRORS IN "SET PHASERS ON STUN"

When reading any of the 20 stories in "Set Phasers on Stun", which appear to be listed in random order, you may take them strictly for entertainment; "What do these stories have to do with me? I'm not an astronaut". However, if you picture yourself in any of these people's shoes, you may act just as they did. To abstract usefulness from these stories, I have identified what is common and what is different among these 20 stories in terms of hazards and scenarios. Table 1.0 summarizes the risks and errors of the 20 stories. The remainder of this paper quantitatively assesses "Never Cry Wolfe" and "Chutes and Ladders".

Commonalties

Most of the stories have one or more fatalities as the hazard consequence.

Most of the scenarios involve either action or planning errors that involve 2 to 4 simultaneous latent and initiating failures. Most of the latent errors involve the equipment designer who is omitted in the story. The initiating failures are either by an operator or a mechanic. When the hazard is experienced, the facility is starting-up, running, or idle during a turn-around.

Differences

"Double Vision" and "The Wizards of Wall Street" have hazards but the errors were recovered before they lead to the accident. The latent error that damaged the x-ray and the initiating misdiagnosis of film was corrected by another doctor before the patient was cut open.

"A Momento of Your Service" and "Business in Bhopal" involve both planning and action errors.

Although my knowledge is second hand for all of these stories, I do disagree with "Business in Bhopal". The book describes a non-prior intentional planning error having the mitigating equipment out-of-service during the turnaround and an action error by maintenance initiating the hazard. However, before this book was published, Arthur D. Little, Inc. presented a paper (8) proving that a prior intentional planning error by an employee initiated the hazard.

Result

Just categorizing the human errors helps to see analogies to our own situations.

ACTION ERROR ANALYSIS

The action error story analyzed is "Never Cry Wolfe". For readers who are unfamiliar with the story, a prisoner notices that the guards no longer respond to a motion detector alarm because birds and the wind set it off daily. The prisoner hopes that the guards will not respond to the sensor when it detects her escape. She was right!

When assessing this story, the system is first described as it is intended to operate and then its risks are described. The scenario frequency is quantitatively assessed using fault tree and human error analysis to deliver findings and lessons learned. Finally, changes are recommended, other stories in the book with similar failures are listed, and lessons learned are applied to individuals and industries.

Operating description (how system works)

Figure 1 shows two fences surrounding a jail. Between the fences, an alarm sounds when motion is detected and two or more guards respond to the alarm. The instrument mechanic is not mentioned in the story.

Figure 1: Jail

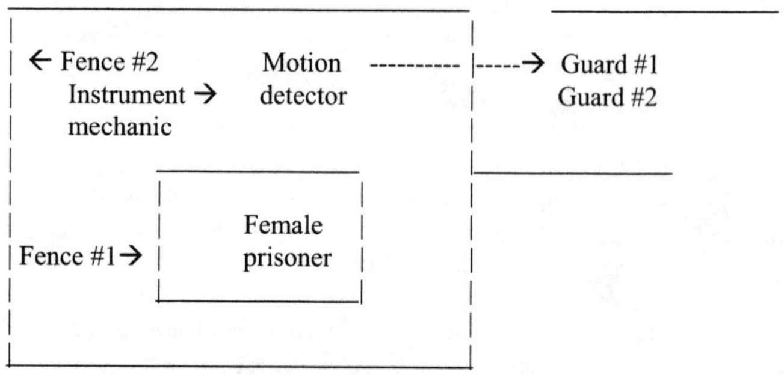

Risk description

Because risk consists of a hazard consequence and a scenario leading to the hazard at some frequency, the risk is described in these terms.

Hazard consequence. Female prisoner escapes.

Scenario (what can go wrong). An instrument mechanic inadvertently sets the motion detector alarm too low. Once a day the alarm detects flying birds and blowing grass. The alarm becomes unimportant to the guards. Thereby, they fail to diagnose the prisoner's escape when she initiates the alarm.

Frequency (how likely). The fault tree and human error analysis in Figure 2 calculates the frequency (f) of the above scenario to be 1.0E-2/year. Both guards fail to diagnose the alarm because it carries little information. The chance that the alarm will fail during the prisoner's escape is 2.7e-4 because the flying birds and blowing grass test it daily. If one of the guards diagnose that a prisoner escaped, the chance that both of them would fail to respond is 9.0e-6. The initiating event is the prisoner jumping the two fences with an assumed frequency of once per year.

The fault tree labels "AND" and "OR" gates with words rather than the standard symbols to overcome a software limitation.

Findings

Simultaneous failures. This story has three simultaneous mitigation failures: the latent error by the instrument mechanic which creates the nuisance alarm and the failure of guard #1 and guard #2 to diagnose the alarm. The nuisance alarm causes the diagnosis of both guards to fail.

Scenario frequency. The calculated failure rate of 1E-2/year does not agree with the prisoner's assessment of 1/year.

Lessons learned

This analysis offers five lessons. One, if you know something has failed, fix it as soon as practical; the longer a failure remains uncorrected, both the cost to correct it and the consequence increase. Because the alarm was not corrected, the alarm was transmitting less information to the guards each day and degrading their decisions. The cost to repair the failure increased from repairing the instrument to also retraining the guards and the cost of the consequence increased from the guards wasting response time to a prisoner escaping. Two, correcting an action error is relatively easy because you know what needs to be corrected. Three, there were three simultaneous failures. However, blame may have only been assigned to the guards because the instrument mechanic's latent error was omitted. Four, adding more guards or retraining them may not improve their decisions because the nuisance alarm was common to all of them. The degree of dependency among the guards was omitted. Five, because the failure frequency does not approximates the actual frequency, the fault tree analysis is incorrect. There is a flaw in the logic that needs to corrected.

Recommendations

Equipment and administrative changes reduce the scenario frequency but equipment changes usually reduce it more than administrative ones.

Equipment improvements. One, resetting the motion detector eliminates the many nuisance alarms. When it is reset, it should be tested. Two, installing a security camera reduces the guards' diagnostic error. However, installing a second motion detector may not reduce the guards' diagnostic error if the instrument mechanic sets both motion detectors too high. Installing an electric fence will kill birds.

Administrative improvements. One, more training may help the instrument mechanic. However, assess the motion detector's design first. Two, hiring more guards will not reduce the diagnostic error. Three, motivating the guards to quickly respond to every alarm or training them further have low benefit to cost ratios.

Analogies and applications

Diagnosis errors appear in other stories in the book and the lessons learned apply to both individuals and industries.

Other stories. In "Set Phasers on Stun", a technician corrects an error but lack of feedback prevents her from diagnosing that equipment had failed. In "Double vision", a defective x-ray is misdiagnosed by one doctor but not by a second doctor. In "An Act of God", a technician misdiagnoses power failures to Manhattan Island.

Applications. Some of the lessons learned apply to individuals. If you do not repair a leaking roof, the cost to repair the damage increases. Fasten your seat belt before starting your car engine rather than afterwards to avoid sounding the alarm. Some of the lessons learned apply to industries. If a control panel has a nuisance alarm, either remove it if it is unnecessary or fix it quickly. If an organization has a problem employee, either help them or remove them quickly.

PLANNING ERROR ANALYSIS

The intentional planning error story analyzed is " Chutes and Ladders ". For readers who are unfamiliar with the story, labors are being electrocuted when they raise one end of a long aluminum water line and inadvertently contact an overhead power line. They intentionally raise one end not as part of relocating the pipe but to clear rabbits out of it.

When assessing this story, the system is first described as it is intended to operate and then its risks are described. The scenario frequency is assessed using fault tree and human error analysis to deliver findings and lessons learned. Finally, changes are recommended, other stories in the book with similar failures are listed, and lessons learned are applied to individuals and industries.

Operating description (how system works):

Figure 3 shows that during dry spells, after an area in a field has been watered, farm workers disassemble a long irrigation pipe, lift and carry it to a new area to be watered, and then reassemble it. The pipe designer is not mentioned in the story.

Figure 3: Farm workers move pipes

```
        6 power lines →          ------------------------
38 foot aluminum pipe            |                      |
--------------------------       | ← pole               |
    Farm          Pipe           |             pole →   |
    worker        designer
```

Risk description

Because risk consists of a hazard and a scenario leading to the hazard, the risk is described in terms of hazard and scenario.

Hazard consequence. Farm worker is electrocuted.

Scenario (what can go wrong). A farm worker intentionally lifts one end of a long irrigation pipe into the air, it inadvertently contacts a power line, and the aluminum conducts the electricity.

Frequency (how likely). The fault tree and human error analysis in Figure 4 calculates the frequency (f) of the above scenario to be 0.7/year. About 10% of the 135 pipes moved per year are raised to clear rabbits. Of these, about 5% contact a power line, conduct electricity, and kill the farm worker.

The fault tree labels "AND" and "OR" gates with words rather than the standard symbols to overcome a software limitation.

Findings

Simultaneous failures. This story has three simultaneous failures: two latent error by the pipe designer who specified the pipe's conductive material and sufficient length to contact the power line and the initiating intentional planning error by the farm worker who raised the pipe.

Scenario frequency. The calculated failure rate of 0.7/year appears to be in the range of actual fatalities.

Lessons learned

This analysis offers four lessons. One, correcting a planning error is extremely difficult because it could require changing motivation. The farm worker intentionally raised the irrigation pipe into the air without fear of the electrical hazard. Two, because the failure frequency approximates the actual frequency, fault tree and human error analyses can help to confirm what happened and to assess the value of any recommendations. Investigators initially thought the farm worker's mistake was an action error. It took a big effort to assess that the farm worker's mistake as an intentional planning error. Three, to what extent is an engineer accountable for how a product can be misused? Because the failure was very difficult to identify, the pipe designer did not know how the pipe would be misused. Four, if you are associated with someone making an intentional planning error, because you can not correct them, your only alternatives may be to remove yourself from the situation or to wait for external forces to change the situation.

Recommendations

Equipment and administrative changes reduce the scenario frequency but equipment changes usually reduce it more than administrative ones.

Equipment improvements. One, changing the pipe's material of construction from conductive to non-conductive eliminates the risk. Because this will be expensive, it is unlikely to be done. Two, install a temporary barrier (e.g. screen) at each end of the pipe to prevent rabbits from entering the pipe.

Administrative improvements. One, hiring guards to watch the farm workers will reduce the risk but at a high expense. Two, training the farm workers will not reduce the risk at a high expense.

Analogies and applications

Intentional planning errors with no prior intent appear in other stories in the book and the lessons learned apply to both individuals and industries.

Other stories. In "A Momento of your Service", a ship captain intentionally decides to sail between rocks versus around them. In "Leap of Faith", an airplane pilot intentionally decides to bypass the "emergency shutdown device". In "Silent Warning", a company intentionally decides to ship grain that has been treated with toxic alkylmercury.

Applications. Some of the lessons learned apply to individuals. For example, intentionally not wearing a seat belt even though the consequence is known. Some of the lessons learned apply to industries. A budget is intentionally reduced to delay maintenance even though the consequence is known. Industry in a state that intentionally raises taxes without cutting spending may eventually leave the state.

CONCLUSION

More training is usually an excuse not to improve on underlying endemic (organizational) failures.

Designers and maintenance need to work with operators. Because two or more failures lead to a hazard, the initiating errors by operators were set up by a designer's latent error.

Your third alternative recommended is usually better than the first two. Whenever you recommend "more training", think again. A deeper analysis may reveal a better and less expensive alternative.

Fault tree and human error analysis can be used to confirm the experienced frequency and rank recommended improvements.

REFERENCES

1. Casey, Steven M., "Set Phasers on Stun", Aegean Publishing Co., 1998.
2. McCormick, Norman J., "Reliability and Risk Analysis", Academic Press, Inc., 1981, pages 231, 235
3. Swain, A.D and Guttmann, H.E., "Handbook of Human Reliability Analysis with Emphasis on Nuclear Power Plant Applications", NUREG/CR-1278, August 1983, pages 11-51, 12-20,
4. Norman, Donald, "The Design of Everyday Things", Doubleday, 1988.
5. Reason, James, "Human Error", Cambridge University Press, 1990.
6. Best, John B., "Cognitive Psychology", West Publishing Co., 1992.
7. Sanders and McCormick, "Human Factors in Engineering and Design", McGraw Hill, 1993, page 49.
8. Kalelkar, Ashok S., "Investigation of Large-magnitude Incidents: Bhopal as a Case Study", The Institution of Chemical Engineers Conference on Preventing Major Chemical Accidents, London, England, May 1988, paper.

Figure 2: Action error fault tree

AND | f= 1.0 E-2/yr

Prisoner jumps fences after three false alarms
AND | f= 1.0 E-2/yr

- Guards fail to mitigate 1st false alarm (see F2a) 0.02
- Guards fail to mitigate 2nd false alarm (see F2b) 0.505
- Guards fail to mitigate 3rd false alarm (see F2c) 0.9901
- Prisoner jumps fences f= 1/yr

Guards fail to mitigate 4th alarm
OR | 1.0

- Alarm fails when set low (note 2) 2.7e-4
- Guards fail to diagnose in 5 min
  AND | 1.0
  - Guard #1 fails to diagnose nuisance alarm (note 3) 1.0
  - Guard #2 fails to diagnose nuisance alarm (note 3) 1.0
- Guards fail to respond (note 1) 9.0e-6

NOTES:

1. Two guards fail to respond after diagnosing that prisoner escaped = 9.0e-6 = 0.003 x 0.003. (3)
2. Because the instrument detected the slightest motion, the instrument mechanic inadvertently set the motion detector too low and failed to test it. However, after the instrument started operating, it was tested daily when a bird would trip it.
   Latent error = 2.7e-4 = 0.33 x 8.2e-4 = 0.33 x 0.3/year x (1 day test interval x year/365 day)
   0.33 is chance of correct setting without a test
   0.3/year is failure rate of motion detector
3. Frequent sounding alarm carries very little information using information theory (7) to both guards.
   Alarm = 0.9992 x log (1/0.9992) / log 2 = 0.9992 x 0.001 = 1.0e-3 bits
   No alarm = 0.0008 x log (1/0.0008) / log 2 = 0.0008 x 10.300 = 8.4e-3 bits
   9.4e-3 bits

F2a

Operators fail to mitigate 1st alarm
AND | 0.02

- 1st bird trips alarm 1.0
- Mitigation fails
  OR |
  - Alarm fails (Note 2) 2.7E-4
  - Decision fails **0.01**
  - Response fails 0.01
- No prisoner found 1.0

F2c

Operators fail to mitigate 3rd alarm
AND | 0.9901

- 3rd bird trips alarm 1.0
- Mitigation fails
  OR |
  - Alarm fails (Note 2) 2.7E-4
  - Decision fails **0.99**
  - Response fails 0.01
- No prisoner found 1.0

F2b

Operators fail to mitigate 2nd alarm
AND | 0.505

- 2nd bird trips alarm 1.0
- Mitigation fails
  OR |
  - Alarm fails (Note 2) 2.7E-4
  - Decision fails **0.5**
  - Response fails 0.01
- No prisoner found 1.0

Figure 4: Planning error fault tree

```
                                   AND      | f= 0.7/yr
          ---------------------------------------------------------------------------
          |                                              |                      |
     Pipe rises vertically                        Pipe contacts           Aluminum pipe
     into the air                                 a power line            conducts electricity
        (initiator)                               (latent error)          (latent error)
     AND   | f= 14/yr                              AND | 0.05                 1.0
        --------------------                     -----------------------
        |                 |                      |                  |
  | Move  |        Walk off pipe               Power line        Pipe length
  |_pipe__|        (planning error)            in the area       sufficient
  (note 1)        (intentional -                 0.1               0.5
  f= 135/yr        no prior intent)
                 AND | 0.1
                 ----------------------
                 |                    |
           Walk-off pipe          Fail to
             (note 2)           fear hazard
               0.1                  1.0
```

Notes:

1. 135/yr = 0.3 dry spell x 10 pipes/2 day x 3 month crops grow/year x 30 day/month
2. To remove rabbits in a pipe, farm worker "walks-off" pipe until it stands-up.

Table 1: Summary of Book "Set Phasers on Stun"

Story	Title	Hazard	Scenario	Scenario mode	Simultaneous failures	Latent person	Initiating person	Initiating error
1	Set Phasers on Stun	3 doses of inadvertently high radiation kills patient	Machine can not correct input err, technician corrects input err but fails to diagnose machine responses	Startup	Three: designer, tech-in, tech-diag	Designer	Technician	Action (diagnostic)
2	Return from Salyut	N2 suffocates 3 Russian astronauts	Wrong valve installed, system fails, astronaut fails to close valve before space capsule depressurizes	Running	Two: designer, system	Designer	Astronaut	Action (response)
3	Rental car	Japanese planes damage a U.S. plane	Designer revises cockpit, Japanese attack, pilot fails to fly new plane	Startup	Three: designer, attack, pilot	Designer	Pilot	Action (input)
4	A momento of your service	Major oil spill from grounded ship	Control designed with 3 options, captain selects wrong path, captain moves lever to wrong position	Running	Three: designer, capt-dec, capt-resp	Designer	Captain	Planning (intentional-not prior) Action (response)
5	Chutes and ladders	Person electrocuted	Pipe material conductive and labor contacts it to power line	Turn-around	Three: material, contact, laborer	Pipe designer	Laborer	Planning (intentional-not prior)
6	Business in Bhopal	Multiple fatalities from toxic release	Mitigation out-of-service when excess water enters reactor	Turn-around	Two: mgt, operator	Mgt	Mechanic	Action (response) Planning (intentional-prior)
7	Never cry wolfe	Inmate escapes prison	Mitigation too sensitive, guards fail to diagnose, inmate escapes	Running	Three: mechanic, guard #1, guard #2	Mechanic, guard #1, guard #2	In-mate	Action (diagnose)
8	Leap of faith	Several people die when Airbus 320 crashes	Pilot bypasses ESD before unusual maneuver	Run-demo	Two: ESD, pilot	Pilot	Pilot	Planning (intentional-not prior)
9	The wizards of wall street	No hazard - Brokerage recovered lost money	Computer designed to enter $ or shares, clerk inadvertently sold shares versus $, checker bypassed	Running	Three: design, clerk, checker	Designer	Clerk	Action (response)
10	Genie in the bottle	Explosion and radiation kills 3 mechanics	Reactor designed to operate on one tube, written instructions fail to describe procedure, mechanic inadvertently pulls reactor tubes out too far	Turn-around	Two: designer, mechanic	Designer	Mechanic	Action (response)
11	Silent warning	Multiple fatalities when poisoned with alkyl mercury	Grain treated with alkyl mercury to inhibit spoilage, red color and labels fail to warn, bread made from grain, and bread consumed	Running	Three: treatment, red color, label	Treatment	Public	Action (long diagnosis)
12	Zzzs in Zeebrugge	Multiple fatalities when water sinks ferry	Open bow door undetectable, Boatswain fails to close bow door, ferry enters channel	Startup	Three: designer, boatswain, captain	Designer, assist boatswain	Captain	Action (memory)

Table 1: Summary of Book "Set Phasers on Stun" (Continued)

Story	Title	Hazard	Scenario	Scenario mode	Simultaneous failures	Latent person	Initiating person	Initiating error
13	Double vision	None - almost operated on healthy person	Lab tech damages X-ray, doctor & four checkers miss-read x-ray, 2nd doctor catches	Running	Three: lab tech, doctor #1, doctor #2	Lab tech	Doctor	Action (long diagnosis)
14	That newfangled technology	Lead ship hits rocks, 6 trailing ships collide, 23 seamen die, Captain's career ends	Several incorrect bearing readings	Run-test	Three: radioman, subordinate, captain	Radioman	Captain	Action (input)
15	In search of the lost cord	Electricity kills girl	EKG plug inadvertently designed, technician inadvertently connects EKG plug to direct power, completes circuit	Running	Two: designer, tech	Designer	Nurse	Action (input)
16	An act of God	Major looting when power to Manhattan fails for 24 hours	Lightning fails power lines, operator fails to diagnose, backup power fails to start, demand fails to decrease	Running	Four: lightning, backup-1, backup-2, demand	Designer of control room	Operator	Action (diagnosis)
17	Tigershark!	Plane crash kills pilot, damages plane, ends sales	Several multiple g-forces inadvertently fatigues pilot	Run-test	Two: designer, many runs	Designer	Airforce	Planning (problem solving err)
18	The peppermint twist	Several people burned by drinking caustic, lawsuits, business gone	Dishwasher uses special soap in bottle used for drinks, bartender serves it	Running	Four: barkeep-1, barkeep-2, waitress, customers	Barkeep-1	Customer	Action (input)
19	The price of the Amagasaki	Toxic oxygen kills Navy seal	While breathing in a closed system, depth creates oxygen toxic, mission motivates time	Running	Two: designer, navy seal	Designer	Navy seal	Planning (problem solving err)
20	Murphy's law and Newton's law	Rocket inadvertently fires, kills 1 mechanic, injures 2, damages bldg	Volt reducer designed to be removed with batteries, technician replaces batteries but fails to replace volt reducer, test inadvertently ignites rocket	Turn-around	Two: designer, tech	Designer of tester, previous technician	Mechanic	Action (input)

Proceedings of IMECE'03
2003 ASME International Mechanical Engineering Congress
Washington, D.C., November 15–21, 2003

IMECE2003-43413

A STUDY INTO LDPE AS AN UNDERSURFACING MATERIAL FOR INJURY PREVENTION AND RISK MINIMISATION IN CHILDREN'S PLAYGROUNDS

Dr David Eager BE(Hons) PhD GradCertDispRes FIEAust MASME MAAS
Faculty of Engineering
University of Technology Sydney Australia
David.Eager@uts.edu.au

ABSTRACT

Low Density Polyethylene (LDPE) closed-cell foam is used extensively as an impact absorbing material for injury prevention and risk minimisation in a variety of applications, including children's playground undersurfacing, padding for trampoline frames, and other fall zones. This paper presents and analyses the data from numerous impact tests performed on samples of LDPE of select different product thicknesses (10, 20, 30 and 40 mm), nominal Relative Densities (30, 45, 60 and 75 kg/m^3) and drops or *free height of fall* (100 mm steps in heights from 300 to 2100 mm). The impact absorption properties of LDPE are characterized using the Australian and New Zealand Standard AS/NZS 4422: Playground Surfacing – Specifications, Requirements and Test Method. The g_{max} and HIC results are presented both graphically and numerically. This paper also discusses uses and limitations of LDPE with particular emphasis on injury prevention and risk minimisation.

NOMENCLATURE

ASTM	American Society for Testing and Materials
CS5	Standards Australia Consumer Safety Committee CS5
EVA	Cross-linked Ethylene Vinyl Acetate
F08	ASTM Standards Committee F08
g	Acceleration due to gravity at sea level (9.8 m/s^2)
g_{max}	A positive multiple of g that represents the maximum deceleration experienced during an initial impact
LDPE	Low Density Polyethylene
HIC	Head Injury Criteria
UTS:ITL	Impact Testing Laboratory, Faculty of Engineering, University of Technology, Sydney, Australia

BACKGROUND

In industrial safety there is a recognized hierarchy of hazard control measures based on the principle that risks need to be reduced to an acceptable level by engineering means. Risk management within the playground environment is more complex than industrial safety because the primary aim of a playground is to stimulate a child's imagination, provide excitement and adventure and allow scope for children to develop their own ideas of play. Children are not little adults and the intervention strategies intended to protect children may differ from those intended to protect adults. Ideally playgrounds should encourage the development of motor skills and present the child with manageable challenges. A playground should also include elements that encourage the development a child's creative, cognitive, social and sensual skills too. Risk taking should be taken as a given with child development. A well designed playground is actually encouraging the child to take risks, but in a semi-controlled environment that protects a child from hazards he or she may be unable to foresee when using playground equipment as intended. A well-designed playground will also help the child to develop a sense of boundaries, a very important and often overlooked life skill. A well-designed playground will also be designed so that risk involved in play is apparent and foreseeable by the child.

When the author was growing up, he walked to school, he rode his bike around the local neighborhood without any fear of abduction, he played in the local natural bushland, and breaking an arm or grazing an elbow was all part of growing up. The free use of the natural environment is no longer easily available to children; instead playgrounds now provide the main opportunities for healthy physical activity. Unfortunately in Australia, Councils are citing the high cost of litigation and insurance as the primary reason for removing playgrounds and/or stimulating playground equipment.

The Australian Bureau of Statistics (ABS) has confirmed that Australian adolescent children are more obese than in previous generations. Approximately 15% of Australian adolescent children are now classified as overweight, while 5% are classified as obese. This obesity is blamed on a number of factors including the removal of stimulating playground equipment and playspace areas. Society has effectively created an environment that makes it harder for children to run around and play. It is well known that the enjoyment of being active, particularly in childhood, is a key factor in becoming and remaining active in later life.

PLAYGROUND SAFETY STANDARDS

As with industrial safety a balance must be found between risk and safety. The encouragement of risk-taking in children needs to occur without the child sustaining an injury that causes death or leaves the child permanently disabled or disfigured.

The author holds the honorary position as the Institution of Engineers, Australia representative on the Australian Standards Committee CS5 Children's Playgrounds (CS5 Committee). The CS5 Committee is responsible for the preparation of Australian Standards AS 1924, AS 2155, AS 2555, AS/NZS 4422 [1], AS/NZS 4486 and AS 4685 (draft only). The author is also on the ASTM International Technical Committee F08 Sports Equipment and Facilities (F08 Committee). This Committee is responsible for several ASTM Standards including F 355 [2] and F 1292 [3].

In 1976 Standards Australia at the request of the National Safety Council of Australia published the first Australian Standard for children's playground safety; namely Australian Standard AS 1924:1976 Playground Equipment for Parks, Schools and Domestic Use [4]. Since the publication of this original playground safety standard, additional standards and revisions have been published taking into account new developments within the playground industry including: sun protection, softfall surfacing, playground maintenance, and the banning of carcinogenic timber additives. Twenty years have now passed since AS 1924:1981 was last updated. AS 1924:1981 has now been totally rewritten by the CS5 Committee over the past two years using EN 1176 Playground Equipment [5] as the base document. It is envisaged that this new playground safety standard will be published as Australian Standard AS 4685: Playground Equipment Parts 1–6 [6] later this year.

THE MAXIMUM FREE HEIGHT OF FALL DEBATE

One of the major decisions that needed to be made by the CS5 Committee when writing the new playground standard was the issue of maximum free height of fall. An issue that was raised within the CS5 Committee was the necessity and practicality of lowering the maximum free height of fall from the existing 2.5 m to 1.5 m based on a study and extensive research by Chambers *et al.* [7] that concluded that the severity of injuries increased greatly for falls greater than 1.5 m. The CS5 Committee was torn between the Chambers study and other studies.

Chambers *et al.* [7] have been quoted extensively in the literature concerning the maximum free height of fall. Chambers *et al.* claim that falls in excess of 1.5 m increase the risk of injury 4.1 times that of falls from *1.5 m or less.* They also claim that an estimated 45% reduction in children attending hospital emergency departments could be achieved if the maximum fall height was lowered to 1.5 m. The data provided by European studies where 3.0 m [5] is the maximum free height of fall do not support the Chambers study.

A study by Nixon [11] of 335 children who were admitted to hospital emergency after falling from playground equipment noted that the most common distance fallen was *0.5 m or less* (40% of children). This study did not confirm a different pattern from falls *greater than 1.5 m* and *less than 1.5 m* as published by Chalmers *et al.*

Statistics in isolation can create a distorted impression of the actual problem so it is important to look at the resulting playground equipment if the Chambers *et al.* recommendations were implemented. Common items of playground equipment such as monkey bars, flying fox (both usually 2.2 m in height) and many slides would all need to be removed. Swings would also need to be carefully assessed to determine the height that could be achieved while in motion (heights of 2 m are easily achieved with a common swing). If all these pieces of playground equipment were removed and/or not installed in new playgrounds they would need to be substituted with other pieces of exciting and challenging equipment. In some cases substitution may be possible to improve safety without lowering the playground's popularity. A study conducted in Wales [12] replaced monkey bars with a rope climbing frame. This change was made because the fracture rate for monkey bars was twice that of other climbing frames. The substitution did not reduce the playground's popularity and no injuries were reported on the new climbing frame. Unfortunately it is unclear whether the replacement equipment had the same or a lower free height of fall.

It is the author's opinion that the practicality of reducing the maximum free height of fall from 2.5 m to 1.5 m must address safety concerns in conjunction with resulting play value. A boring playground without sufficient stimulation and challenges will not satisfy the developmental needs of a child. It is also the author's opinion that the causal problem of fall injuries is non-compliance with the Standards, particularly AS4422: 1996 (the Australian and New Zealand Standard equivalent to ASTM F1292 or EN1177). This is the Standard that specifies the playground undersurfacing requirements. It is also the author's opinion that both the number and severity of fall injuries would be reduced if Councils (and other organisations) were required to comply with the testing requirements of AS4422: 1996. In simple terms this would ensure that the surface that children fell onto absorbed a controlled amount of impact energy and limited the maximum impact to less than 200 g and a Head Injury Criteria (HIC)[1] to less than 1000 [1].

The author is happy to advise that at the time of writing the CS5 Committee had agreed to continue to use a maximum free height of fall limit of 2.5 m.

THE PLAYGROUND UNDERSURFACING DEBATE

The impact attenuating surfacing beneath playground equipment is known by a number of terms, including: playing surface system, surface system, softfall, impact absorbing playground surfacing, soft surfacing, soft pour, playground surfacing, and undersurfacing. It has been the focus of much attention within the playground industry particularly in recent years. Gone are the days when concrete or asphalt (bituminous material) beneath playground equipment was considered acceptable.

In 1996 an Australia and New Zealand Standard was released that specifically addressed this issue of undersurfacing and impact attenuation. This standard was the Australian and New Zealand Standard AS/NZS 4422:1996 Playground Surfacing – Specifications, Requirements and Test Method [1]. There are similar standards in other countries, for instance ASTM F355 [2] and ASTM F1292 [3] are the Impact Attenuation Standards used in the US, and EN 1177 [8] is the equivalent Impact Attenuation Standard used in Europe. The common thread throughout all of these Standards is the requirement for the entire fall zone[2] (see Figure 1) to comply with a maximum deceleration or no greater than 200 g and a HIC of no greater than 1000.

Traditionally natural materials such as grass, pine-bark nuggets (the broken bark of conifers, grain size 20 mm to 80 mm), wood chips (mechanically broken wood with no wood based materials and without bark and leaf components, grain size 5 mm to 30 mm), sand (washed without silt or clay particles, grain size 0.2 mm to 2 mm) and gravel (round and washed, grain size 2 mm to 8 mm) have been used as a undersurfacing material. A major problem with natural materials is the high ongoing maintenance cost. This cost has two components, namely: the cost of regular topping up to maintain impact absorbing properties; the cost of grooming or racking to ensure that minimum depths are maintained; and the removal of broken glass and used syringes. There are other problems including the supply and use of appropriate high quality materials that exhibit energy absorbing properties. For example, sand that is used as a road-base is not suitable for use as an undersurfacing material. There are also problems associated with needle-stick injuries, splinters, ingestion, choking, and accessibility for people with mobility problems. Ideally all materials should be on-site tested and certified for compliance stating the playground undersurfacing depth and associated maximum free height of fall. Nominally a minimum depth of 200 mm to 300 mm is standard practice.

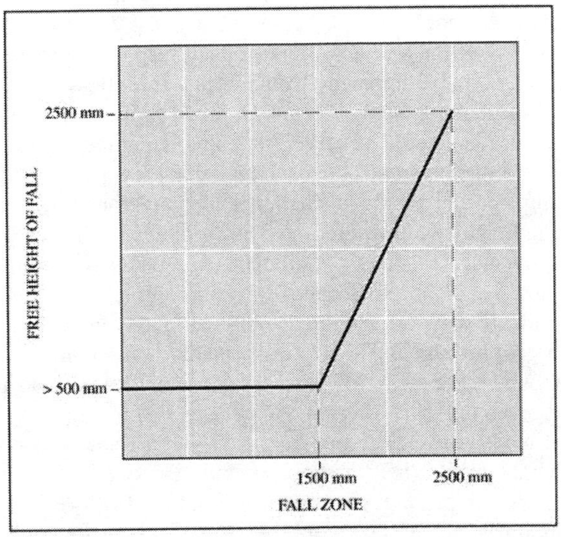

Figure 1: Minimum extent of fall zones [6]

In recent years artificial or synthetic materials have progressively replaced natural materials as the preferred undersurfacing material within playgrounds. These synthetic materials include recycled ground soft-rubber and LDPE. A major reason for the change in usage is the lower ongoing maintenance costs and the perceived reduction in risk as the market perceives the synthetic material as a safer more reliable product that once installed will continue to provide a safe play space for many years.

Until relatively recently all playground undersurfacing energy absorption testing was confined to the laboratory. Recent technological advances have allowed the development of relatively low cost portable impact test rigs that can test in-situ in accordance with the relevant test methods and procedures. Previously all in-situ testing was limited to a visual inspection and measurement of depth (for natural materials).

In a study by Witheaneachi and Meehan [13] 240 Council playgrounds were assessed to determine the extent to which they complied with safety guidelines. Of the 723 pieces of equipment that required undersurfacing only 45.4% had the recommended type of undersurfacing while only 42 of the 723 pieces had undersurfacing to the required depth. When the fall height of the equipment was considered in addition to the

[1] The HIC is an extension of the Wayne State Tolerance Curve (WSTC) and the Gadd Severity Index (SI) derived from motor vehicle accident research conducted on adult cadavers and animals. The HIC is the most widely accepted human head injury measurement. The HIC does not account for differing brain mass, brain structural capacity and skull size [9]. There was no investigation into the response of a child's trauma in the development of the WSTC and thus the HIC. Thus, when the HIC is applied to children falling from play equipment it really is an arbitrary number that allows baseline referencing across the industry. However, injury data do support the use of the 1000 HIC threshold for reducing the statistical prevention of brain injury [10].

[2] The fall zone is the surface beneath playground equipment that may be hit by a child falling from that equipment. The line in Figure 1 represents the minimum extent of the undersurfacing for the corresponding adjacent equipment height.

undersurfacing only 1.8% of the 723 pieces of equipment simultaneously satisfied all the safety requirements.

In a study that was conducted by the Impact Testing Laboratory, Faculty of Engineering, University of Technology Sydney (UTS:ITL) in 2000 [14] into playgrounds and playground fall heights it was concluded that the main issue with playground safety was not whether to lower the fall height from 2.5 m to 1.5 m, but rather how to maintain compliant surfacing beneath the equipment. This study surveyed 25 Councils and the overwhelming response was that the fall height should not be reduced from the existing 2.5 m to 1.5 m. The general opinion given in the Council responses was that a lowering of the fall height would result in a 'loss of many highly used' playgrounds. Comments were varied and included that Councils 'would remove offending equipment and not replace them'. Several Councils argued that they are accused of creating 'boring' and 'unchallenging' playgrounds as the existing Standards have 'sterilized playgrounds so they provide little adventure for kids'. 'Children's wants and requirements needed to be taken into account'. 'At present the Standards are strict and if they were tightened further playgrounds will be hardly worth having'. It was stated that 'homogenous playground equipment, low deck height and a litigious society are preventing the design of really good entertaining playgrounds'. Several responses stated that a reduction in the fall height would 'be covering our kids in cotton wool even more than at present'. It was also feared that lowering the maximum height of fall would lead to climbing on roofs or high sections to encounter a challenge and that 'boring' equipment 'would lead to more aggression, misuse and higher rates of vandalism' of equipment.

The UTS:ITL study also conducted a detailed physical inspection of 48 playgrounds to determine actual compliance with the existing Australian Standards with respect to soft-fall zone (see Figure 1), soft-fall depth and equipment height. We were surprised by the low level of compliance with the existing Australian Standards for both soft-fall zone and soft-fall depth (generally less than 50%).

Figure 2 depicts the protection against falls for various free heights of fall from the playground platform that will be included in the proposed new Australian Playground Equipment Standard AS 4685. It diagrammatically shows the differing requirements for guardrails and barriers as represented by rail and fencing above the platform. It also shows that playground undersurface testing is required above 500 mm. When used in conjunction with Figure 1, one can easily determine the appropriate fall zone for any specific playground.

Unless playground equipment is totally enclosed, children will continue to fall off this equipment. It is now generally accepted that undersurfacing material is a necessary component of good playground design for the purpose of reducing the severity of accidents. As the playground equipment height is increased the need for more effective undersurfacing and or other hazard reduction methods, such as: guardrailing, barriers, and total-enclosure increases.

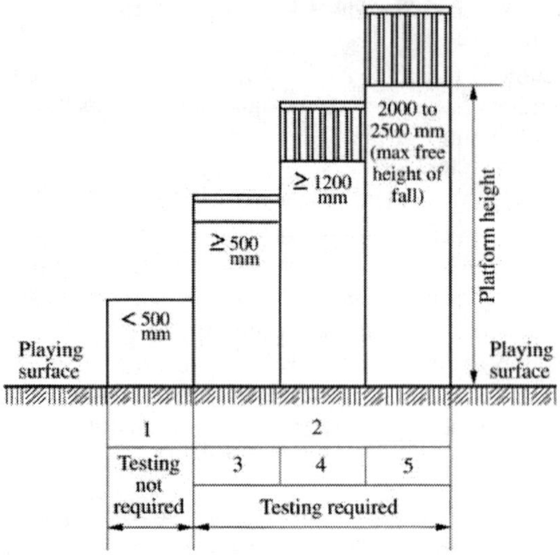

LEGEND
1 = Surfacing required but need not be tested
2 = Tested surfacing required as per AS/NZS 4422
3 = Guardrails required: 700 – 900 mm (height above platform)
4 = Barriers required not less than 900 mm (height above platform)
5 = Barriers required not less than 1200 mm (height above platform)

Figure 2: Protection against falls [6]

The alarmingly low rate of actual playground undersurfacing compliance reported in the literature and the author's personal observations motivated this study.

LDPE IMPACT RESULTS

This study reports some of the resent research undertaken by the author within the UTS:ITL. Over 1,000 individual impact tests were conducted in this study. Figure 3 shows a typical data output from the tri-axial impact test rig. The test rig software was written in LabView® and acquires the three channels of acceleration data. It then performs a vector sum maxima to calculate the g_{max} for each impact test.

The HIC for each impact test was calculated in accordance with AS/NZS 4422:1996. The HIC for each time/acceleration curve is based on the average value of deceleration over the most critical part of the deceleration impulse such that the following function is maximized.

$$\mathrm{HIC} = \left[\left(\frac{\int_{t_1}^{t_2} a \, dt}{t_2 - t_1} \right)^{2.5} (t_2 - t_1) \right]_{max} \quad [1]$$

where:

t_1 is the time, at the start of the impact event, when the deceleration of the headform first equals or exceeds zero;

t_2 is the time, at the completion of an impact event, when the end deceleration of the headform first equals or falls below zero;

a is the deceleration experienced by the headform and expressed in g; and

t_2-t_1 is the duration of impact measured in seconds [1].

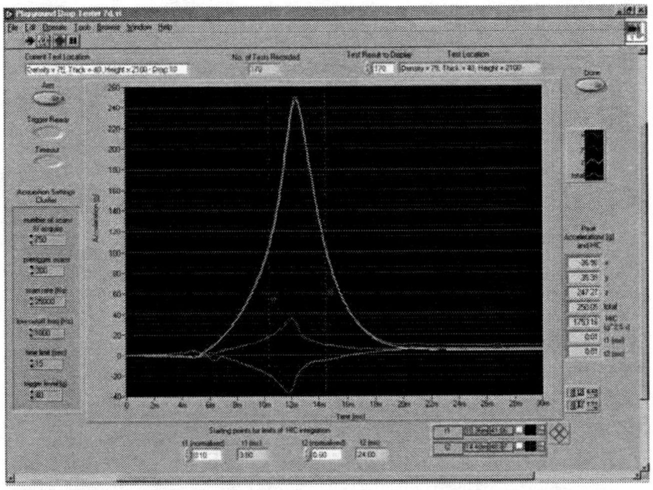

Figure 3: Typical output from tri-axial test rig

The testing was conducted in accordance with requirements of AS/NZS 4422:1996 with one deviation, namely: that the testing methodology was changed so that only one impact location at each specific drop height was used for each test sample and the drop (impact) was repeated ten times in short succession at this one location. AS/NZS 4422:1996 stipulates that three impacts are conducted in the same test position and the greatest value of the three impacts is used.

The pass/fail requirements AS/NZS 4422:1996 are essentially that the g_{max} must not exceed 200g and HIC must not exceed 1000. Both these requirements must be met at the chosen test *free height of fall* or an impact failure is recorded at and above that particular test free height of fall.

The free height of fall was incrementally increased by 100 mm after each set of ten drops at the one location until the test sample failed to meet either the g_{max} or the HIC requirements stipulated within AS/NZS 4422:1996.

AS/NZS 4422:1996 stipulates that the '*greatest values of HIC, or gmax, or both, obtained from the three impacts shall be used for the critical drop height determination*'. The author noted during preliminary testing of LDPE that under most conditions when more than three impact tests were performed the values of g_{max} and HIC continued to increase or deteriorate. It was also noted that even after these samples were subjected to extensive recovery time period they did not return to their pre-test condition. It was hypothesized that choosing the worst of three impacts to obtain the HIC and g_{max} figures did not provide a full or accurate description of the LDPE test results.

Tests were conducted to show the effect of repeated impacts in the same impact location. The variability and ability of the LDPE to absorb successive impacts under differing test conditions can be observed by comparing Figures 4 to 7. An ideal test result would be represented by a horizontal plot. Whereas a test result that continues to deteriorate with repeated impacts would be represented by a plot with a positive gradient.

Figure 4 is a plot of a set of ten impact results obtained for a 900 mm fixed free height of fall from a LDPE test sample with a density of 30 kg/m^3 and a thickness of 30 mm. This plot depicts the typical deterioration in performance that occurs after the initial impact. It was noted that the g_{max} increased from 189 g to 286 g while the HIC increased from 612 to 1072 between the first and second impact. It also confirms that it took several additional impacts for the performance to deteriorate by a similar magnitude. It was observed throughout this study that LDPE never regains the pre-impact performance, no matter how long it is allowed to recover.

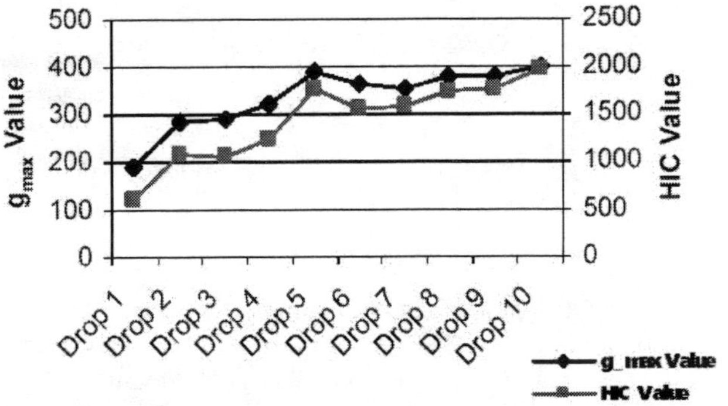

Figure 4: Impact test on a 30 mm thick x 45 kg/m^3 density LDPE sample at a 900 mm fixed free height of fall [15]

Figure 5 is a plot of a set of ten impact results obtained for an 1100 mm fixed free height of fall from a LDPE test sample with a density of 60 kg/m^3 and a thickness of 40 mm. This plot depicts a typical sample when tested within the desired performance range. It was observed that after the initial pre-compression phase that occurred during the first four impacts there was very little drop in measured performance. It appears that a sample with this shaped plot would generally recover to a performance at or near the magnitude of HIC and g_{max} values of the fifth impact. A phenomenon that was given the name *pre-compression* was observed. In short, *pre-compression* is the condition that the test sample returned to days or weeks after the conclusion of the last impact test when the sample was not tested beyond its limiting yield height of fall.

Field testing of playground undersurfacing by the author has confirmed that the effectiveness of LDPE foam products reduces over a long period of time. As a general rule LDPE is not used alone, but as a composite undersurfacing treatment in conjunction with a sand filled artificial grass layer on top. The top layer distributes and lowers the actual impact forces experienced by the LDPE beneath to a level such that walking and running do not noticeably lower the performance. The mechanism of slow deterioration within the microstructure of the LPDM is similar to that within the sole of a typical running shoe that is the gas within the foam is gradually squeezed out and the cross-linking is eventually broken with repeated impacts.

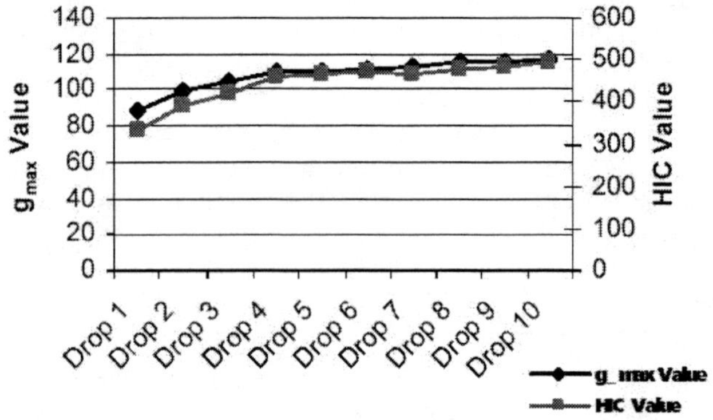

Figure 5: Impact test on a 40 mm thick x 60 kg/m^3 density LDPE sample at an 1100 mm fixed free height of fall [15]

Figure 6 is a plot of a set of ten impact results obtained for an 1100 mm fixed free height of fall from a LDPE test sample with a density of 30 kg/m^3 and a thickness of 40 mm. This plot is a good example of a set of test conditions that continue to deteriorate with successive impacts. With the exception of the tenth and final impact both the HIC and g_{max} values were noted to increase with each successive impact. The lower results obtained for the tenth impact were most probably due to the headform impacting the test sample in a slightly different location to the previous nine impacts. The only difference between the tests that are depicted in Figures 5 and 6 is the LDPE test sample density. As a rule-of-thumb, within the range of samples tested, the amount of permanent damage decreased with increasing sample density.

Figure 7 is a plot of a set of ten impact results obtained for a 400 mm fixed free height of fall from a LDPE test sample with a density of 45 kg/m^3 and a thickness of 30 mm. This plot is a good example of a set of test conditions that after the initial *pre-compression* phase exhibited little deterioration with successive impacts. This sample performed in a consistent manner producing similar results impact after impact. It is also worth noting that at a 400 mm free height of fall this sample complied with the HIC and g_{max} requirements of AS/NZS 4422:1996. This plot is also a good example of a sample that has not been subjected to impacts beyond its yield. In an ideal world this is the characteristic curve that one would like to achieve at the maximum free height of fall for undersurfacing installed within a playground.

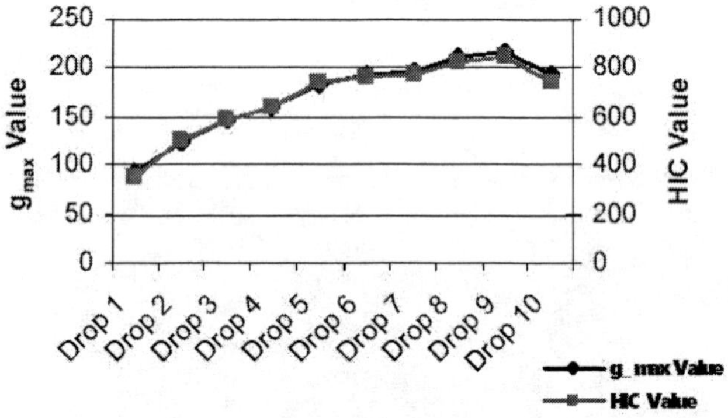

Figure 6: Impact test on a 40 mm thick x 30 kg/m^3 density LDPE sample at an 1100 mm fixed free height of fall [15]

It was noted that the LDPE test samples exhibited visual signs of either temporary or permanent deformation. Permanent deformation was noted at free heights of fall where test sample did not have the ability to absorb successive impacts. This behaviour was more apparent with lower density test samples where the LDPE was noted to tear and/or puncture. It was also noted that failure of the HIC and g_{max} requirements of AS/NZS 4422:1996 was accompanied by a *bottoming out* as the headform compressed the LDPE to an extent that it struck the reinforced concrete beneath the test sample with a thud. By the end of this study one could audibly predict a pass or fail result from the noise emitted from the headform impact.

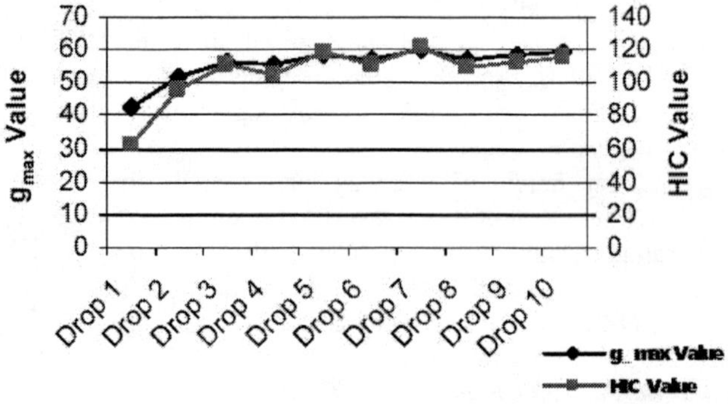

Figure 7: Impact test on a 30 mm thick x 45 kg/m^3 density LDPE sample at a 400 mm fixed free height of fall [15]

Temporary deformation was evident in test samples when the free height of fall was within the test sample's ability to absorb successive impacts. It was observed that when the test samples were impacted within their limit to successive impact at that specific free height of fall that the test sample would uncompress after an impact leaving little evidence of any impact except for a slight bruising at the point of impact.

To analyze the effect that the sample thickness had on the g_{max} and HIC the average test results were plotted to obtain a family of constant density graphs. Each data point on each of the curves within Figure 8 to 11 represents the average of ten consecutive impact tests at the same impact location.

Figure 8 is a plot of the impact test results obtained from LDPE with a density of 30 kg/m^3 for thickness range 20 mm to 40 mm at a 300 mm to 1200 mm free height of fall. This is the first graph in the family of constant density graphs. As was predicted, this plot shows a general trend that the impact performance increases with increasing sample thickness.

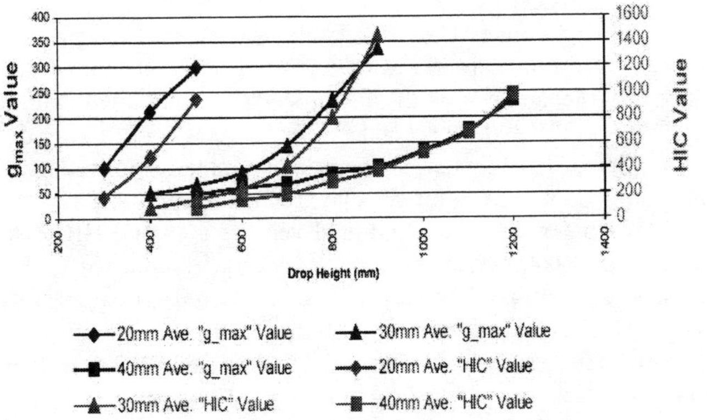

Figure 8: Impact test on 30 kg/m^3 density LDPE samples for the thickness range 20 mm to 40 mm at a 300 mm to 1200 mm free height of fall [15]

Figure 9 is a plot of the impact test results obtained from LDPE with a density of 45 kg/m^3 for thickness range 20 mm to 40 mm at a 300 mm to 1500 mm free height of fall.

Figure 10 is a plot of the impact test results obtained from LDPE with a density of 60 kg/m^3 for thickness range 20 mm to 40 mm at a 300 mm to 1800 mm free height of fall.

Figure 11 is a plot of the impact test results obtained from LDPE with a density of 75 kg/m^3 for thickness range 20 mm to 40 mm at a 300 mm to 2100 mm free height of fall.

It was observed that as the LDPE sample thickness increased, the rate of improvement in performance also increased (rate of increase in the g_{max} and HIC lower). This was primarily due to the fact that the thicker test sample had a greater ability to absorb successive impacts.

As expected, at a common free height of fall the thicker test samples produced lower g_{max} and HIC values than thinner test samples. The effects of LDPE sample thickness was more pronounced as the free height of fall was increased.

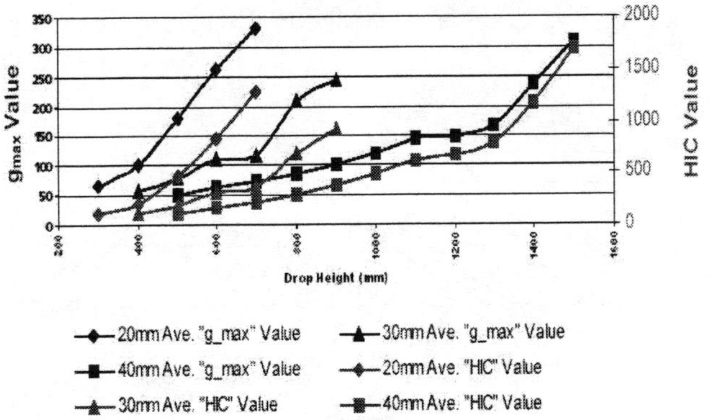

Figure 9: Impact test on 45 kg/m^3 density LDPE samples for the thickness range 20 mm to 40 mm at a 300 mm to 1500 mm free height of fall [15]

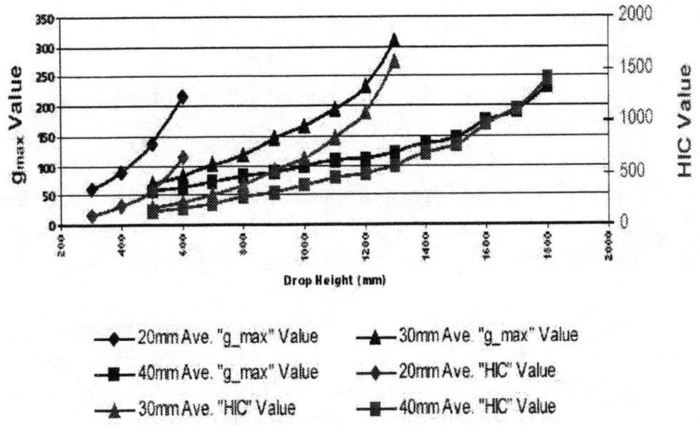

Figure 10: Impact test on 60 kg/m^3 density LDPE samples for the thickness range 20 mm to 40 mm at a 300 mm to 1800 mm free height of fall [15]

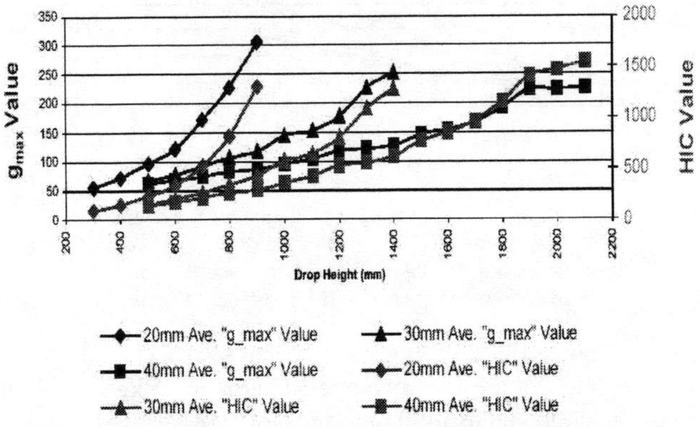

Figure 11: Impact test on 75 kg/m^3 density LDPE samples for the thickness range 20 mm to 40 mm at a 300 mm to 2100 mm free height of fall [15]

Figure 12 is a plot of the impact test results obtained from LDPE with a thickness of 20 mm for the density range 30 kg/m^3 to 75 kg/m^3 at a 300 mm to 900 mm free height of fall.

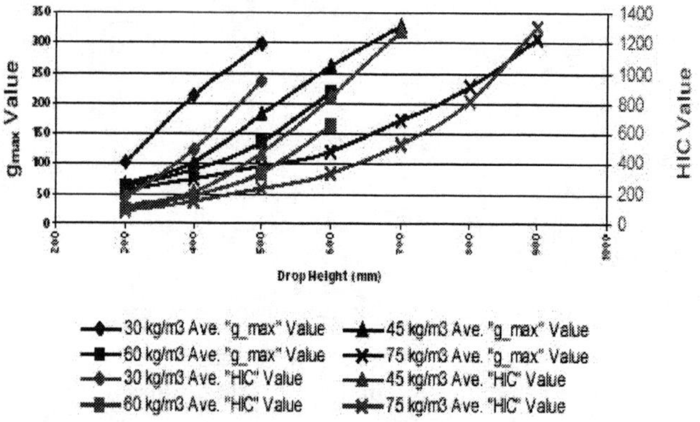

Figure 12: Impact test on 20 mm thick LDPE samples for the density range 30 kg/m^3 to 75 kg/m^3 at a 300 mm to 900 mm free height of fall [15]

Figure 13 is a plot of the impact test results obtained from LDPE with a thickness of 30 mm for the density range 30 kg/m^3 to 75 kg/m^3 at a 400 mm to 1400 mm free height of fall.

Figure 14 is a plot of the impact test results obtained from LDPE with a thickness of 40 mm for the density range 30 kg/m^3 to 75 kg/m^3 at a 500 mm to 2100 mm free height of fall.

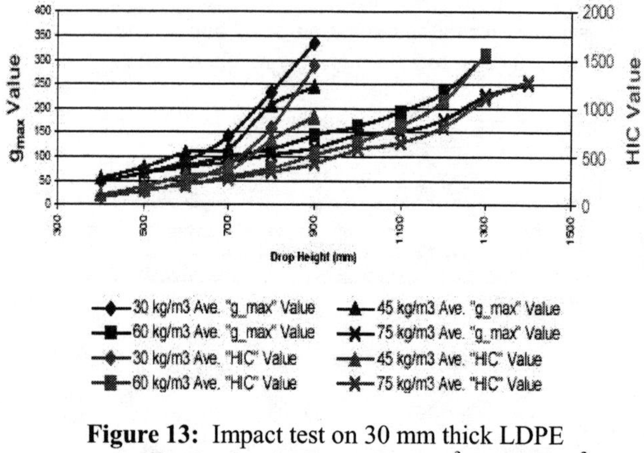

Figure 13: Impact test on 30 mm thick LDPE samples for the density range 30 kg/m^3 to 75 kg/m^3 at a 400 mm to 1400 mm free height of fall [15]

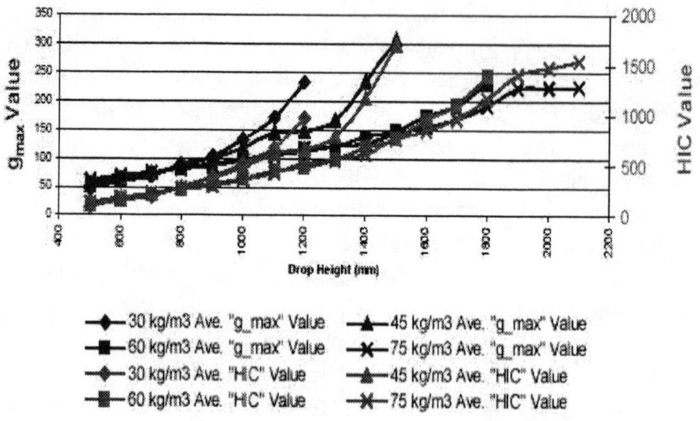

Figure 14: Impact test on 40 mm thick LDPE samples for the density range 30 kg/m^3 to 75 kg/m^3 at a 500 mm to 2100 mm free height of fall [15]

CONCLUSIONS

The primary objective of this study was to characterize the impact absorption properties of LDPE closed-cell foam when tested in accordance with the procedures contained within AS/NZS 4422:1996. A secondary objective was to assess the suitability of LDPE as an undersurfacing material for reducing head injuries within children's playgrounds.

This paper presented and analysed the g_{max} and HIC data from over 1,000 impact tests performed on samples of LDPE closed-cell foam. These tests were conducted on samples of various thicknesses, namely: 10 mm, 20 mm, 30 mm and 40mm. The tests were also conducted on samples of various nominal Relative Densities, namely: 30 kg/m^3, 45 kg/m^3, 60 kg/m^3 and 75 kg/m^3. The test height of fall varied from 300 to 2100 mm.

It was observed that both the g_{max} and HIC increased with increasing free height of fall for constant sample thickness or material density. Within the range of samples tested a 15 kg/m^3 increase in density approximated a 100 mm to 200 mm increase in free height of fall, and a 10 mm increase in sample thickness approximated a 350 mm to 550 mm increase in free height of fall.

The critical free height of fall obtained ranged from less than 300 mm for the 10 mm x 30 kg/m^3 sample which was g_{max} limited, to a maximum at 1720 mm for the 40 mm x 75 kg/m^3 sample which was HIC limited.

It was also observed that in the more severe impact test regime LDPE did not regain the pre-impact performance, no matter how long it is left to recover.

A phenomenon that was given the name *pre-compression* was observed. This condition is described in detail within the results section of this paper. In short, it is the condition that the test sample returns to days or weeks after the conclusion of the last impact test when the sample is not tested beyond its limiting yield height of fall.

It was concluded that LDPE closed-cell foam was suitable for use as impact absorbing material in playground areas of low to medium impact but would be unsuitable for use in areas of high impact as it becomes permanently damaged and loses the ability to fully recover.

It is generally known that cross-linked EVA foams provide superior recovery after repeated compression than LDPE foams but require higher density foam to achieve a similar impact performance[3]. Further research is warranted in the use of EVA foams for potential usage in areas of high impact requirements and/or heavy traffic.

ACKNOWLEDGMENTS

The author would like to acknowledge the support provided by the UTS:ITL, in particular the support provided by Mr Chris Chapman, Mr Adrian Willingham, Mr Lambert Seeto and Ms Sarah Howarth.

The author would also like to acknowledge the in-kind support by Thermoplastic Foam Industries Pty Ltd which provided test samples of various densities and thicknesses to UTS:ITL.

Finally the author would like to acknowledge the intellectual support over many years by various members of the Australian Standards Committee CS5, in particular Ms Margaret Cavanagh of Kidsafe NSW, Ms Barbara Champion of PRAV, Mr Ian Scott of Kidsafe Australia, and Mr Don Wark of PlayWorks Pty Ltd.

REFERENCES

[1] Australian and New Zealand Standard AS/NZS 4422:1996 Playground Surfacing – Specifications, Requirements and Test Method, Standards Australia, Australia, 1996.

[2] ASTM F355 – 01 Standard Test Method for Shock-Absorbing Properties of Playing Surface Systems and Materials, American Society for Testing and Material, USA, 2001.

[3] ASTM F1292 – 99 Standard Specification for Impact Attenuation of Surface Systems Under and Around Playground Equipment, American Society for Testing and Material, USA, 1999.

[4] Australian Standard AS 1924:1976 Playground Equipment for Parks, Schools and Domestic Use, Standards Australia, Australia, 1976.

[5] EN 1176: 1998, Playground Equipment, European Committee for Standardization, Brussels 1998.

[6] Australian Standard AS 4486:2003 (draft) Playground Equipment, Standards Australia, Australia, to be published.

[7] Chamber D.J., Marshall S.W., Langley J.D., Evans M.J., Brunton C.R., Kelly A.M., Pickering A.F. '*Height and surfacing as risk factors for injury in falls from playground equipment: A case-control study*' Injury Prevention 2:98-104, New Zealand 1996.

[8] EN 1177: 1997 Impact Playground Surfacing – Safety Requirements and Test Methods, European Committee for Standardization, Brussels 1997.

[9] Goldsmith W. '*Current controversies in the stipulation of head injury criteria*' Journal of Biomechanics 14(12): 883-84, 1981.

[10] Willingham A.F. 'A preliminary study into children's head injury and playground surface coverings' BE Thesis, University of Technology Sydney, Australia 2001.

[11] Nixon J. '*Playground surfaces*' 2nd Australian Playground Safety Conference (Kidsafe), 109-111, Brisbane 1998.

[12] Silbert J.R., Mott A., Rolfe K., James R., Evans R., Kemp A., Dunstan F.D.J. '*Preventing injuries in public playgrounds through partnership between health services and local authority*' British Medical Journal 318:1595, UK 1999.

[13] Witheaneachi D., Meehan T. 'Council playgrounds in New South Wales: Compliance with safety guidelines' Australian and New Zealand Journal of Public Health 21(6):577-580, Australia 1997.

[14] Howarth S.A. '*A preliminary investigation into playgrounds and their fall heights*' BE Thesis, University of Technology Sydney, Australia 2001.

[15] Seeto L. 'Impact absorption performance of polyethylene foam' BE Thesis, University of Technology Sydney, Australia 2003.

[3] The author was unable to identify any references on EVA recovery as research into the properties of this material appears to be in-house. Notwithstanding, preliminary laboratory tests have indicated EVA has superior recovery properties to that of LDPE.

Proceedings of IMECE'03
2003 ASME International Mechanical Engineering Congress
Washington, D.C., November 15–21, 2003

IMECE2003-43444

VALIDITY OF THE USE OF GENERIC FAILURE FREQUENCIES IN A QRA

Herlinde I. Beerens
National Institute for Public Health and the Environment,
P.O. Box 1, 3720 BA Bilthoven,
The Netherlands. Tel: +31-30 274 45 79,
Fax: 31-30 274 44 42, e-mail:herlinde.beerens@rivm.nl

P. A.M. Uijt de Haag
National Institute for Public Health and the Environment,
P.O. Box 1, 3720 BA Bilthoven,
The Netherlands. Tel: +31-30 274 37 13,
Fax: 31-30 274 44 42, e-mail:paul.uijt.de.haag@rivm.nl

J.G. Post
National Institute for Public Health and the Environment,
P.O. Box 1, 3720 BA Bilthoven,
The Netherlands. Tel: +31-30 274 35 96,
Fax: 31-30 274 44 42, e-mail:jos.post@rivm.nl

B.J.M. Ale
National Institute for Public Health and the Environment,
P.O. Box 1, 3720 BA Bilthoven,
The Netherlands. Tel: +31-30 274 22 45,
Fax: 31-30 274 44 42, e-mail:ben.ale@rivm.nl

ABSTRACT

In QRA calculations the frequency of an accident scenario is assessed by a generic failure frequency approach, defined in the Dutch Purple Book. In this paper we will examine the origin of certain failure frequencies in the Purple Book. The credibility and validity of the failure frequencies are reviewed here by means of a literature search. A study to revise and update these failure frequencies has been initiated.

Keywords: risk assessment, failure frequency, validity

INTRODUCTION

The assessment of risk is an activity of great importance. Responsible industrial companies engage in improving the safety of their processes to prevent incidents that may cause a serious threat to life and the health of workers or to the environment. Risk assessment assists in making cost-effective choices.

With the implementation of the Seveso directive, the drafting of an external safety report became obligatory in the Netherlands in the early nineties. The Quantitative Risk Assessment (QRA) for major hazards forms part of this report. Within the scope of regulatory decision-making on major industrial hazards, the results of a QRA form one of the main concerns for identifying the risk of use, handling, transport and storage of hazardous materials. QRAs provide the competent authorities with relevant information for defining risk acceptability related to developments on-site, around the establishment or on the transport route. Since these decisions made by the authorities strongly depend on the outcome of the QRA, it is necessary to have a reliable risk assessment.

Scenarios and the corresponding failure frequencies play a prominent role in risk calculations. The accuracy of the calculations is determined by the quality of the base data used. The frequency of accidents is mainly determined by multiplying the frequency of release of hazardous substances by the probability of ignition (for flammables) and the probability of exposure of persons. The risk assessments should be in such a way compatible as to have two independent analyses performed on the same plant yield the same result and the same risk value. Therefore it is necessary to standardise the terms by which risk is expressed as much as possible. Risk can be described by a combination of two attributes: the extent of the consequences and the frequency at which these consequences occur. The accident frequency data used for a risk assessment are critical to the final result. The sensitivity of the assumed failure rate data on the calculated risk can be illustrated by the large variation in the extent of the hazard zone (in which the individual risk exceeds 10^{-6} per year) for variations in values for rupture and large leak frequencies of vessels within the present bounds of uncertainty. This underlines the necessity for credible and valid failure frequencies.

In the QRA calculations the frequency of an accident scenario is assessed by a generic failure frequency approach. These generic failure frequencies are defined in the Dutch Purple Book [1] for a number of standard installations. The credibility and validity of these failure frequencies is reviewed here by means of a literature search. The question also arises how applicable and valid these failure frequencies are in current-day practice, so a study was initiated in order to revise, validate and update these failure frequency data.

HISTORY OF THE PRESSURE VESSEL FAILURE FREQUENCIES IN THE PURPLE BOOK

This section discusses the historical background of the failure data for pressure vessels in the Purple Book. More specifically, the origin of the failure frequencies is investigated, concentrating on the historical data sources used in the COVO study [2]. Of the six industrial objects analysed in the COVO study, emphasis is given to the Akzo chlorine study. The historical outline followed in the literature survey is depicted in Fig. 1.

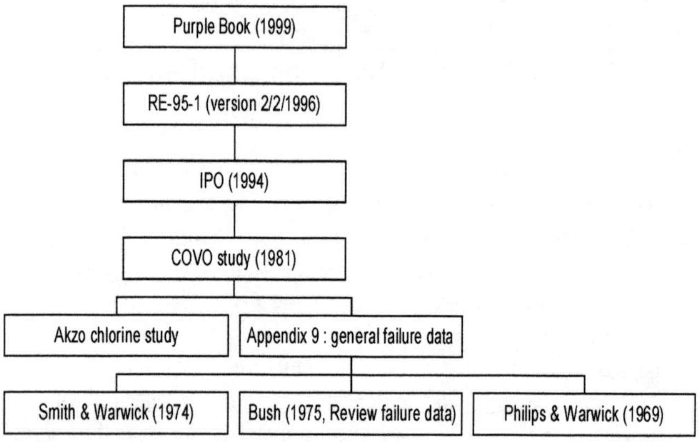

Figure 1: Historical outline started from the COVO study and the references cited therein.

The COVO study (1981)

Part of the COVO study comprised the Akzo chlorine study, in which two different approaches were used for the determination of the failure frequencies. The catastrophic failure of a tank was derived from fault-tree analysis, whereas the other failure types were based on historical data. The inclusion of failure causes like human error, corrosion and fatigue due to vibrations in the fault-tree complicates the comparison with the Purple Book data, in which these failure types are excluded. The leakages in vessels and pipelines are based on the historical data, incorporated in Appendix 9 of the COVO study. Only one frequency is given for leakages and is defined as "serious leakage" for vessels and "significant leakage" for pipelines. Because of the lack of an exact definition, the failure causes which are included as well as the hole size must be estimated to make a comparison with the Purple Book data possible.

In the references cited in the COVO study, failures for pressure vessels were classified in two categories: potentially and catastrophic failure (Philips & Warwick, Smith & Warwick) [3,4] and noncritical and (potentially) disruptive failure (Bush) [5]. Contrarily, for vessels, the COVO study makes a distinction between catastrophic rupture and serious leakage. Although the COVO study refers to these preceding studies, there is no clear link between the different types of classifications. The question posed is to what extent *potential* failure must be taken into account in a failure frequency study.

Furthermore, the values collected by Philips & Warwick, Smith & Warwick and Bush are based mainly on data for steam generation systems and a very limited number of process type vessels. The failures quoted by these UKAEA studies involved only some 10-25% actual catastrophic rupture of vessels. In order to obtain an estimate of the base failure rate for pressure vessels, in the COVO study the range quoted in the above-mentioned references is divided by a factor 10. However there is no good basis given for this conversion of data.

Bush [5] reviews different data sources from various countries. An additional limitation in the examination of these data is the use of upper limits of confidence levels based on zero leakage. As a result of the unclear indication of the confidence level used for certain data, some differences in the reported failure frequencies cannot be explained. Some data sets are basically subsets of larger extrapolated data sets, which as a consequence can not additionally contribute to the determination of an appropriate upperbound value for failure probability. Moreover, these historical data sets do not give enough information for the assignment of a leak to the vessel or to the pipelines connected to the vessels.

In some data sets defects found during periodic inspection of pressure vessels have been compiled and failure rates derived for the defects that are, in general, pertinent to pressure vessels. One cannot exclude the notion that many of the "failures" cited were probably generated during the fabrication stage and remained virtually unchanged until detected during a periodic inspection. These types of failures did not, however, directly interfere with functional use and do not result in emission. In order to prevent confusion, attention should be given to imprecise definitions and inconsistent use of terms such as "failure statistics" and "catastrophic failure".

IPO (1994)[6]

The first document in which standardised failure frequencies were published by the Dutch authorities was the IPO document. The frequency of catastrophic failure for a pressure vessel in the IPO is in good accordance with the value in the COVO study (10^{-6} per year). Catastrophic failure is modelled as instantaneous release (5×10^{-7} per year), release in 10 min (2.5×10^{-7} per year), and release through a diameter of 50 mm (2.5 x

10^{-7}per year). For leakages the comparison is more complicated. Whereas in the COVO study only "serious leakage" (10^{-5} per year) is defined, in the IPO-study two types of leakages are considered: leakage from a hole with a diameter of 50 mm (10^{-5} per connection) and leakage from a hole with an effective diameter of 10 mm (10^{-5}per year). In fact, the frequency of 10^{-5} per year is transformed into $(n+1)$ 10^{-5} per year, in which n is the number of connections. Another point of discussion is that the serious leakage defined in the COVO study as a leak of 50 mm on an arbitrary position on the tank, is transformed in the IPO into a leakage from a hole of 50 mm on the connection of the pipelines. On the other hand, in the COVO study there is a separate scenario defined for the rupture of connections on the tank.

Re-95-1 [7]

The IPO document was amended in the RE-95-1 document. In the IPO document, a leak from a hole with an effective diameter of 50 mm (with a failure frequency of 10^{-5}per year) is defined. This Loss of Containment event (LOC) was meant to cover the rupture of the pipe connected to the vessel; however, rupture of pipes is already covered by the LOCs of pipes. To avoid double counting, this LOC is left out and only the class of leakage with a hole size of 10 mm is left. One could ask whether this omission can be justified. This consideration is based on the fact that the frequency of 10^{-5} per year is considered as the frequency for serious leakage and therefore probably has a hole size larger than 10 mm. As a consequence, the deletion of the 50 mm category may be an underestimation.
The three scenarios for catastrophic failure, defined in the IPO document, were reduced to two scenarios: instantaneous release (5 x 10^{-7} per year) and release of the complete inventory from a hole with an effective diameter of 50 mm or, if the duration exceeds ten minutes, a continuous release with a duration of ten minutes (5 x 10^{-7} per year).

Purple Book (1999) [1]

The Purple Book is based on the IPO and the RE-95-1. In the RE-95-1, the LOC of 10^{-5} per year is defined as a release from a hole with an effective diameter of 50 mm or a continuous release in ten minutes. For simplicity, this LOC is in the Purple Book defined as a release of the complete inventory in ten minutes at a constant rate of release. In other words, the catastrophic failure is defined in the Purple Book as an instantaneous release or a release in ten minutes. The total history of the vessel failure frequencies is depicted in Fig. 2.

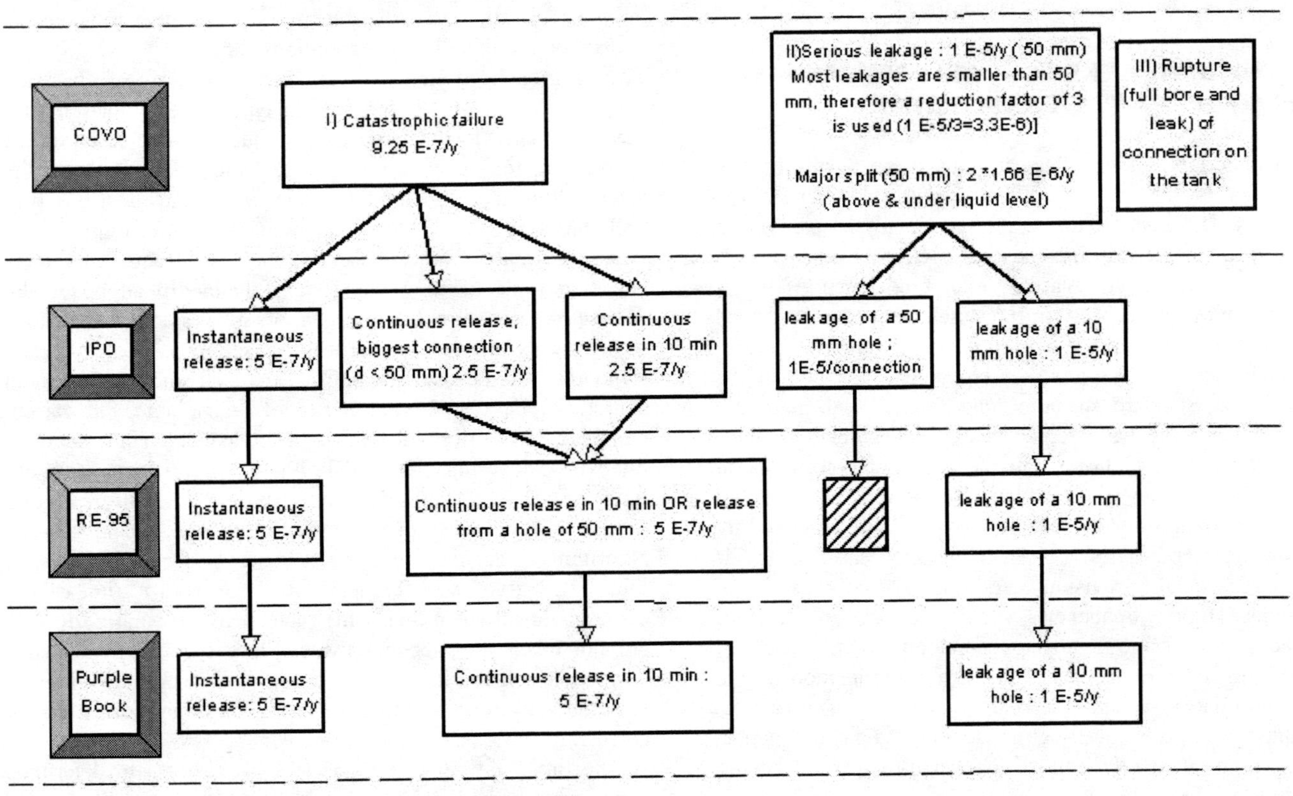

Figure 2: History of vessel failure frequency data.

This study pointed out that it is rather difficult to judge the reliability of the sources. Both the link between the IPO and the COVO study, as well as the derivation of the failure frequencies from the references of the COVO study, are not evident. This can be attributed to various aspects:

- There is no univocal terminology used for the classification of the failures, which complicates the comparison of the data. Moreover the expression "catastrophic failure" has different meanings in the different reports, which shows the difficulties in the application of uniform criteria.
- Besides real failure, a lot of potential failure frequency data were considered. Failures during inspection and testing (pressure) are also included. This contradicts the Purple Book, where these failures are not included.
- The information in the data sets is often limited and does not always give an indication of the kind of failure causes included and the confidence upper boundary of the failure numbers. Furthermore, the underlying assumptions and relationships to certain causes are not unambiguously described. Some of the studies are not representative, because they are based on an extremely small data set.
- Rough estimates are made in order to convert data from one type of industry (steam generators), into data for another type of industry.

THE CURRENT GENERIC DATA SET AND THE NECESSITY FOR UPDATES AND REVISION

An important source of uncertainty in the results of risk assessments is the use of different data sets for failure frequencies. Different performers of risk analyses use varied failure data sets and methods of selecting the failure frequencies, which, in most cases, is expert judgement.

The problem of repeatability in risk assessments can be solved to a certain level by giving a fixed standard value for release frequencies for different types of equipment, as done in the Purple Book. Standard scenarios and failure frequencies are described here too. These frequencies are set as default values; it is recognised that these data will only be seen as an approximation for any specific plant. Deviation is possible, but only on the condition that it can be motivated and approved by the competent authorities. Lower frequencies can be used if there are special provisions additional to the standard provisions. Higher frequencies should be used if standard provisions are missing or under uncommon circumstances. In fact, the use of generic data requires a determination of the conditions under which the data would be appropriate, something which is not well-defined to date. As a consequence, data are used outside the context for which they are originally collected.

Besides the uncertainties mentioned in the previous section, another disadvantage of the current set of standard failure frequencies, which should not be underestimated, is the fact that they are based on outdated numbers. Most data were published around 1970, which means that most of the data are at least 25 years old. It could be argued that because of the changes of design rules, materials and experience the failure frequencies have changed since then. However, at the moment there is no real proof of this from the data found in literature. It is unknown to what extent these data are representative for the current standard in the chemical industry.

The failure frequencies are based on a limited number of characteristics. For vessels, there is only a distinction between storage vessels and process vessels, and for pipes only the length and the diameter is relevant. As a consequence, special provisions directed to integrity of the installation and significant for safety are not included in the failure numbers. Moreover, review studies have been published that show a tendency for some systems towards higher failure frequencies than the ones reported in the Purple Book [8,9]. These figures clearly underline the need to have a better understanding of the actual failure data and the underlying assumptions.

To meet this objective a study to revise and update the standard failure data has been started [10]. One of the goals of the study is to throw some light on the way in which equipment-release frequencies vary according to the standards of design, construction, operations and maintenance and to the actual operating conditions. The intention is also to give some guidance on the way the release and accident frequencies depend on the underlying assumptions.

The approach used involves the selection of baseline frequencies for different equipment types in combination with modification factors. The baseline frequencies can be corrected by these factors to obtain a more realistic risk estimate of the actual frequency. The failure frequencies obtained in this way will be very useful for non-standard applications. When choosing modification factors for release frequencies it is necessary to decide in relation to what the modification is made. A baseline of design, construction, operations and maintenance standards is needed. In order to determine the reference standards, it is necessary to look at the data sources. The main sources of data in this study are of recent date and varied: company data from the oil, petrochemical and chemical industries covering the mid-1970s to the present day are included. As model for failure rate prediction a fault-tree-based model will be used in combination with a checklist-based algorithm. Algorithms will be made only for the causes of failure that arise most frequently. For the modification factors care should be taken that some factors, like special corrosion, are not taken into account twice. More specifically, when a correction is made for the application of the equipment item, it is not always necessary to provide an additional correction for corrosion.

At the time of writing work is still in progress to determine appropriate baseline values, modification factors and to apply the above-mentioned approach [10].

CONCLUSION

This study shows the difficulty of tracing back failure data due to the lack of univocal terminology.
The failure data of pressure vessels in the Purple Book have been copied from the IPO. The unclear links between the IPO and the COVO study, as well as between the references of the COVO study and the COVO study itself, seriously hamper the validation of failure data in the Purple Book. To obtain validated failure data we initiated a study based on recent data and a fault-tree-based model with algorithms and modification factors. This initiative should ultimately lead to failure data that can be easily applied in current-day practice and gives opportunity to obtain appropriate data for non-standard applications.

REFERENCES

[1] Committee for the Prevention of Disasters (CPR), 1999, "*Guideline for Quantitative Risk Assessment-"Purple Book*" CPR18E, SDU, The Hague.
[2] COVO Commission, 1981, "*Risk analysis of six potentially hazardous industrial objects in the Rijnmond area, a pilot study. A report to the Rijnmond public authority*", Central Environmental Control Agency, Schiedam, The Netherlands.
[3] Philips C.A.G., Warwick R.G., 1969, "A survey of defects in pressure vessels built to high standards of construction and its relevance to nuclear primary circuit envelopes." Authority Health and safety branch, Risley, Warrington, Lancashire, UK. AHSB (S) R162, UKAEA Report.
[4] Smith T.A., Warwick R.G., 1974, "The second survey of defects in pressure vessels built to high standards of construction and its relevance to nuclear primary circuits." United Kingdom atomic energy authority, Warrington, SRD-R-30, UKAEA Report.
[5] Bush S.H., 1975, "Pressure vessel reliability", Journal of Pressure Vessel Technology, pp 54-69.
[6] IPO, 1994, "*Guidelines for the preparation of off-site safety industrial sites (in Dutch).*" Report IPO Project A-73. The Hague.
[7] RE-95-1, Version 2-2-1996, KO-95, KO-96, KO-100 performed by TKO working group.
[8] Logtenberg, M.Th., 1998, "Derivation of failure frequencies for LOC cases". TNO, Apeldoorn, The Netherlands.
[9] Hoorelbeke P., Van Cayseele, Cnudde, 1994, "Handboek kanscijfers ten behoeve van het opstellen van een veiligheidsrapport." AMINAL Dienst gevaarlijke stoffen en risicobeheer, Brussels.
[10] Taylor J.R. 2002, "Hazardous materials release and accident frequencies for process plant", Draft version.

Proceedings of IMECE'03
2003 ASME International Mechanical Engineering Congress
Washington, D.C., November 15–21, 2003

IMECE2003-43945

SIMULATION-BASED PARAMETRIC RELIABILITY MODELING USING COVARIATE THEORY

Mr. Jon M. Wallace
Graduate Research Assistant, Aerospace Systems Design Laboratory, School of Aerospace Engineering, Georgia Institute of Technology

Dr. Dimitri N. Mavris
Associate Professor, Aerospace Systems Design Laboratory
School of Aerospace Engineering, Georgia Institute of Technology

ABSTRACT

Reliability analysis methods used for preliminary safety assessments of complex systems typically assume a predetermined and invariant set of input variable statistical properties. However, during the product development phase of the system and especially during in service operation, the characteristics of the random variables can themselves be subject to variation. Thus, the resulting failure probability distribution can vary greatly from early predictions. The objective of this paper is to explore a technique used to create a general parametric failure probability distribution as a function of key input variables. This technique is constructed around covariate theory which is the basis of the familiar Accelerated Life Testing and Proportional Hazards Modeling approaches. Where these approaches have traditionally been used with physical experiments, they are applied within this study to Monte Carlo simulation data generated using an available component modeling and simulation environment of a gas turbine airfoil limited by a single failure mode. Necessary modifications to the traditional form of the covariate approach are identified for application to controlled Monte Carlo simulation data. Implications to potential safety improvements early on in the product development phase are discussed.

KEYWORDS
Engine Reliability, Accelerated Life Testing, Failure Probability, Monte Carlo Simulation

INTRODUCTION

Reliability is an important customer metric for expensive and complex vehicle system designs. This is especially true for systems that require extended and manned flight operations. Yet, accurate system reliability assessment traditionally occurs far downstream of the conceptual design process when critical design decisions have already been made. Recent advances in multi-disciplinary analyses, design, optimization, probabilistic, and computational methods and capabilities now provide a medium that is much more conducive to considering reliability earlier in the design process of complex vehicles.

The objective of this paper is to demonstrate use of covariate theory to create a parametric representation of a component's probabilistic failure distribution as a function of several driving explanatory variables. Where response surface and other regression methods have been used to create a deterministic approximations of a given response, covariate theory can be applied to provide an approximation of the lifetime distribution not only as a function of the traditional index, time, but also as a function of additional variables that may affect the properties of the response distribution.

NOMENCLATURE

A_h	blade hub cross-sectional area
F_c	centrifugal force
$N_{fatigue}$	fatigue life
N	number of Monte Carlo simulations
r_t	blade tip radius
r_h	blade hub radius
T_c	internal blade cooling temperature
T_g	external turbine gas temperature
T_m	blade metal temperature
ΔT_a	ambient temperature change
ε	Monte Carlo simulation error
η_c	cooling effectiveness
η_f	fan efficiency
η_{hpc}	high pressure compressor efficiency
η_{cu}	combustor efficiency
η_{hpt}	high pressure turbine efficiency
η_{lpc}	low pressure compressor efficiency
η_{lpt}	low pressure turbine efficiency
ω	rotor speed
ρ	material density
σ_a	centrifugal stress amplitude
σ_c	centrifugal stress
σ_f',b	fatigue life fitting constants
σ_m	mean centrifugal stress

BACKGROUND

Parametric Distribution Determination

Before a covariate model of the suspect lifetime response can be created, the baseline distribution must either be specified or determined, for instance, through experimentation. Two methods are used with experimental lifetime data to statistically guide the process of selecting an appropriate distribution to represent the data: The Kolmogorov-Smirnov and Anderson-Darling EDF based methods[1,2,3,4,5]. EDF stands for Empirical Distribution Function which is a non-parametric determination of the cumulative failure, or survival, function and is the basis for many distribution test methods. However, unlike the traditional application of these distribution test methods where the data is generated through physical testing, it is generated within this study using computational simulation.

The Kolmogorov-Smirnov (K-S) goodness-of-fit test is useful in determining whether or not a given data set follows a hypothesized, continuous distribution. The K-S test is easy to implement and is completely general for any suspect parametric distribution when the empirical distribution is available. The K-S test statistic is defined as the maximum difference between the empirical cumulative distribution function, $\hat{F}(x)$ and the hypothesized fitted cumulative distribution function, $F_o(x)$. The null and alternative hypotheses for this test are:

$$H_o : F(x) = F_o(x) \quad H_1 : F(x) \neq F_o(x) \tag{1}$$

The test statistic for a complete uncensored data set is

$$D_n = \sup_x \left| \hat{F}(x) - F_o(x) \right| \tag{2}$$

where *sup* is an abbreviation for supremum. Thus, the test statistic is computed by looping over the entire data set. Larger values of D_n indicate a poorer fit of the hypothesized distribution. The test statistic formula is modified to a slightly different form for computational efficiency given as

$$D_n = \max\left\{ \max_{i=1,2,\ldots,n}\left(\frac{i}{n} - F_o\left(x_{(i)}\right)\right), \max_{i=1,2,\ldots,n}\left(F_o\left(x_{(i)}\right) - \frac{i-1}{n}\right)\right\} \tag{3}$$

where i is the ith failure, $x_{(i)}$ is the ith sequential lifetime or response value and n is the total number of realizations (samples) of the random variable. Algorithmically, one would fail to reject the null hypothesis and accept the hypothesized distribution with significant evidence if the computed K-S test statistic is less than the critical value of the statistics. The critical value is independent of the hypothesized distribution which is an appealing characteristic of the method. However, the K-S goodness of fit test is only valid for fully specified distributions. Therefore, if the same data for the goodness-of-fit test was used to estimate the parameters of the probability distribution, then the K-S statistic method cannot be used. Due primarily to this limitation, the Anderson-Darling test statistic is recommended.

The Anderson-Darling (A-D) test is a modification of the K-S test that weights the tails of the distribution resulting in a more sensitive test for goodness of fit of the hypothesized distribution. Contrary to the K-S statistic, the A-D test uses the hypothesized distribution to calculate the critical value of the test statistic. Therein lies the disadvantage. Although superior to the K-S statistic, the A-D method requires critical values to be computed unique to each distribution. Further, tables of critical values found in the literature have only been created for a few distributions such as the normal, lognormal, exponential, Weibull, extreme value type I, and logistic distributions. Fortunately, these distributions are some of the most widely used in statistics, especially in reliability modeling.

The A-D test statistic is defined as

$$A^2 = -n - S \tag{4}$$

where

$$S = \sum_{i=1}^{n} \frac{(2i-1)}{n}\left[\ln\left(F_o\left(x_{(i)}\right)\right) + \ln\left(1 - F_o\left(x_{(n+1-i)}\right)\right)\right] \tag{5}$$

In order to compare the A-D test statistic to its critical value, it must be modified by a factor which is dependent on the number of samples as well as the specific case encountered. For example, assume the case of testing the goodness-of-fit of a normal distribution for a given data set. If the location and scale (mean and variance) are to be estimated from the same data as that used for the test, then the appropriate modified A-D test statistic will be

$$A^2\left(1 + 4/n - 25/n^2\right) \tag{6}$$

The A-D method without using the critical value estimate has been applied in the literature to several distributions simultaneously. In this application, the distribution with the smallest A-D test statistic value would be the most appropriate distribution of the set to model the data. Once the most ideal distribution is selected, the critical value could then be calculated to statistically test for a specified significance level whether or not the distribution is truly representative of the data set. See the work by Stephens for a list of modifying factors as well as a complete account of widely excepted goodness-of-fit test techniques[1,2,3,4,5].

Covariate Models

Covariate models are used to represent the effect of suspect treatments in a probabilistic lifetime distribution or reliability model. A covariate is defined as a treatment or explanatory variable that influences the failure time of the component or item. A vector of covariates, *z*, is chosen with each entry of the vector representing a unique explanatory variable. Typical covariates include those that represent mechanical forces, material properties, and environmental variables. There are two rather popular approaches for linking these covariates to the probability function. The first method, known as Accelerated Life Testing (ALT), is based on modifying the survivor function. The premise of the second approach, called Proportional Hazard Model (PHM), is to modify the hazard rate function to include the covariates.

In either method, the covariates are modeled by modifying the random variable with a 'link' function, $\psi(z)$, which is a function of the covariates. Usually, the link function chosen is of a log-linear form given as

$$\psi(z) = e^{\beta' z} \tag{7}$$

where β is an n by 1 vector of regression coefficients corresponding to each of the n covariates in the vector, z. The appealing characteristic of this link function compared to others that have been proposed is that the exponential form is highly compatible with most traditional parametric distributions in that the model retains the necessary properties of a probability distribution. Also, the log-linear function is asymptotically stable across large ranges of the regression coefficients.

The resulting survivor function in the ALT approach becomes

$$S(t) = S_o(t\psi(z)) \tag{8}$$

where S_o is the baseline survivor distribution function determined at the nominal values of each of the potential covariates (i.e. z_i=0). This formulation imposes certain requirements on the link function. Namely, the function must be equal to unity at the nominal covariate setting, z=0, and must be positive for any and all values of z. In essence, the ALT approach uses the link function to modify the time axis. The form of the log-linear link function used in both of the two approaches allows for the use of the familiar applied statistical model theory to determine the respective covariate regression coefficients, β_i. However, rather than use least squares regression, as is typical for statistical linear models and response surface equations, covariate models use maximum likelihood estimation (MLE) to estimate the covariate coefficients. The coefficients are treated as parameters of a multivariate distribution to be estimated using MLE. As a consequence, statistical significance tests of each coefficient in addition to whole model tests are readily computed resulting in a rapid importance ranking and error estimation of each coefficient.

MODELING AND SIMULATION ENVIRONMENT

Propulsion System Description

The propulsion system selected for this study is the CFM56 separate flow turbofan engine shown in Figure 1. The CFM56 is one of the most successful engine programs powering medium range transport aircraft such as the B737. The overall pressure and bypass ratios of the CFM56 are roughly 30:1 and 6:1, respectively. The maximum thrust at the sea-level static condition is around 20,000 lb_f.

The vehicle mission used in this study is typical of a B737 aircraft; although, it has been simplified to facilitate the demonstration of the proposed method. Only the cruise condition is considered for this study. The take-off and landing segments are modeled as discontinuous jumps in engine rotor speed between the shut-down and cruise segments of the operating profile. The cruise segment occurs at an altitude of 35,000 feet and a speed of Mach 0.745 as is typical for the B737-400. The mission parameters identified for this study are the cruise Altitude and Mach number.

Figure 1: CFM56 Separate Flow Turbofan Engine Cutaway.

Multidisciplinary Turbine Blade Failure Analysis

The safe, reliable operation of the entire aircraft is highly dependent on the propulsion system utilized. One of the most critical components affecting the safety and reliability of turbine engines is the turbine blade. Failure of a turbine airfoil could cause a high-energy part to be released from the rotor system destroying the entire engine through a cascade of subsequent events. Such a situation could cause a catastrophic condition for the entire vehicle. Therefore, the propulsion system reliability is determined, within this study, by conducting a reliability assessment of a single turbine airfoil. There are numerous other failure conditions that can occur which would also affect system reliability and safety[6]; but, modeling and analyzing multiple critical-to-reliability components and failure mechanisms is not the focus of this study.

Turbine blade analysis requires the involvement of many disciplines and the use of advanced computational and experimental techniques which are beyond the scope of this study. Therefore, a first-order, integrated turbine blade multi-physics environment was constructed. The physical analysis structure, depicted in Figure 2, begins with system operational and ambient conditions providing input to a thermo-dynamic cycle model. This cycle model then is used to solve for the engine station state conditions as well as the mechanical speed of the rotors. A thermo-mechanical analysis is then conducted at the part level to determine the thermal and mechanical state of the part as required for the failure analysis.

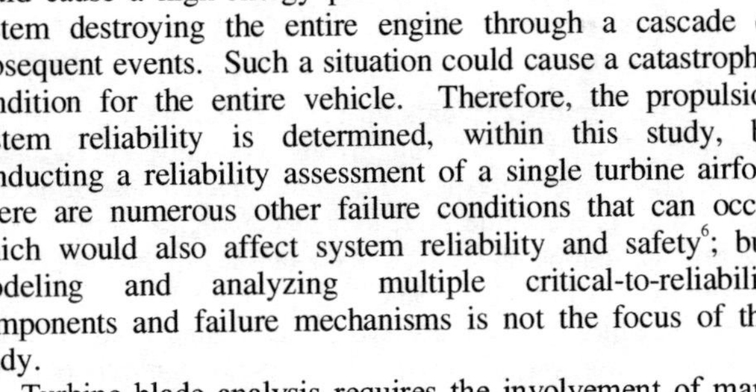

Figure 2: Turbine Blade Life Analysis Structure Matrix.

For the thermo-dynamic cycle analysis, a baseline NEPP thermodynamic cycle model of the CFM56 turbofan engine was utilized to compute the 1-D steady-state engine station and component properties[7,8]. The thermodynamic cycle parameters identified for this study are given in Table 1. Notice that three operational parameters representing the cruise flight condition are chosen along with the efficiency of each of the major cycle components. The required output provided by the cycle analysis is gross thrust, specific fuel consumption, high-pressure rotor speed, and turbine entrance and compressor exit temperatures. The gross thrust and specific fuel consumption responses were chosen as top-level metrics for comparison to the reliability prediction, which is also considered a top-level system metric. The remaining cycle responses are required as intermediate input for the subsequent thermo-mechanical and failure analyses.

After the core flow temperature and shaft speed are determined from the cycle analysis, a thermo-mechanical analysis is performed. The input for the thermo-mechanical analysis is given in Table 2. The thermal solid analysis required for determining the blade metal temperature is conducted using the definition of an empirical cooling effectiveness[9] given as

$$T_m = T_g - \eta_c \cdot (T_g - T_c) \quad (9)$$

This simple formula accounts for aerothermal and heat transfer behavior through this empirical cooling effectiveness parameter. A baseline expected cooling effectiveness of 0.63 is used, which is the value used to size the engine cycle model. The flow gas temperature, T_g, and coolant flow temperature, T_c, are simply the turbine entrance temperature and compressor exit temperature, respectively; which, are output from the cycle analysis. Radial variation of the temperature is not modeled in this analysis.

Although complex multi-axial states of stress exist in rotor airfoils, only the uni-axial bulk radial stress is considered here. The mechanical analysis for determining the bulk radial stress can be accomplished using Newton's second law applied to a rotating body. The centrifugal force produced anywhere in the airfoil can be determining by calculating the product of the mass of the supported material at the radius of interest and the square of the component rotational speed. Dividing this quantity by the airfoil cross-sectional area gives the bulk centrifugal stress at this radius. The maximum bulk centrifugal stress therefore occurs at the hub of the airfoil and is given as

$$\sigma_c = \frac{F_c}{A_h} = \rho\omega^2 (r_t - r_h) r_h \quad (10)$$

for a non-tapered geometry. The baseline material density and annulus area are provided as input. The rotor speed is calculated using the cycle model described earlier. With the thermo-mechanical conditions solved for, a failure analysis can then be conducted.

Table 1: Thermodynamic Cycle Input Values

Paramete	Categor	Unit	Baseline
Altitude	Operational	feet	3500
Mac	Operational	-	0.74
ΔT	Operational	°F	0
η_f	Component	-	0.88
η_{lpc}	Component	-	0.90
η_{hpc}	Component	-	0.87
η_{cu}	Component	-	0.99
η_{hpt}	Component	-	0.93
η_{lpt}	Component	-	0.93

Table 2: Thermo-Mechanical Input Values

Parameter	Category	Units	Baseline Value
η_c	Aerothermal	--	0.630
ρ	Material	lbm/in^3	0.300
r_T	Geometric	in	15.0
r_H	Geometric	in	13.0

Table 3: Failure Input Values

Parameter	Category	Units	Baseline Value
σ_f	Material	ksi	280
b	Material	--	-0.14

The failure condition, fatigue, can be highly involved but simplified here to facilitate the current study. Assuming that linear-elastic conditions exist, the straightforward stress-life, S-N, approach[10] can be utilized to estimate the number of cycles to failure given completely reversed, constant stress cycle amplitude. Morrow suggests a variation to the S-N approach to compensate for non-zero mean stress, which is given as[11]

$$N_{fatigue} = 0.5\left[\frac{\sigma_a}{\sigma_f'\left(1 - \dfrac{\sigma_m}{\sigma_f'}\right)}\right]^{\frac{1}{b}} \quad (11)$$

where σ_a, σ_m, and σ_f are the stress amplitude, mean stress, and fatigue strength coefficient of the blade.

At this point, it is necessary to make another simplifying assumption with regards to the fatigue analysis. The Morrow low cycle fatigue function assumes that the cyclic stress amplitude is constant. Consequently, the stress cycle generated within the context of this problem is defined as the centrifugal stress cycling between the zero stress condition before starting the engine and the stress state at the cruise condition. Realistically, the stress-state is highly complex both temporally and spatially due to the varying operation conditions of the vehicle throughout its mission. Furthermore, material property and geometry non-linearity may also exist, which further complicate the analysis. Several experts within the area of mechanics of materials have focused considerable effort towards improving the prediction of the complex stress-strain state. For the purposes of demonstrating the proposed reliability method, a simple failure function is considered.

RESULTS

Identification of Baseline Distribution

The fatigue life data from 10,000 simulations of the multi-disciplinary turbine blade failure model was used to determine an appropriate parametric distribution to represent the distribution of the turbine blade fatigue life. The Anderson-Darling Test statistic is used to compare how well each distribution fits the given data. The smaller the A-D test statistic the better the fit of the parametric distribution. As shown in Figure 3, only the lognormal distribution properly represents the empirical data as it exhibited the smallest A-D statistic and visually captures the empirical data as shown by its probability plot. The other candidate distributions poorly model the data in the tail regions of the distributions. Thus, the popular Weibull and normal distributions are inferior to the lognormal distribution for this data set. Therefore, the baseline distribution chosen for this problem is the lognormal distribution. The parameter estimates were calculated using the maximum likelihood method and found to be 14.4912 and 1.6990 for the location and scale, respectively. The selection of the lognormal distribution can be validated using the A-D goodness-of-fit test provided earlier. Using equation (4) and (5) the adjusted A-D statistic is computed as 0.4571. The critical A-D value can be found using tables published by Stephens[1] for a normal distribution. Remember that the logarithm of the random variable, Fatigue life, follows a normal distribution. The critical value at a significance level of $\alpha=0.05$ for a large number of simulations (i.e. n>>100) and for the case where the distribution parameters are *estimated* from the data is found to be 0.787. Therefore, with significant evidence the null hypothesis that the data follows a lognormal distribution with the computed parameter estimates is accepted and our selection of the appropriate distribution can be supported statistically. One can infer from this finding that the fatigue life, given normally distributed input, follows a lognormal distribution. This result can also be proven using analytical statistics.

The number of simulations used to generate the data set was originally chosen qualitatively as a balance between at least a moderate sample size, recommended for Monte Carlo simulations, and a sample size that can be processed efficiently to identify the appropriate life response distribution. However, it is well known that low probability predictions require a large sample size to achieve an acceptable accuracy when using the Monte Carlo simulation method. Shooman's formula[12] for estimating the error (%) of probability predictions from Monte Carlo data is useful for estimate the required number of Monte Carlo simulations. The formula is given as

$$\varepsilon = \sqrt{\frac{(1-P_f)}{N \cdot P_f}} x200 \quad (12)$$

where N is the number of simulations, and P_f is the probability level. Solving for N gives

$$N = \frac{1-P_f}{P_f}\left(\frac{\varepsilon}{200}\right)^{-2} \quad (13)$$

It is assumed that the lowest anticipated probability of failure level is 0.001 and an acceptable error is 5%. Thus, 1.5984E6 simulations are required to achieve this. Evaluating the location and scale parameter using the increased sample size had only a slight affect on the location and scale. They were recomputed as 14.5017 and 1.7047, respectively.

Should the required number of simulations be prohibitive to execute, there are several methods under the class of variance reduction techniques[13] that can be employed to improve the sampling points within the low probability regime. In this case, a mixed approach can be applied whereby a manageable number of Monte Carlo simulations can be executed then augmented with further variance reduction samples to improve the low probability prediction capability.

Predictions can now be made for the failure probability as a function of time. Using the lognormal parametric distribution, failure probabilities of as low as 1% and 0.1% can be expected at 37641 and 10236 cycles, respectively, with an error of at most 5% for the 0.1% probability level. Also, the predicted fatigue life at 0.1% can be converted to years by assuming that on average the aircraft takeoff and landing cycle occurs four times daily resulting in a 0.1% failure probability at 7.01 years of operation. Stated in another way, one can expect 1 in 1,000 turbine blades to have failed after 7 years. However, these predictions are valid only for the given mean values of such variables as Altitude and Mach number. If for example the mean cruise Altitude or Mach number were to be different, then this process would need to be repeated. The generation of a predictive model of the distribution function as affected by the values of suspect influential variables, known as covariates, is now demonstrated.

Covariate Model of Component Failure Distribution

The previous section determined that the baseline fatigue life distribution is best represented with a two-parameter lognormal distribution. As discussed earlier, one can use either the PHM or ALT methods to parameterize the failure distribution to so that it is a function of multiple explanatory (covariate) variables. This essentially is a parametric way to create a multivariate distribution function utilizing the baseline life distribution. The PHM method is advantageous because it can be used in a semi-parametric sense to prevent erroneously specifying an incorrect baseline distribution. However, our baseline analysis and subsequent A-D goodness-of-fit hypothesis test showed that the lognormal distribution function accurately modeled the fatigue life distribution. Therefore, the ALT method was selected.

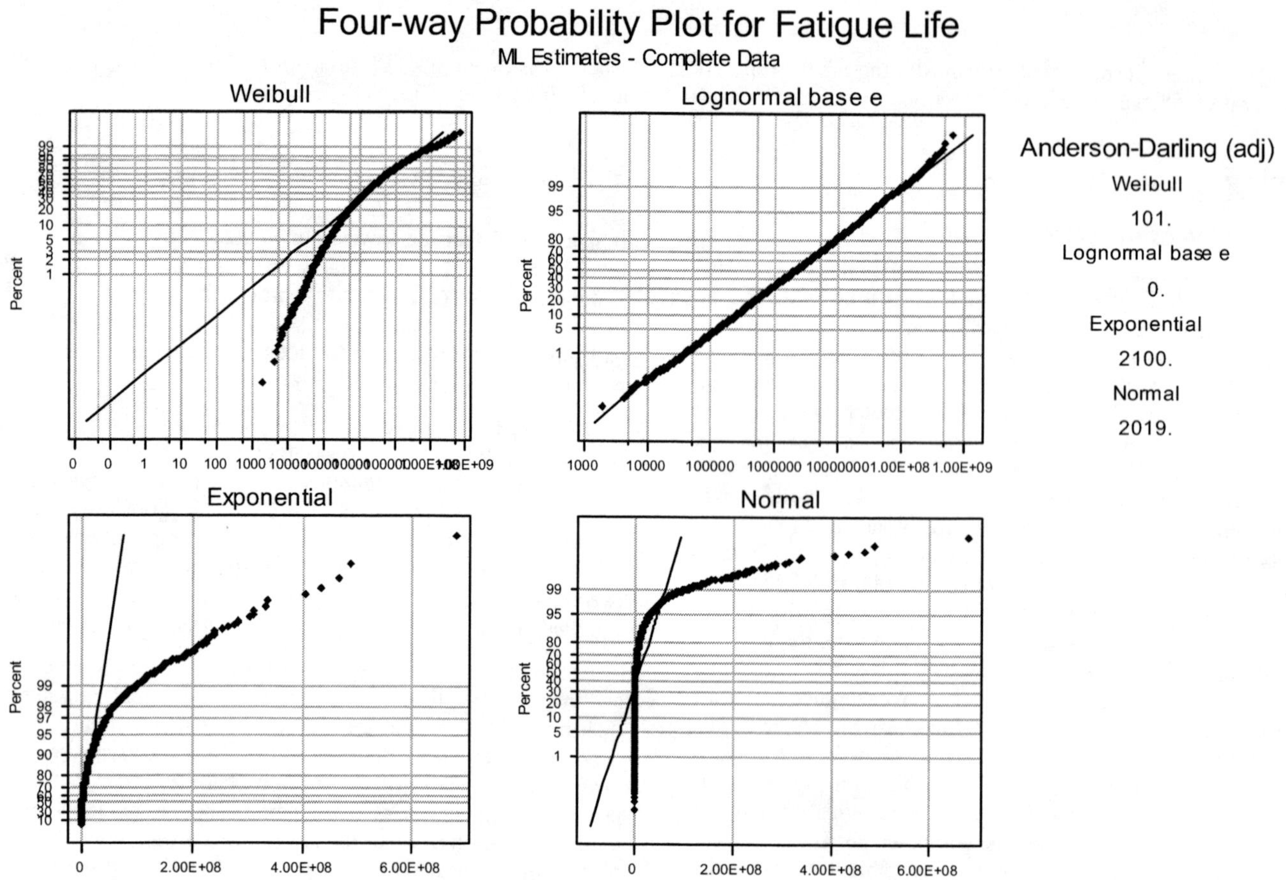

Figure 3: Probability plots of four hypothesized failure distributions

The ALT method requires the baseline survivor function which is determined in this case from the probability density function (PDF). The probability density function for the lognormal distribution is

$$f(x,\theta)=\frac{1}{\theta_2\sqrt{2\pi}}e^{\frac{-(\ln x-\theta_1)^2}{2\theta_2}} \tag{14}$$

where θ_1 and θ_2 are the location and scale of the distribution, respectively, corresponding to the mean and variance of the natural logarithm of the fatigue life. Integrating over x gives the cumulative density function

$$F(x,\theta)=\int_0^x \frac{1}{\theta_2\sqrt{2\pi}}e^{\frac{-(\ln \tau-\theta_1)^2}{2\theta_2}}\,d\tau \tag{15}$$

Since $\ln(x)$ is normally distributed, the cumulative density function (CDF) from equation (15) can be determined alternatively using

$$F(x,\theta)=P\left(Z\le\frac{\ln x-\theta_1}{\theta_2}\right)=\Phi\left(\frac{\ln x-\theta_1}{\theta_2}\right) \tag{16}$$

which is easily found using any standard normal table. The survivor function is the complement to the CDF given as

$$S(x,\theta)=1-F(x,\theta) \tag{17}$$

The log-quadratic link function is used to account for the covariate parameters in the ALT model. A quadratic, polynomial function is assumed for the exponent of the link function and is given as

$$g(\beta z)=\beta_1 z_1+\beta_2 z_2+\beta_3 z_1 z_2+\beta_4 z_1^2+\beta_5 z_2^2 \tag{19}$$

which upon substitution into the log-quadratic link function gives

$$\psi(z)=e^{g(\beta z)}=e^{\beta_1 z_1+\beta_2 z_2+\beta_3 z_1 z_2+\beta_4 z_1^2+\beta_5 z_2^2} \tag{20}$$

The PDF of the ALT model then is expressed as

$$f(x,\theta,\beta,z)=\frac{1}{\theta_2\sqrt{2\pi}}e^{\frac{-(\ln(x\cdot\psi(z))-\theta_1)^2}{2\theta_2}} \tag{22}$$

The ALT survivor function then becomes

$$S(x,\theta,\beta,z)=1-\Phi\left(\frac{\ln(x\psi(z))-\theta_1}{\theta_2}\right) \tag{23}$$

The objective then is to acquire samples of the life at various values of the *covariate vector*, z, and use the maximum likelihood method to estimate the *coefficient vector* $\boldsymbol{\beta}$, as well as the vector of baseline distribution parameters, $\boldsymbol{\theta}$.

Model Generation

As with any exercise of determining an appropriate model, a sample set which is indicative of the space to be modeled should be pursued. Determining the appropriate covariate model sample set, however, can be challenging. Meeker and Escobar recommend several considerations when designing the test matrix14. The considerations relevant to this problem are now discussed.

The coefficients of the ALT model are to be determined during the modeling process. Therefore, the accuracy of these coefficients is dependent on this process, especially the number of simulations at each vector value of the covariate variables. An initial estimate of the number of required simulations at each level of the covariates is helpful as a guide to designing the test matrix. Meeker and Eskobar recommend that the designer estimate the coefficient values from experience or available data. Then these temporary coefficient values should be utilized with the ALT model to allow the designer to rapidly explore both the levels of the covariate variables as well as the number of required simulations. Such an initial estimate can be used to evaluate different test schemes such as the spacing of covariate vector values as well as the number of simulations allocated to the various cases. Equal spacing and allocation of test points is the typical starting point. The authors of this paper also feel that under a limit of the number of experiments to run, one should allocate the number of experiments appropriately to each design of experiments case so that the variation of the initial maximum likelihood parameter estimates are minimized. For instance, if certain combinations of the covariates produce a lower probability level then one should allocate more experiments to these cases so that the accuracy at this probability level is improved. Also, one could alternatively use the classical power of test method to statistically determine the required number of total experiments[15]. For this study, a simple design of experiments test matrix with equally spaced covariate values was found to be sufficient.

With these considerations in mind, the design of experiments method was employed. To be able to model quadratic behavior, a 3-level full-factorial design was created using the two covariate variables, Altitude (z_1) and Mach number (z_2). The 3-level design was chosen here to be able to capture quadratic behavior should it exist. Also, the small number of candidate variables permitted a full-factorial design to be utilized. However, should a larger number of variables and levels be pursued a fractional factorial DOE, such as the popular Box-Benkin or Central Composite Design, is recommended.

The ranges of the two covariates are given in Table 4. Nine different DOE cases, as shown in Table 5, each with a unique value of Altitude and Mach number are required for the full-factorial design. However, the number of simulations to be executed for each combination of covariate variable values is somewhat subjective. Within this study the number of simulations for each case was chosen based on a fraction, say 1%, of the number suggested using Shooman's model (equation 13). This is an initial value which can be modified based on the adequacy of the model to be discussed later. An initial value of 10,000 simulations was chosen.

Table 4: Covariate Variable Ranges

Covariate	Minimum (-1)	Centerpoint (0)	Maximum (+1)
Altitude	34000	35000	36000
Mach	0.725	0.745	0.765

Table 5: Coded DOE Case Values

Case	Altitude	Mach
1	-1	-1
2	-1	0
3	-1	1
4	0	-1
5	0	0
6	0	1
7	1	-1
8	1	0
9	1	1

The results of the MLE fit of the covariate coefficients to 10,000 Monte Carlo simulations generated for each case are summarized in Table 6. The whole model test verifies that the covariates considered provide an exceptional model for representing the failure distribution while the standard error for each of the covariate coefficient estimates is minimal. However, the likelihood ratio tests for each coefficient show that the interaction and quadratic terms can be neglected in the final ALT model. Based on the model statistics produced by the maximum likelihood estimation approach, the number of case simulations was deemed to be sufficient.

Table 6: Parametric Survival (ALT) Fit Results

Parametric Survival Fit

Log-Normal Distribution

Whole Model Test

Model	-LogLikelihood	ChiSquare	DF	Prob>Chisq
Difference	7428.57462	14857.15	5	0.0000
Full	175722.538			
Reduced	183151.113			

Parameter Estimates

Term	Estimate	Std Error
Intercept	14.5049955	0.0127079
Altitude	0.814098	0.0069604
Mach	-0.3457485	0.0069604
Mach*Altitude	0.00711812	0.0085247
Altitude*Altitude	-0.0242726	0.0120558
Mach*Mach	0.01575175	0.0120558
Sigma	1.70494715	0.0040186

Effect Likelihood Ratio Tests

Source	Nparm	DF	L-R ChiSquare	Prob>ChiSq
Altitude	1	1	12734.8699	0.0000
Mach	1	1	2434.24084	0.0000
Mach*Altitude	1	1	0.69721637	0.4037
Altitude*Altitude	1	1	4.05351982	0.0441
Mach*Mach	1	1	1.70712408	0.1914

Validation of Covariate Model

As with any model, one should determine whether it is an adequate representation of the actual physical phenomena being modeled. However, with a probabilistic response this quickly becomes prohibitive for physical validation and more practical using computational simulation. Large simulation Monte Carlo analyses can be used to validate the model accuracy by verifying the probability or life prediction at various values of Altitude and Mach number. However, the choice of covariate value cases from which to evaluate the model or even how many cases to pursue is somewhat arbitrary. In the interest of computational efficiency the minimal number of cases necessary to assess the accuracy of the model is sought. For this study a random sample of validation cases was generated numbering roughly around half of the number of original cases used to create the model. The metric to be used to assess the accuracy of the model is the percent error of the probability predicted by the covariate model for three probability levels: 0.01, 0.1, and 0.50. Four cases were randomly generated to validate the model and are given in Table 7. The last three columns give the percent error of the model in predicting the life as compared to that which was computed using large sample Monte Carlo simulation. Notice that the maximum error across all of these points is 4.7% for the 1% probability of failure point, which is well within the 5% error originally specified for the baseline distribution probability prediction. Should this percent error be deemed to high, further simulation points could be run within this region of the probability space to improve the model. Therefore, the covariate model proved to be accurate across the probability space under consideration.

Table 7: Percent Error of Coded Validation Cases

			Probability Level (%)		
Case	Altitude	Mach	1	10	50
10	0.2076	0.4936	0.6	0.5	0.3
11	-0.4556	-0.1098	1.9	0.8	0.6
12	-0.6024	0.8636	4.0	1.4	1.7
13	-0.9695	-0.068	4.7	2.1	1.0

Parametric Life Prediction

Now that the ALT model has been created and validated, several exercises are possible. For instance, the life corresponding to a certain probability of failure, as shown by Figure 4, can be predicted for any combination of Altitude and Mach number within the specified ranges considered. Thus, an optimal setting of these two flight condition parameters could be searched that would maximize the life for a given acceptable probability of failure. Conversely, the probability of failure as a function of Altitude and Mach number for a given life, as shown by Figure 5, can be assessed. Further, the ALT model could serve as a robustness predictor enabling rapid identification of robust points from which to operate the aircraft. Additionally, the ALT model process could be repeated for design variables such as blade length and stress concentration (geometric concentration factor) thus creating a truly probabilistic design tool.

Another approach to create a parametric response function of reliability, a nondeterministic quantity, was pursued by the authors in an earlier study where the Response Surface Method (RSM) was utilized[16]. The response was reliability and the functional relationship was determined through a least squares solution of a quadratic formula relating flight and usage conditions and the reliability of a gas turbine component. Although the RSM produced an accurate response function of reliability, it was necessary to run at least 1,000,000 simulations per case to achieve such accuracy. With the covariate model explored within the current study, a comparable accuracy was achieved with much less simulations. Additionally, the RSM method does not provide statistical properties of the model as a function of the number of simulations executed. Thus, this potentially useful information is lost. With the covariate model approach, each simulation point is input into the maximum likelihood calculation and therefore used in determining the statistical properties of the assumed functional model. For example, the expected value and variation of each ALT coefficient is a direct function of the number of simulations executed as well as the assumed baseline probabilistic distribution. This is an important property of the covariate model approach which is necessary for creating a truly parametric life distribution function.

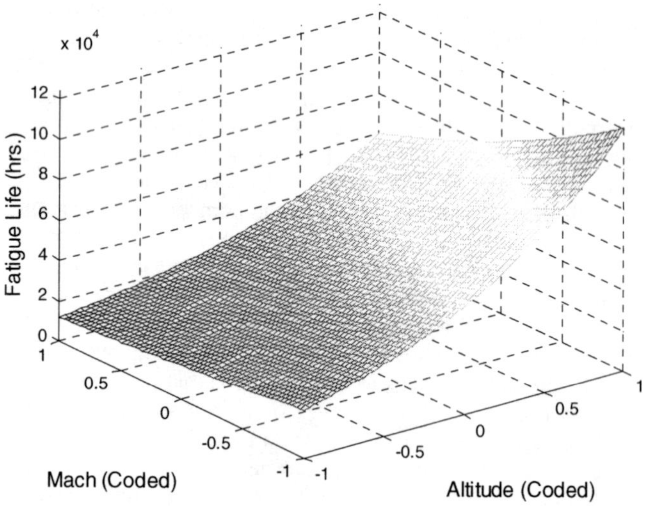

Figure 4: Life prediction using covariate model (P_f=1%).

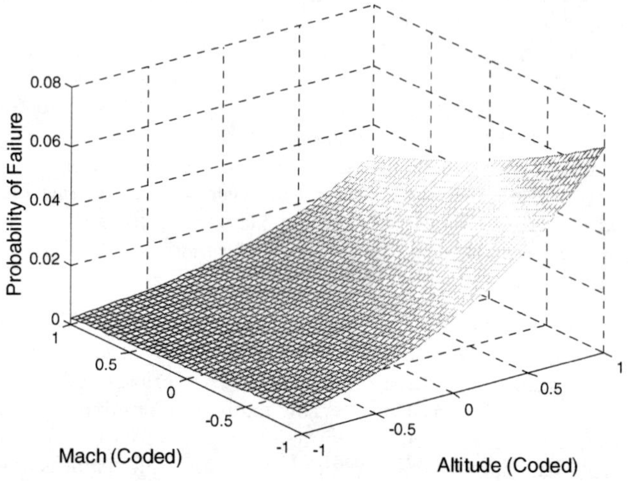

Figure 5: Probability prediction using covariate model (Life=50,000 hrs).

CONCLUSIONS

The covariate modeling method, popular in reliability testing, has been applied to simulation generated fatigue failure lives from an integrated gas turbine blade modeling and simulation environment. A multivariate probabilistic distribution model of the turbine fatigue life as a function of time, flight altitude and speed was produced using this method. Preliminary investigation of the fatigue failure behavior revealed lognormal behavior which was quantified using statistical hypothesis testing. General recommendations for implementing covariate models with simulation-based data inferred from the study include combining design of experiments and Shooman's law of Monte Carlo simulation error to assist in creating the fatigue life sampling test matrix. A random case generation procedure was executed to validate the accuracy of the model.

The model provides a functional relationship of the fatigue failure probabilistic distribution with respect to top-level system parameters and thus is a useful tool for conducting preliminary reliability design, optimization, and even prediction activities. Future work is anticipated to include developing a method of considering multiple failure modes simultaneously as well as a more refined and detailed procedure for utilizing the covariate model method with simulation results. A potential further advancement of this method could be formulation of the model generation steps to accept data generated from alternative probabilistic methods.

ACKNOWLEDGMENTS

This work is being supported by the N.A.S.A. University Research Engineering Technology Institute (URETI) for Aero-Propulsion and Power Technology (APT). Dr. Bryce Roth of the Aerospace Systems Design Laboratory at Georgia Tech is to be thanked for his assistance in the final preparation of this paper.

REFERENCES

1. Stephens, M. A. (1974). EDF Statistics for Goodness of Fit and Some Comparisons, Journal of the American Statistical Association, Vol. 69, pp. 730-737.
2. Stephens, M. A. (1976). Asymptotic Results for Goodness-of-Fit Statistics with Unknown Parameters, Annals of Statistics, Vol. 4, pp. 357-369.
3. Stephens, M. A. (1977). Goodness of Fit for the Extreme Value Distribution, Biometrika, Vol. 64, pp. 583-588.
4. Stephens, M. A. (1977). Goodness of Fit with Special Reference to Tests for Exponentiality, Technical Report No. 262, Department of Statistics, Stanford University, Stanford, CA.
5. Stephens, M. A. (1979). Tests of Fit for the Logistic Distribution Based on the Empirical Distribution Function, Biometrika, Vol. 66, pp. 591-595.
6. Cowles, B.A., 1996. "High Cycle Fatigue in Aircraft Gas Turbines-an Industry Perspective," International Journal of Fracture, Vol. 80, pp. 147-163.
7. CFM56 Thermodynamic 1-D Cycle Model, NASA Glenn, September 18, 2000.
8. NEPP-Navy/NASA Engine Program (NNEP89) Engine Cycle Simulation Computer Code.
9. Hill, P. and Peterson, C., 1992. Mechanics and Thermodynamics of Propulsion, 2nd Edition, Addison-Wesley Publishing Company, Inc., New York, NY.
10. Basquin, O., 1910. "The Exponential Law of Endurance Tests," Proceedings of the American Society for Testing and Materials Vol. 10, pp. 625-630
11. Morrow, J., 1968. Fatigue Design Handbook-Advances in Engineering, Vol. 4, Sec. 3.2, pp. 21-29. Warrendale, PA: Society of Automotive Engineers.
12. Shooman, Martin, 1968, Probabilistic Reliability, an Engineering Approach, McGraw-Hill, New York, NY.
13. Haldar, A. and Mahadevan, S., 2000, Probability, Reliability, and Statistical Methods in Engineering Design. New York, NY: John Wiley and Sons.
14. Meeker, W.Q., and Escobar, L.A., 1998, Statistical Methods for Reliability Data, John Wiley & Sons, Inc, New York.
15. Sethuraman, J., and Singpurwalla, N.D., 1982, "Testing of Hypotheses for Distributions in Accelerated Life Tests," Journal of American Statistical Association, vol. 77, no. 377, March, pp. 204-208.
16. Wallace, J.M., and Mavris, D.M., 2002, "Propulsion System Reliability Prediction and Optimization Using a Multi-Physics Environment," AIAA/ISSMO Symposium on Multidisciplinary Analysis and Optimization, Atlanta, GA, September 4-6.

Proceedings of IMECE'03
2003 ASME International Mechanical Engineering Congress
Washington, D.C., November 15–21, 2003

IMECE2003-42391

IMPACT OF PROGNOSIS ON ASSET LIFE EXTENSION AND READINESS

Yevgeny Macheret
Institute for Defense Analyses
4850 Mark Center Drive
Alexandria, Virginia 22311-1882, USA
Tel: 703-845-2261
Fax: 703-931-7792
e-mail: ymachere@ida.org

Leo Christodoulou
Defense Advanced Research Projects Agency
3701 N. Fairfax Drive
Arlington, Virginia 22210, USA
Tel: 703-696-2374
Fax: 571-218-4553
e-mail: lchristodoulou@darpa.mil

ABSTRACT

Fatigue response of structural components is determined by environmental conditions, material microstructure, and loading history. Variation of these factors results in significant scatter in fatigue-crack growth rates and component life. In this paper, the impact of prognosis capability on asset life extension and readiness is evaluated. Fatigue-crack growth data on aluminum samples under controlled spectrum loading are used to describe the statistics of the crack-size distribution. Several sensors with different probability of detection (POD) characteristics are considered for detecting cracks of critical size, and the effect of the POD on the component life extension is evaluated. Although the crack-detection capability leads to the asset life extension, it is not sufficient to maintain required mission readiness. On the other hand, the prognosis capability, which is based on the knowledge of the component's current damage state, damage evolution laws, and upcoming mission loading, allows required mission readiness to be maintained.

Keywords: component life, fatigue crack, failure response, prognosis.

NOMENCLATURE

a	crack length
a_0	initial crack length
a_{cr}	critical crack length
GAU	standard Gaussian distribution
P	cumulative probability distribution function
f	density probability distribution function
Tm	mission time
Ts	average service life
POD	probability of detection
σ	standard deviation

1. INTRODUCTION

Current time-based maintenance of operational assets requires performing periodic inspections or terminating service life of components at a predetermined number of hours of operation. As common experience shows, the service life of components under fatigue loading has a considerable degree of uncertainty, making an accurate prediction of the service life of a specific component difficult if not impossible. In addition, in many cases it is impossible to detect the damage before it becomes critical. As a result, to limit the risk of a component failure to an acceptable level, the service is typically terminated at some time very early in the life of the component, resulting in a significant underutilization of its service life.

A current DARPA prognosis program is focused on developing technologies to predict the remaining life of critical components in military assets, giving the military commanders a capability to extend the assets' service life and increase asset readiness. A recent simulation of the prognosis capability demonstrated significant benefits of the prognosis approach in terms of general trends by comparing time-based and prognosis-based maintenance procedures [1]. The authors of this study performed the calculations assuming standard statistical distributions (Gaussian distributions with specified mean and standard deviation) of the component life and initial damage-state conditions. Those distributions were not representative of common materials and components under typical loading conditions; instead, they were chosen to simulate general trends.

More accurate realistic estimates of prognosis benefits for typical materials and loading conditions are important, however, because the operational and economic advantages will be the driving force for transitioning this capability to military applications. In addition, since the prognosis approach includes use of damage-detection sensors, a realistic analysis should include the effect of the sensors' detection characteristics, such as probability of detection (POD), on the service life extension.

In this paper, we evaluate the impact of prognosis capability on asset life extension and readiness. Fatigue crack growth data on a crack emanating from a rivet hole in a common aluminum alloy, Al 7475-T3, under a controlled loading spectrum are used to describe the statistics of the crack-size distribution. The service life in terms of flight hours is then calculated based on a small number representing acceptable risk of failure. Next, several sensors with different POD characteristics are considered for detecting cracks, and the effect of the POD on the component life extension is evaluated. A summary of the results is presented in the conclusion.

2. LIFE EXTENSION AND MISSION READINESS ANALYSIS

2.1 Experimental Observations

Experimental observations show that fatigue-crack propagation behavior in structural components involves considerable statistical variability, which is caused by the variation in material's microstructure, loading history, and environmental conditions. The fatigue-crack growth models that have been proposed over the years can be described by the following general formulation: $da/dt = F$(loading, a, material properties, environmental conditions). If the applied loading and environmental conditions are kept constant, then the variability in the fatigue response is often described by various statistical distributions. When the Paris fatigue-crack growth equation is used, for example, Wiebull distribution for the initial flaw size and log-normal distribution for fatigue-crack growth proportionality coefficient are often used [2].

A convenient approach to describe fatigue-crack growth rate under spectrum loading is reported in [3]. To account for statistical variability of the crack growth rate, the authors propose the following formulation: $da/dt = X(t)F$(loading, a), where X is a stochastic variable with a median value of one. The authors consider a fatigue crack emanating from a rivet hole in Al 7475-T73 under bomber spectrum loading. They show that the data could be accurately described by $da/dt = X\,Ca^n$, where X is a time independent log-normal random variable: Log $X = GAU(0,\sigma_Z)$. The values of the parameters in this formulation are found from the experimental data: $\sigma_Z = 0.109$, $C = 0.000284$, and $n = 1.014$. The initial crack size is assumed to be 0.004 inches. The details of the specimen geometry and spectrum loading are described in [4].

2.2. Cumulative Failure Distribution Calculations

2.2.1 Analytical formulation

The crack-growth rate,

$$\dot{a} = XCa^n \, , \tag{1}$$

is integrated to obtain the crack length as a function of time,

$$a(t) = \frac{a_0}{\left(1 - XmtCa_0^m\right)^{1/m}} \, , \tag{2}$$

where $m = n - 1$.

The stochastic variable X is introduced to describe the crack-size distribution at a given time, t, as shown by

$$X(x = a) = \frac{a_0^{-m} - x^{-m}}{mCt} \, . \tag{3}$$

The log(X) is then a standard Gaussian distribution, and

$$f(v) = f\left[\log\left(\frac{a_0^{-m} - x^{-m}}{mCt}\right)\right] = GAU(0,\sigma_z) = \frac{1}{\sqrt{2\pi\sigma_z}} e^{\frac{-v^2}{2\sigma_z^2}} \, , \tag{4}$$

with zero mean and standard deviation, σ_z, determined from the experimental data.

If the component failure is defined as the fraction of components that will have a crack that exceeds a defined critical length, then the cumulative probability of failure is calculated by integrating the crack-size density distribution function from the critical crack length to infinity:

$$P(a \geq a_{cr}) = \int_{a_{cr}}^{\infty} f(v(a))da \, . \tag{5}$$

When crack-detection sensors with a specified POD are used, some cracks are detected and removed (according to the sensor's POD characteristics), and the cumulative probability of failure is

$$P(a \geq a_{cr}) = \int_{a_{cr}}^{\infty} f(v(a))[1 - POD(a)]da \, . \tag{6}$$

In this paper, the value of the critical crack length is chosen to be 0.07 inch. This selection was motivated by the analysis of the cause of the two C–130 failures in the summer of 2002 [5]. After the investigation, the FAA issued airworthiness directive No. 2002-19-14, requiring C–130 operators to "inspect fastener holes in the wing center section…for cracks as small as 0.1 inch or perhaps smaller." As an acceptable risk of failure, a common number of 0.1 percent (i.e., failure of 1 component out of 1,000) is chosen for the analysis.

2.2.2 Service life without damage-detection sensors

Figure 1 shows the results of the cumulative probability of failure calculations with no damage sensors. At 4,910 flight hours, 1 out of 1,000 components is expected to fail (i.e., a crack length in this component is expected to reach 0.07 inches). At this time, that component will have to be inspected or replaced and the system taken out of service. Note the significant underutilization of the component's service life.

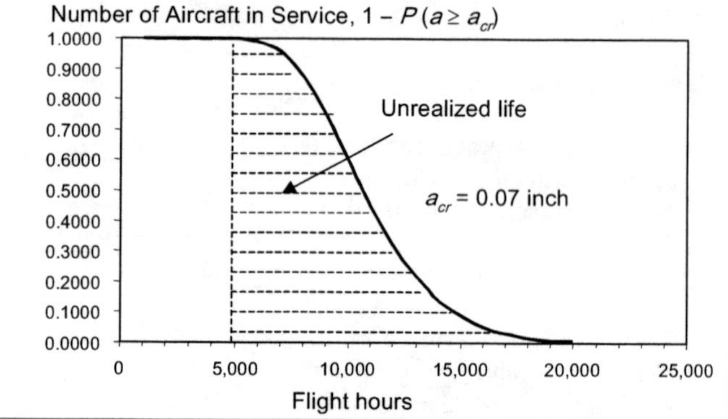

Figure 1. Cumulative probability of failure calculations with no damage sensors

size, but it does not allow for predicting whether the component under consideration will survive the upcoming mission.

This can be restated as follows. The sensors do allow for the extension of the service life, that is, operating the asset without increasing the failure risk. (If, in fact, a sensor is capable of detecting all cracks before they become critical, the component can be operated without failure.) Using the component in the upcoming mission, therefore, will not result in the failure of the component, since the crack will be detected and the component removed before the crack reaches the critical size. Removing the component, however, means that the asset will not be operational (i.e., it will not be mission capable). In other words, although the sensors allow service-life extension and prevent the component failure, they do not reduce the risk of mission failure.

2.2.3 Life extension with damage-detection sensors

Many sensors are available for detecting fatigue damage in structural components. Figure 2 shows the POD characteristics of three novel magneto-optical damage sensors [6]. Note that the sensors allow some of the cracks to grow beyond the critical crack length. To evaluate the effect of the POD on service life extension, we introduce a fictitious sensor (with the POD characteristic shown in black dashed line) having the capability to detect all the cracks before they reach critical size.

The calculations of the cumulative service life distribution with damage detection sensors are performed according to Eq. (6), and the results are given in Fig. 3. Shown in this figure by arrows are the times at which the component service life has to be terminated. At each termination point the risk of failure is equal to 0.001 (i.e., a fatigue crack reaches 0.007 inches in 1 component out of 1,000). Note that the POD has a significant effect on the service life extension. Utilizing sensor 1, for example, results in the component life extension from 4,910 (no sensors) to 9,500 flight hours, an increase in service life of over 90 percent.

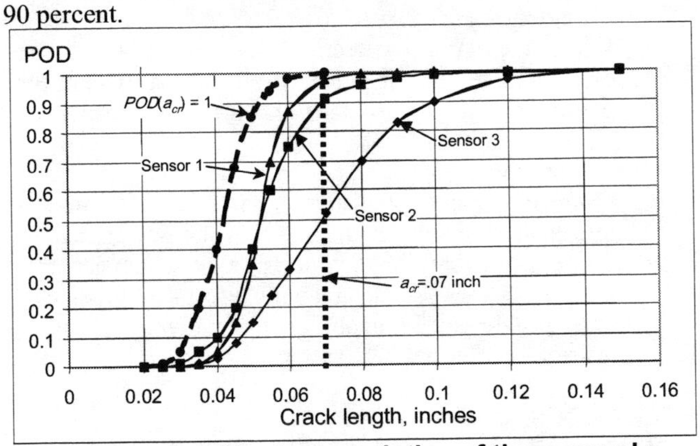

Figure 2. POD characteristics of three novel magneto-optical damage sensors

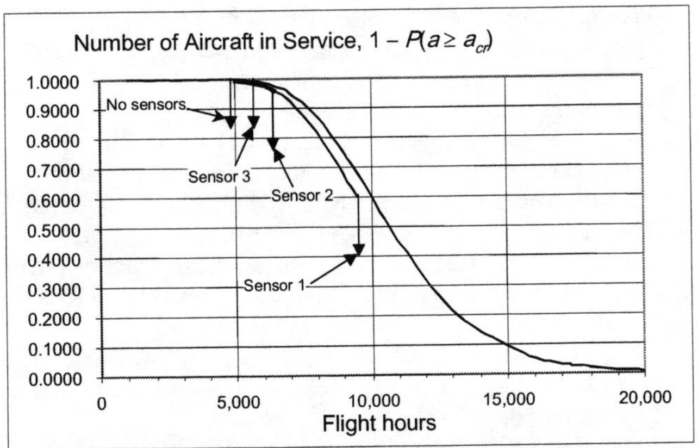

Figure 3. Flight hours at which the component service life has to be terminated

3. MISSION READINESS ANALYSIS

While the damage sensors extend the component service life by detecting cracks, by themselves they do not provide data regarding the remaining service life. In other words, this damage-detection capability only indicates a flaw of a certain size, but it does not allow for predicting whether the component under consideration will survive the upcoming mission.

A problem of calculating the mission-failure risk can be stated in terms of determining the probability that the upcoming mission will be successful, calculated as a function of the length of the upcoming mission and the initial damage state of the component.

The results of such calculations can be presented in terms of mission readiness (*MR*) as a function of the upcoming mission length and average asset service life. *MR* is defined as the ratio of the number of assets that are capable of performing the upcoming mission to the total number of assets. In this paper, *MR* is calculated as the ratio of the number of assets with crack size, after the mission is completed, below that of critical, to the total number of assets. In the calculations, it is assumed that a sensor capable of detecting all cracks before they reach critical size is available. (The POD characteristic of such a sensor is given in Fig. 2 by a black dashed line.)

Figure 4 shows the results. Note that as the average asset service life, *Ts*, and mission length, *Tm*, increase, *MR* decreases, as expected. For example, at *Ts* = 4,500 flight hrs and mission length *Tm* = 1,000 hrs, the risk of failure is almost 4 percent. That means that there is a 4-percent chance that this particular mission will fail. The significance of this event depends on the nature of the particular mission. One can easily visualize missions for which even a 1-percent failure rate is unacceptable.

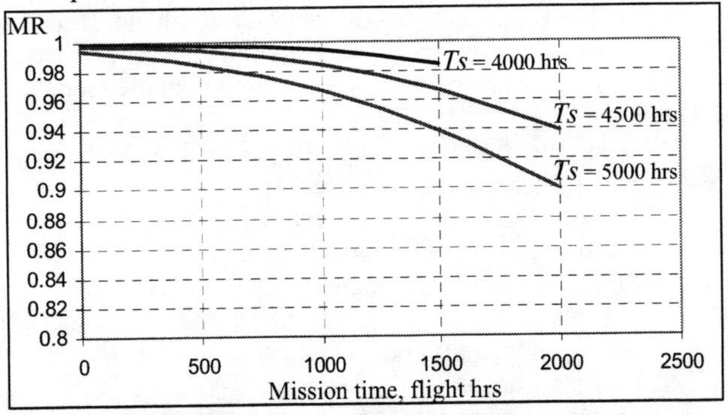

Figure 4. Flight hours at which the component service life has to be terminated

4. PROGNOSIS APPROACH

The prognosis approach is focused on developing a capability to predict the remaining service life of the asset under consideration. Such a capability makes it possible to determine whether the asset will be able to perform the upcoming mission (i.e., prediction of mission readiness).

Figure 5 shows the essential components of the prognosis approach: state awareness, mission history and material property information database, damage evolution modeling, future mission profile, and decision algorithm capability.

The damage prognosis operation can be described as follows. Interrogating the system (if possible, real time, online) yields information on the system's state awareness, which includes system-performance indicators: speed, acceleration, temperature, component loads, etc., as well as possible component damage characteristics (i.e., fatigue, wear, corrosion, etc.).

The database module includes information on the mission history and maintenance and the component's material properties. This information is used as input to the damage-evolution models for evaluating the current damage state.

The loading conditions of the upcoming mission (speed, duration, altitude, etc.) are "translated" in the component loading by means the structural analysis.

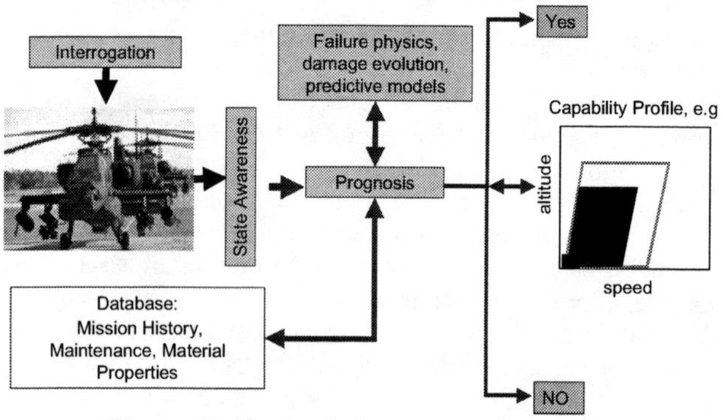

Figure 5. Essential components of the prognosis approach

The prognosis algorithm utilizes knowledge of the current damage state, the upcoming loading conditions, and the damage evolution models to estimate the remaining life of the asset. If the predicted damage exceeds a critical value during the upcoming mission, the mission is rejected (other missions for which the damage remains below critical may be recommended).

The online damage-detection sensors supply information on a component's damage state to validate the predictive models and reduce the uncertainty associated with the modeling predictions.

To make the prognosis capability possible, technology development efforts must include the following:

- Sensors that monitor system performance indicators and detect damage in components;
- Structural analysis models that translate mission profiles and system performance indicators (speed, acceleration, altitude, temperature) into the components loads;
- Databases capable of storing and retrieving the information on mission history, material properties, and components damage state during the service;
- Physics of failure models capable of predicting damage evolution of the specific component (characterized by the microstructure and loading history) for a given upcoming loading conditions.

To summarize, the prognosis capability will allow predicting, for the loading conditions of the upcoming missions, the number or duration of future missions that the asset is capable of, thus *providing a capability to maintain required mission readiness.*

5. CONCLUSIONS

As a result of this study, the following conclusions are reached.

- For a common, time-based maintenance approach, a substantial variation in fatigue crack growth rate in Al 7475-T73 results in significant underutilization of components.
- Use of damage-detection sensors results in asset life extension. The POD characteristics of the damage sensors have a considerable effect on magnitude of the life extension.
- While the damage-detection capability extends the asset service life, it is by itself not sufficient for predicting the remaining life of the asset and its mission readiness.
- A prognosis capability, which includes state awareness, mission history and material property information database, damage evolution modeling, future mission profile, and decision algorithm capability, allows the asset's remaining life to be predicted, thus providing a capability to ma*intain required mission readiness.*

REFERENCES

1. Nicoll, J.F., Cartier, J.F., 2001, DARPA Prognostics "Seedlings Program" meeting, Institute for Defense Analyses, Alexandria, Va.
2. Tryon, R.G., Cruse, T.R., and Mahadevan, S., 1996, "Development of a Reliability-Based Fatigue Life Model for Gas Turbine Engine Structures," *Engineering Fracture Mechanics,* **53**, No. 5, pp. 807–828.
3. Yang, J.N., Manning, S.D., Hsi, W.H., and Rudd, J.L., 1987, "Probabilistic Fracture Mechanics and Reliability," J.W. Provan, ed., Martinus Nijhoff Publishers, Dordrecht, The Netherlands, ICBN 90-247-3334-0, pp. 171–211.
4. Naronha, P.J., et al., 1978, "Fastener Hole Quality," Air Force Flight Dynamic Laboratory, AFFDL-TR-206, Vol. II.
5. Dornheim, M.A., 2002, "Metal Fagigue Cited in Fire-fighter Crashes," *Aviation Week and Space Technology,* September 30, p. 35.
6. Shih, W.C.L., and Fitzpatrick, G.L., "Magneto-optic Imaging Technology: A new Trend for Aircraft Inspection," *The AMPTIAC Quarterly,* **6**, No. 3, p. 17.

Proceedings of IMECE'03
2003 ASME International Mechanical Engineering Congress
Washington, D.C., November 15–21, 2003

IMECE2003-42749

RISK-BASED ANALYSIS AND DECISION MAKING IN MULTI-DISCIPLINARY ENVIRONMENTS

Martin S. Feather **Steven L. Cornford** **Kelly Moran**

Jet Propulsion Laboratory, California Institute of Technology

ABSTRACT

A risk-based decision-making process conceived of and developed at JPL and NASA, has been used to help plan and guide novel technology applications for use on spacecraft. These applications exemplify key challenges inherent in multi-disciplinary design of novel technologies deployed in mission-critical settings:

1) Cross-disciplinary concerns are numerous (e.g., spacecraft involve navigation, propulsion, telecommunications). These concerns are cross-coupled and interact in multiple ways (e.g., electromagnetic interference, heat transfer).

2) Time and budget pressures constrain development, operational resources constrain the resulting system (e.g., mass, volume, power).

3) Spacecraft are critical systems that must operate correctly the first time in only partially understood environments, with no chance for repair.

4) Past experience provides only a partial guide: New mission concepts are enhanced and enabled by new technologies, for which past experience is lacking.

The decision-making process rests on quantitative assessments of the relationships between three classes of information – objectives (the things the system is to accomplish and constraints on its operation and development), risks (whose occurrence detracts from objectives), and mitigations (options for reducing the likelihood and/or severity of risks). The process successfully guides experts to pool their knowledge, using custom-built software to support information gathering and decision-making.

Keywords: Risk, Decision-making, Design, Novel Technology

INTRODUCTION

NASA's Mission statement reads: "*To understand and protect our home planet. To explore the Universe and search for life. To inspire the next generation of explorers . . . as only NASA can.*". In April 2002 the NASA Administrator stated: "*... In broad terms, our mandate is to pioneer the future . . . to push the envelope . . . to do what has never been done before. ...*" [1]. These quotes emphasize the novel aspects of NASA's activities. Spacecraft design involves multiple disciplines (e.g., interplanetary navigation, spacecraft propulsion, telecommunications, properties of materials in deep space environments). Spacecraft development and operation are driven by budget and time concerns (e.g., planetary configurations dictate launch windows). Spacecraft operate remotely from earth; so third-party repair is almost always impossible. Their purpose is to yield information of distant, unknown environments, meaning they often must operate in those same unknown environments. The aim to improve the quantity and quality of the science information they gather and in return drives the use of novel technologies in novel ways.

At first glance, these would seem to imply that spacecraft development and operation has little in common with earthly activities. Certainly the specifics of the information involved (e.g., behavior of materials in a deep space environment; interplanetary navigation) are particular to space exploration. However, the fundamental challenges that underlie NASA's activities are familiar to a wide range of endeavors – the challenges stem from the need to do multi-disciplinary design of novel technologies intended for deployment in critical settings. They are:

- Multiple disciplines are involved. No single individual has depth of knowledge in every one of these disciplines. Furthermore, these disciplines are cross-coupled and interact in multiple ways. As a consequence, there is the need to pool the knowledge from multiple discipline experts, and to conduct decision-making taking the sum total of this knowledge into account.
- Time and budget pressures constrain development (an almost universal phenomenon, e.g., development of almost

any product these days is subject to time to market needs, and aspects of competitiveness). Limited operational resources constrain many systems (e.g., think of power, mass and volume constraints on hand-held devices).

- Depending upon the purposes for which systems are employed, the consequences of failure during their operation could range from a minor nuisance to catastrophic. Spacecraft may be irreparably lost if things go too badly wrong. On earth, the systems themselves may be more repairable, but the not necessarily the consequences of their operational failure.
- Spacecraft must operate in a harsh and often unpredictable physical environment. While more may be known about the physical operating environment for Earth-bound systems, considerable uncertainty arises from the unpredictability of the ways in which users will make use of systems. In the extreme, a deployed system will induce a change in the environment around it, leading to an unpredictable feedback loop (e.g., the phenomenon of "E-type Software" described by Lehman [2]).
- Past experience provides only a partial guide as to how to develop the new system. For NASA's missions, new mission concepts are enhanced and enabled by use of new technologies, and/or of established technologies applied in novel ways. Past experience is lacking in either case. The same is true for almost any novel solution to a real world problem.

In response to these challenges, at JPL and NASA we have been developing and applying a risk-based approach to analysis and decision making for novel system applications. This paper summarizes the approach, and indicates how it addresses the challenges listed above.

NOMENCLATURE

Defect Detection and Prevention (DDP) – a risk-based decision-making process conceived of and developed at JPL and NASA.

Objectives – the things the system/technology is to accomplish and constraints on its operation and development.

Risks – the things whose occurrence would detract from attainment of Objectives.

Mitigations – options for reducing the likelihood and/or severity of Risks.

A RISK-BASED ANALYSIS AND DECISION-MAKING APPROACH

The focus of this paper is on a risk-based analysis and decision-making approach developed and applied at JPL and NASA. The approach is called "Defect Detection and Prevention (DDP)". The name reflects its origins as a structured method for planning the quality assurance of hardware systems [3]. Since then its scope has expanded to also encompass decision-making earlier in the development lifecycle, and to be applicable to software, hardware and systems [4].

The approach has been generalized to aid in the decision-making during the early phases of advanced technology and system development. Decisions made in these early phases are important because they have the most leverage to influence the development to follow. However, decision-making during these phases is challenging because information on which to base those decisions is incomplete and uncertain, and in the case of advanced technologies and systems, there is little past experience from which to extrapolate.

DDP's Simple yet Quantitative and Detailed Risk Model

DDP offers a conceptually simple yet quantitative, detailed model of risk as foundation for reasoning:

- The simple conceptual core of the DDP risk model rests on three sets of information: *objectives* – the things we want the system/technology to achieve, *risks* – the things that, should they occur, detract somewhat from the attainment of objectives, and *mitigations* – the things that we could choose to do to reduce risks (by reducing their likelihood of occurrence and/or their severity should they occur).
- The quantitative treatment derives from information on how much each risk (should it occur) detracts from attainment of each objective, and by how much each mitigation (should it be applied) reduces each Risk.
- The detail arises from the ability to populate the DDP model with system/technology specific objectives, risks and mitigations. These are open-ended sets of information, and the level of detail to which these are populated is not pre-ordained, but rather is determined in the course of the DDP application process, to the extent necessary to support decision-making.

Note that our use of the word "risk" may be non-standard – we have sometimes referred to these as "risk elements" or "failure modes". The DDP model of information does accommodate the elements of the traditional definition of risk, as follows: each DDP risk item has a likelihood (the calculation of which takes into account the effect of mitigations); each DDP risk item has a consequence, namely it detracts from attainment of objectives. The usual definition of risk as "probability * severity" is a measure we calculate from the DDP information, explained further in the next section.

The combination of these aspects yields an approach that can encompass a wide range of concerns – many problems can be cast in terms of objectives (what do we want), Risks (what can get in the way) and mitigations (what can we do about it). Their quantitative treatment allows reasoning about their net combined effect. For example, the biggest risks are those with the greatest sum total adverse impact on attainment of objectives. The ability to incorporate problem-specific detail permits the representation and reasoning over system, technology and application-specific nuances.

The later sections will examine the ways in which the DDP information model helps address the key challenges of multi-disciplinary decision making. In preparation, we now expand upon the description of the DDP model.

Details of DDP's Risk Model

In more detail, DDP deals with the following three sets of information:

- "**Objectives**" – these encompass all of the things the system/technology is to achieve, how and under what conditions it is to be operate, and how it is to be developed. In our area of spacecraft technologies, typical examples include science return objectives (e.g., quantity and resolution of data), aspects of its operational environment (e.g., ambient temperature, operating temperature, available electrical power), and nature of its development (e.g., must be ready for launch 2 years from now). Since

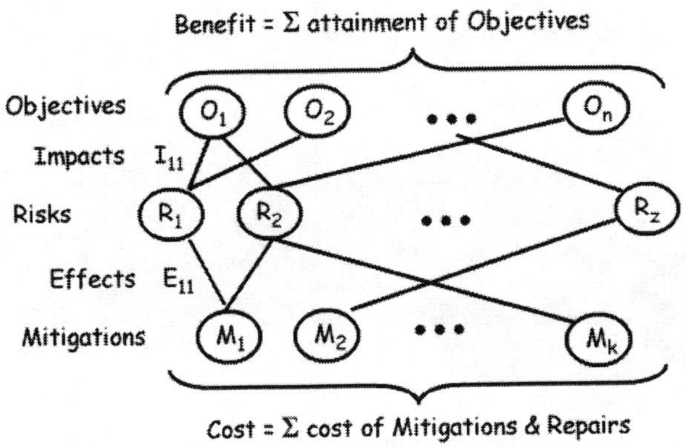

Figure 1. Topology of DDP information model

Figure 2. Topology of data in a completed DDP model

not all objectives are equally important, DDP allows users to assign "Weights" (numerical values of a unitless scale) to reflect their relative importance. For example, minimal success criteria ("must-have" capabilities) can be assigned high weights, while optional criteria ("nice to have" capabilities) can be assigned lesser weights.

- "**Risks**" – broadly speaking, all the problems that, should they occur, will adversely impact attainment of objectives. For hardware, these cover the gamut of failure modes of design, assembly (won't fit, too heavy, etc.) and operation (cracks, breaks, short-circuits etc.). For software, these include development risks (ambiguous requirements, flawed designs, poor contract arrangements, lack of configuration management) and risks in the developed software (erroneous results, deadlocks, timing errors, poor performance, mis-use of computing resources). Human and organizational concerns can also be incorporated here (e.g., key personnel unavailable; vendor goes out of business).
- "**Mitigations**" – the entire range of actions and options to reduce risks, including preventative measures (e.g., training of personnel, use of design standards), alleviations (e.g., hardware redundancy; software bounds checking that make a software component more resilient to erroneous inputs), and detections (e.g., design reviews, code inspections, analyses, tests of all kinds) that can potentially detect problems ahead of their manifestation in use, and so allow for their correction. There are almost always resource costs associated with mitigations (e.g., budget and schedule; for spacecraft hardware, power, volume and mass are perennial concerns; for spacecraft software, CPU and memory utilization), and one of the primary purposes of DDP is to help guide the selection of mitigations that reduce risk in a cost-effective manner.

In DDP, information in these three categories is *quantitatively* linked. Risks are quantitatively related to objectives to indicate *how much* a risk, should it occur, adversely impact an objective's attainment. Mitigations are quantitatively related to risks to indicate *how much* a mitigation, should it be applied, will reduce a risk (by preventing it from occurring in the first place, by detecting it and so allowing for its repair, or by reducing its impact on objectives).

The overall structure of a DDP information model is relatively simple – see Fig. 1. However in practice there can be numerous items and links among them. For example, the topology of the data in a recently completed DDP study (of a technology intended for spacecraft application) is shown in Fig. 2. This comprises 29 objectives, 58 risks, 36 mitigations, and some 900 links among them.

MULTI-DISCIPLINARY CONCERNS

The single most important aspect of the DDP approach is that it supports multiple experts pool their knowledge and allows them to take the sum total of their pooled knowledge into account as they make decisions. This helps address two key concerns of multi-disciplinary studies, namely:

- No single individual has depth of knowledge in every one of the disciplines, and
- Disciplines are cross-coupled and interact in multiple ways.

DDP enables pooling of multiple experts' knowledge by utilizing:

- a structured group process to elicit in-depth information from experts,
- software support to perform calculations and search over the entirety of the gathered information (e.g., calculate aggregate risk information, search for near-optimal solutions), and
- cogent visualizations to present the information back to the experts and so retain their engagement in a risk-informed decision-making process.

The DDP process is conducted in a group setting in which all the experts relevant to the technology/system under consideration are simultaneously involved. It is crucial the experts' combined areas of expertise span the concerns of the study, and that those experts are able and willing to contribute their expertise. In this respect DDP is akin to traditional risk assessment methods, which gather their risk information from representatives of all relevant disciplines. However, DDP differs by seeking more fundamental kinds of information than do typical risk assessment methods. This provides opportunities for experts to contribute specific and detailed knowledge. The DDP software tool performs the calculations to pool their separately contributed items, and present that information back to the experts.

DDP's use of fundamental risk information

Most risk assessment methods take as starting point a technology/system *incorporating* all the risk reducing measures to be applied to make that technology/system reliable. They focus on the risks *remaining* in that technology/system. In

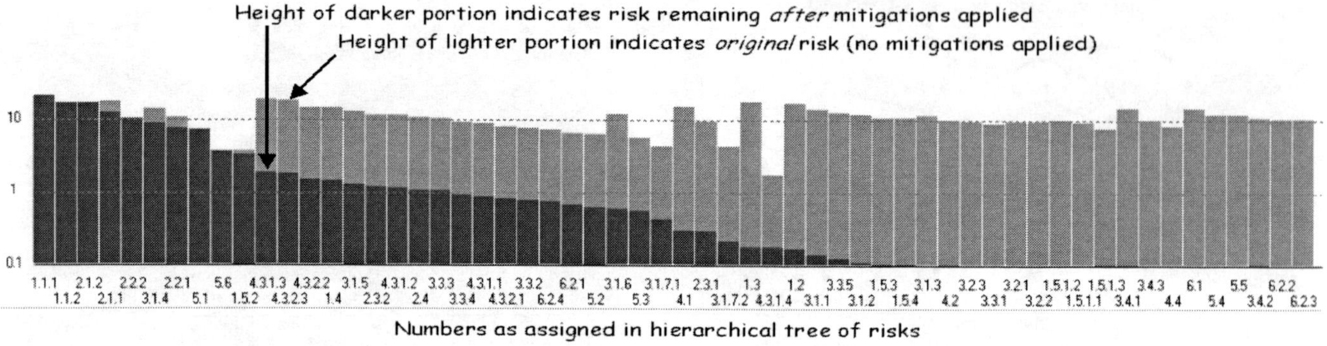

Figure 3. Graphical presentation of risks sorted into descending order of remaining risk levels

particular, these methods ask experts to provide, for each risk, estimates of the remaining likelihood and impact of that risk. The product of these is the usual definition of risk (sometimes referred to as "risk exposure"), formula (1):

$$\text{risk} = \text{likelihood} \times \text{severity (a.k.a. consequence)} \quad (1)$$

In contrast, DDP takes as starting point a technology/system *separate* from the risk reducing measures expected to be applied. Those are represented separately, explicitly accounting for their risk reducing effects. The risks remaining in the design are *calculated* from the a-priori risks, and the risk reductions effects of the mitigations. In detail, DDP asks experts to provide, for each risk, estimates of:

- what would be its likelihood of occurrence were nothing done to prevent it, and
- what adverse impacts it would have on objectives should it occur, and
- what likelihood- and severity- reducing effects each of the mitigations would have on that risk were they to be applied.

The first two are combined to calculate the *unmitigated* contribution of a risk, using the formula (2):

$$\begin{aligned}&\text{Unmitigated-Risk}(R) = \\ &\text{A-Priori-Likelihood}(R) \times \\ &\Sigma\ (O \in \text{Objectives}) : \text{Impact}(R, O) \times \text{Weight}(O)\end{aligned} \quad (2)$$

where A-Priori-Likelihood(R) is risk R's likelihood were nothing done to prevent it, Impact(R, O) is the proportion of the objective O that would be lost were R to occur, and Weight(O) is the weighting (importance) ascribed to objective O. The summation of Impact × Weight over all objectives adds up the risk's total severity were it to occur.

Note that the information on a risk's impacts on objectives is explicit and separate from information on the objectives' weights. This gives an opportunity to different discipline experts to contribute their own knowledge. Typically the mission scientists will be the ones who ascribe relative weights to the various science objectives (e.g., the relative importance of measuring the water content in soil vs. the importance of measuring atmospheric composition). The inventors of a novel piece of technology may be the most knowledgeable in estimating the adverse effect of a particular type of failure. The discipline engineers may be the most knowledgeable in estimating the a-priori likelihood of a given fault (e.g., single event upsets caused by radiation).

DDP then goes on to *calculate* the risk remaining after taking into account the risk-reducing effects of mitigations, using formula (3):

$$\begin{aligned}&\text{Mitigated-Risk}(R) = \\ &\text{Unmitigated-Risk}(R) \times \\ &\Pi\ (M \in \text{Mitigations}) : (1 - \text{Effect}(M, R))\end{aligned} \quad (3)$$

where Effect (M, R) is the proportional risk reduction that Mitigation M has on Risk R. The product of (1 – Effect) over all mitigations is DDP's formula for calculating the combined effect of multiple mitigations against the same risk. The intuition is that a mitigation acts like a "filter", reducing a risk by some proportion; multiple mitigations act like filters in series. For example, a mitigation with an effect of 0.8 against a risk reduces that risk by that proportion, leaving (1 – 0.8) = 0.2 of the risk remaining; a subsequent mitigation with an effect of 0.7 against that same risk reduces the 0.2 that's left of that risk further, leaving 0.2 × (1 – 0.7) = 0.2 × 0.3 = 0.06 of the risk remaining. For a more detailed discussion of DDP's risk model the reader is referred to [5].

Note that the information on a mitigation's effect at reducing a risk is explicit and separate from information on the risk's impact. This provides another opportunity to different discipline experts to contribute their own knowledge. For example, the testing and quality assurance experts may have the best knowledge of how effective a given practice is at detecting problems which can then be repaired (e.g., formal inspections of software requirements may uncover defects which can quickly and easily be corrected while early in the software development process).

DDP's calculation and presentation of aggregate risk

Custom software has been developed to support all the steps of the DDP process. In particular, DDP supports pooling the fundamental risk data that the experts provided, and presenting the aggregate risk information back to those experts via cogent visualizations.

As described in the previous section, the sum total impact of a risk on objectives is *calculated* from the objectives' relative weights, how much the risk adversely impacts those objectives (taking into account the mitigations that reduce its impact), and how likely it is to occur (taking into account the mitigations that reduce its likelihood). This is done for each of the risks.

As example is shown in Fig. 3, where both the Unmitigated-Risk and the Mitigated-Risk values have been calculated for each of the 58 risks, and the results plotted in a bar chart. This is a screenshot taken from the DDP software operating on actual data taken from one of the DDP conducted risk studies. The data shows risks at an intermediate stage, when only some of the mitigations have yet been identified; by

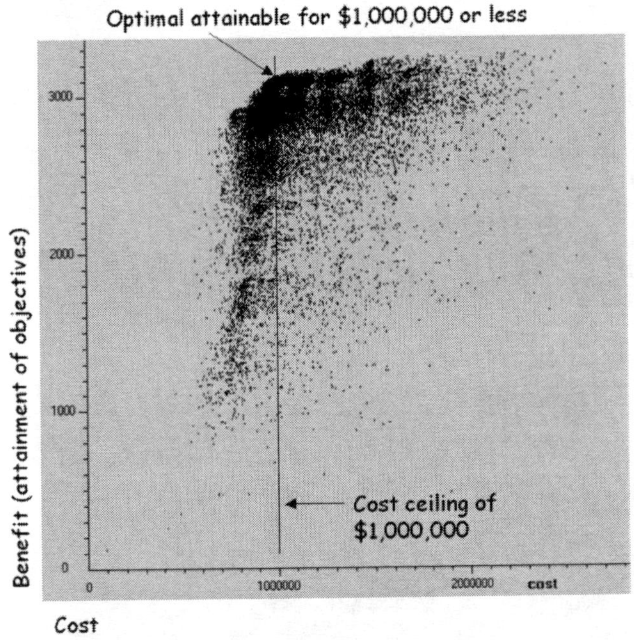

Figure 4. Search for an optimal cost-bounded solution

the conclusion of the study risks had been reduced much further than this chart shows.

DDP also calculates the sum total risk against each objective, which when presented in a similar bar chart allows experts to see the status on an objective-by-objective basis.

Finally, DDP includes a capability to do straightforward sensitivity analysis of the quantitative estimates that populate a DDP model. For example, given a selection of mitigations, one-by-one each effectiveness value connecting a mitigation to a risk is perturbed (we offer the user a choice of several ways to do this, e.g., proportional reduction, reduction to zero), and the benefit is recalculated. The results are then ranked in decreasing order of the change to the benefit value.

CONSTRAINED DEVELOPMENT

Time and budget pressures constrain the development of systems and technologies, and limited operational resources constrain systems themselves. The fundamental observation is that it takes an expenditure of resources to apply the mitigations that reduce risks, and hence the selection of those mitigations must be driven by consideration of both the benefits (risks reduced) and costs (resources consumed). Decision-making involving cost/benefit tradeoffs in this constrained setting is facilitated in DDP by:

- Capturing the cost information in the DDP data model
- Automatic calculation of total costs and benefits for given selections of mitigations
- Heuristic search to locate near-optimal cost/benefit solutions
- Calculation and presentation of the cost/benefit tradeoff space

The DDP date model includes a place to record resource cost information associated with mitigations (of course, this requires that the experts provide this information). For a given selection of mitigations, the DDP software computes the sum of their costs. When experts make decisions as to which mitigations to apply, they take into account both their costs, and their benefits (namely, the increase in attainment of objectives that results from the effect of mitigations at reducing risks). The DDP model also accommodates another important source of cost, namely the cost of *repair* of problems detected during development. For example, a fatal problem discovered at test time will necessitate some kind of repair, part replacement, rebuild etc. In the DDP model information on such costs is associated with the risks themselves – the cost of repairing the problem (risk) depends on what it is, not how it was discovered (i.e., which mitigation detected it). Furthermore, DDP accommodates the escalation of repair costs that occurs when repairs are made later rather than sooner.

For non-trivial DDP applications, the search space of possible mitigation selections is huge. For example, in the application whose data is pictured in Figure 1, there are 36 individual mitigations, so the number of possible selections of such is 2^{36} (more than 10^{10}). In other DDP applications we have seen even greater numbers of mitigations, with correspondingly larger search spaces. Furthermore, selecting an optimal set of mitigations for a given cost ceiling is non-trivial, because of the interconnectedness seen in Figure 2. For example, the simple strategy of selecting mitigations one by one, at each step picking from the as-yet unselected still affordable mitigations the one that achieves the most risk reduction per unit cost, is *not* guaranteed to yield the optimal solution.

These considerations make cost-effectively selecting mitigations a considerable challenge. In response, DDP uses the heuristic search technique of simulated annealing to locate near-optimal selections. Figure 4 shows an example of DDP's search for a near-optimal solution in a study with 58 mitigations: the search has been directed to look for maximal

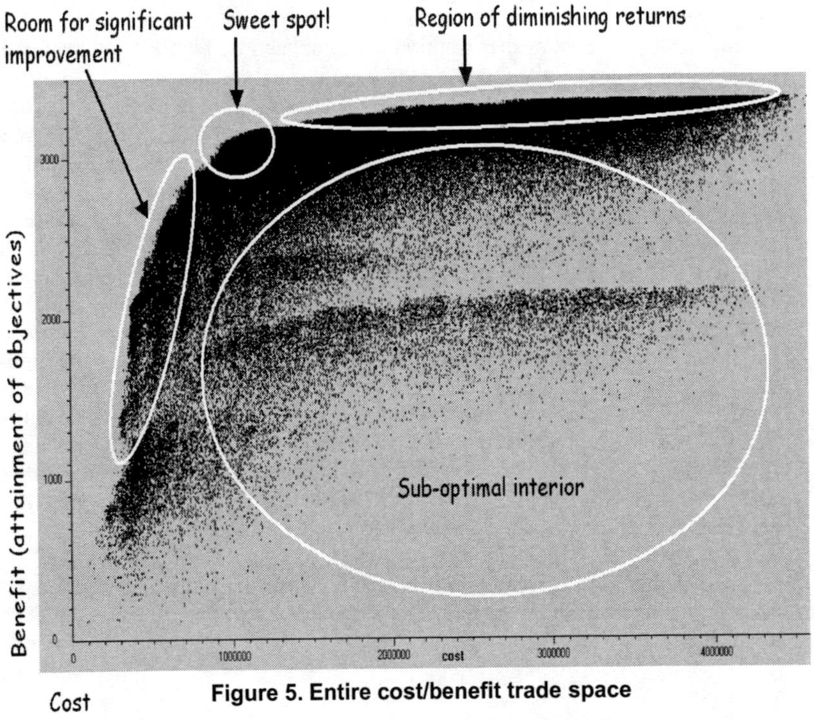

Figure 5. Entire cost/benefit trade space

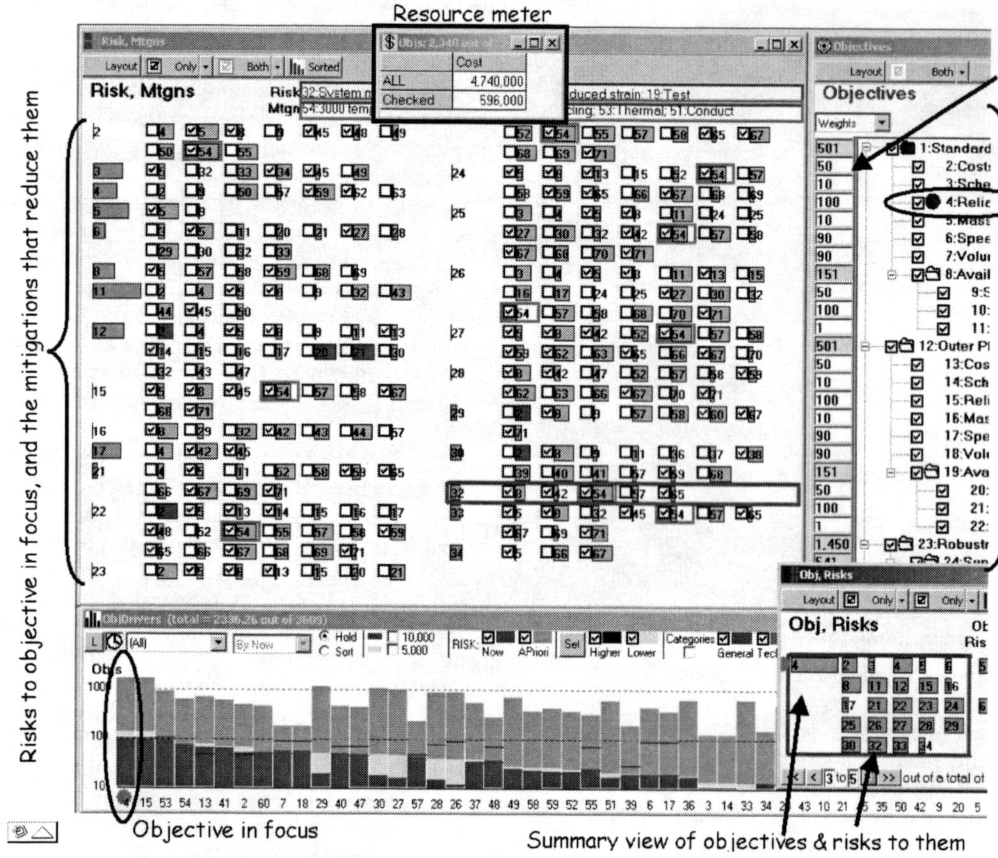

Figure 6. Screenshot of DDP in use to examine detailed risk information

attainment of objectives while costing no more than $1 million. The black cloud on the figure is composed of thousands of points, each one corresponding to a distinct selection of mitigations. The location of each point on the horizontal axis is determined by the cost of that solution (as calculated by DDP), and location on the vertical axis by the benefit, i.e., attainment of objectives (again, as calculated by DDP). The cost ceiling of $1 million is indicated by the vertical line, about a third of the way from the left. All the points on or to the left of the line cost less than or equal to that cost ceiling, so the optimal is the point highest up in that region, the vicinity of which is pointed to.

In addition to simulated annealing, we have also explored the use of other heuristic search techniques for finding near-optimal solutions. Further information on this work is reported in [6].

From an amalgamation of a series of such searches DDP can reveal the entire cost/benefit trade space, as shown in Figure 5. The upper left boundary of the widespread black cloud of points represents the optimal boundary (a.k.a. the "Pareto front"). Experts can look at this overall space to help decide the region of spending they would prefer. Depending on how risk averse they are, they may wish to extend partway to the right of what is here labeled as the "sweet spot" – the purpose of this kind of search and plot is to make the human decision makers aware of the nature of the overall cost/benefit trade space.

CRITICAL NATURE, UNKNOWN/UNPREDICTABLE ENVIRONMENT, AND LACK OF EXPERIENCE BASE

Spacecraft are critical systems that must operate correctly in only partially understood environments with no chance for repair. Furthermore, spacecraft often employ new technologies to yield better science return, etc. Past experience provides only a partial guide in these cases.

When knowledge and experience is lacking, no process can claim to yield perfect answers. The advantage that DDP conveys is that it encourages a more detailed decomposition of the problem, so that well understood aspects can be separated from the less understood aspects. Because of the critical systems nature of our studies, where aspects are poorly understood, we err on the side of caution. We use a pessimistic estimate of the impact of a risk on an objective, and the pessimistic estimate of the effectiveness of a mitigation at reducing a risk. We are often find it advantageous to decompose an item (e.g., a risk) into subcases, those that we do understand well, and can accurately score the impacts of, etc., and those that we understand less well, for which we adopt the pessimistic posture.

The overall purpose of gathering the information within DDP is to help guide decision-making. It may well turn out that even with pessimistic assumptions, a number of well-understood risks still emerge as more serious than certain less understood risks. For example, compare risks that relate to the challenging time of planetary landing phase with risks that relate to operation of a rover on the planetary surface. We might have a much greater understanding of the relevant aspects of former environment (e.g., atmospheric density) than of the latter (e.g., composition of the soil over which a rover will move). However, a failure during landing is almost certainly going to be fatal, while a failure to move up a slope on the surface may simply bring the rover to a halt. This may direct our decision-making as to where to appropriately expend our mitigation resources. When risks do show up as significant, the DDP tool gives the experts the ability to scrutinize those risks in detail, by looking at the objectives they impact, and the mitigations that could be applied to reduce those risks. If it turns out that risks cannot be satisfactorily reduced with the resources available, DDP can help experts understand how to "descope" their mission (i.e., select which objectives to abandon so that resources can be better directed to quell the risks impacting the remaining objectives, and not squandered n what turns out to be an especially challenging objective, say).

The key to all these is the simultaneous ability to view the "big picture" of risks, objectives' attainment, cost of selected mitigations, along with the ability to drill down in detail into the data that underlies these calculations. DDP offers a number of cogent visualizations that support such investigations. To convey a feel for DDP's capabilities in this regard, Figure 6 shows (most of) a screenshot of DDP in use to examine detailed risk information. Within this screen are the following elements:

- The bar chart in the lower left pane is showing the status of risk against each of the objectives (akin to the bar chart of risks seen earlier in Fig. 3, but here each bar corresponds to an objective, and the heights correspond to various measures of risk to that objective).
- The hierarchy of objectives is shown in the upper right pane (in this image we have truncated the right of the screenshot to hide the actual names of objectives, since this is data of a proprietary nature). One of the objectives has been brought "in focus", meaning the tool highlights it, and displays information particular to that objective. Alongside each objective is its user-assigned weight.
- The fragment of the pane in the lower right shows a summary view of the risks on that objective, where each risk is portrayed as a small rectangle whose width is proportional to the risk's impact on that objective.
- The large pane in the upper left, filled with a multitude of rectangles and checkboxes, shows in greater detail the risks that impact the in focus objective. Alongside each risk are tiny checkboxes coupled with rectangles. Each represents a mitigation which (if applied) reduces the risk it is listed alongside. The checkbox is checked if and only if the mitigation is currently selected for application. The user can click to toggle this checked status, causing DDP to recalculate risks, costs, etc., in response.
- The resource meter is shown floating over the top middle of the screen. This keeps a running total of the resource costs of the checked (i.e., selected for application) mitigations.

At first sight this appears a dauntingly complex display, but in practice it serves to convey the risk-related information to the experts in such a way as to aid them in their decision making. The responsiveness of the DDP software helps in this regard – when the user clicks a checkbox to toggle the status of a mitigation, complete recalculation and display update is fast. For a typical DDP dataset with a hundred or so in total of objectives, risks and mitigations, and on the order of a thousand links among them, this takes on the order of one second, operating on a modern-day PC laptop. Thus it is possible for experts to quickly try "what if" experiments – e.g., "what would be the change to cost and risk if we turned off that mitigation and turned on that other one?" The bar chart displays can help in these "what if" explorations, by visual presentation of the *changes* to risks from some baseline. In fact this is just barely discernable on the barchart in Fig. 6: some of the bars have three shades of gray, the lightest indicating how much the risk against that objective has dropped compared to a previously set baseline.

SUMMARY AND CONCLUSIONS

DDP has now been applied to make risk-informed decisions in over 20 different studies of advanced technologies intended for spacecraft use. The nature of the decisions has varied. In some cases the primary outcome has been selection of a suite of mitigations that in concert will cost-effectively reduce risk to sufficiently low levels. The ability to search and reveal the cost/benefit tradeoff space has been of assistance here. In other cases the primary outcome has been selection from among major design alternatives, using DDP to study the risks of each and indicate the cost of mitigating the risks in each option. In almost all cases the process leads to clarification of requirements. In one of the studies, a particularly problematic requirement was identified. It was found to be problematic because the cumulative risk information contributed by the engineering experts showed the requirement to be significantly at risk, and furthermore showed that sufficient mitigation of those risks would be very expensive in terms of schedule and budget. The net result was that the mission scientists were motivated to rethink their objectives, and make the problematic one a much lower priority.

We have not performed any formal experiments to measure the overall effectiveness of DDP applications. We do know the cost – predominantly the time of the experts involved in populating the study with data and making decisions. For example, a typical DDP application, with 15 participants in all four of the 4-hour sessions, consumes 240 hours (6 work weeks) of time. We require the participation of experts. Such people are always in demand, so this is by no means a trivial investment of time. Estimating the benefit of DDP applications is much more subjective. We are encouraged as to the validity and utility of the findings that DDP yields for the following reasons:

1. Most of the results calculated by DDP, based on the information gathered from the multiple discipline experts, match those experts' overall intuitions. For example, DDP's list of most significant risks that remain despite application of mitigations is usually in agreement with what the experts tell us they would have expected. Overall this suggests that the detailed information we are gathering from those experts, and the ways we combine that information in DDP's risk calculations, is reasonably valid.
2. In almost every case study there is some result calculated by DDP that is a "surprise". That is, it does not match the experts' intuitions (e.g., a risk shows up as more significant than they would have anticipated). Furthermore, when the experts look at the detail underpinning that result (and the capability of DDP to let them explore the details as well as

see the big picture is crucial in this regard), they concede that the "surprise" is a genuine finding. Overall this suggests that the approach we follow is capable of findings that would be overlooked by even a well-qualified set of experts.

3. The decisions arrived at with the aid of DDP have led to much clarified in problem statements (objectives), better substantiated cost estimates, well-organized rationales for development plans, and, in some cases, very significant cost savings by stimulating design revisions motivated by the realization that resources were being mis-applied to (over)address a relatively tiny risk area, while other risk areas were in more dire need of attention. Perhaps the most important aspect shared by these decisions is that they are being made relatively early in the technology or system lifecycle, when the costs of revising choices are still relatively small. In contrast, fundamental design changes later in development incur much higher costs.
4. DDP is not a mandated activity, yet is seeing increasing use. The earlier DDP applications were supported by research funding (as was the development of the DDP process and support software itself). Specifically, the discipline experts' time was covered by this funding. More recently, there has been a trend to the use of DDP paid for predominantly by the technology/system under scrutiny. While we continue to use research funds to expand DDP's capabilities, the bulk of the cost of performing the risk studies themselves, namely the cost of the discipline experts' time, is no longer being borne by the research funding. This, plus the increasing frequency of DDP based studies, is indicative of an expanding acceptance of the net value of the DDP process.

Related work

Early decision-making is often assisted by *qualitative* decision support methods. For example Quality Function Deployment (QFD) has been used in a wide variety of settings [7] DDP's effect and impact matrices are reminiscent of the Relationship Matrix used in many forms in QFD. DDP is distinguished by its foundation upon a *quantitative* risk model, which gives meaning to DDP's cost and benefit calculations.

As a design matures there are other decision support techniques that better capitalize upon knowledge of design details. For example, probabilistic risk assessment techniques (e.g., fault tree analysis, Bayesian methods) compute overall system reliability from design knowledge of how the system is composed of those components, and estimates of individual component reliabilities. The origins of these approaches lie in applications to assess risk in the nuclear power industry [8], with its need to estimate the probability of catastrophic failure (e.g., meltdown) from knowledge of the power system's design, and reliability measures for the components used in that design. Fault Tree Analysis [9] is now applied to a wide variety of systems, both hardware and software (e.g., software fault tree analysis [10]) including some NASA missions and their hardware and software components [11]. In contrast, DDP aims to fill the niche of early decision making for advanced technology and system development. We are currently exploring means to connect DDP and PRA techniques [12].

ACKNOWLEDGMENTS

The research described in this paper was carried out at the Jet Propulsion Laboratory, California Institute of Technology, under a contract with the National Aeronautics and Space Administration. Reference herein to any specific commercial product, process, or service by trade name, trademark, manufacturer, or otherwise, does not constitute or imply its endorsement by the United States Government or the Jet Propulsion Laboratory, California Institute of Technology.

REFERENCES

[1] O'Keefe, S. "*Pioneering the Future*", address by the NASA Administrator, Maxwell School of Citizenship & Public Affairs, Syracuse University, April 12, 2002 ftp://ftp.hq.nasa.gov/pub/pao/okeefe/2002/pioneering_the_future.pdf

[2] Lehman, M.M. "Some Characteristics of S-type and E-type Software" in *Proceedings of the Second FEAST Workshop*, Dept. of Computing, Imperial College of Science Technology and Medicine, London, UK. http://www.doc.ic.ac.uk/~mml/feast2/papers/pdf/2feastwks.pdf

[3] Cornford, S.L. "Managing Risk as a Resource using the Defect Detection and Prevention process", *Proceedings, 4th International Conference on Probabilistic Safety Assessment and Management*, 13-18 September 1998, New York City, NY, International Association for Probabilistic Safety Assessment and Management.

[4] S.L. Cornford, M.S. Feather & K.A. Hicks. "DDP – A tool for life-cycle risk management", *IEEE Aerospace Conference*, Big Sky, Montana, Mar 2001, pp. 441-451.

[5] Feather, M.S. & Cornford, S.L.. "Quantitative risk-based requirements reasoning", to appear in *Requirements Engineering* (Springer), published online 25 February 2003, DOI 10.1007/s00766-002-0160-y. Available from: http://eis.jpl.nasa.gov/~mfeather/Publications

[6] Cornford, S.L., Feather, M.S., Dunphy, J.R., Salcedo, J. & Menzies, M. "Optimizing Spacecraft Design – Optimization Engine Development: Progress and Plans", Proceedings, 2003 IEEE Aerospace Conference, Big Sky, Montana, March 2003.

[7]. Akao, Y. 1990 "*Quality Function Deployment*", Productivity Press, Cambridge, Massachusetts.

[8] *Reactor Safety Study*, Report WASH-1400, Nuclear Regulatory Commission, 1975.

[9] Vesely, W.E., Goldberg, F.F., Roberts, N.H. & Haasl, D.F., "*Fault Tree Handbook*", U.S. Nuclear Regulatory Commission NUREG-0492, 1981.

[10] N.G. Leveson. "*Safeware: system safety and computers*". Addison Wesley, Reading, MA, 1995.

[11] *Probabilistic Risk Assessment Procedures Guide for NASA Managers and Practitioners, version 1.1*, prepared for the Office of Safety and Mission Assurance, NASA HQ, Washington, DC, August 2002.

[12] S.L. Cornford, T. Paulos, L. Meshkat & M.S. Feather. "Towards More Accurate Life Cycle Risk Management Through Integration of DDP and PRA". *IEEE Aerospace Conference*, Big Sky MT, Mar 2003.

Proceedings of IMECE'03
2003 ASME International Mechanical Engineering Congress
Washington, D.C., November 15–21, 2003

IMECE2003-55496

THE VALUE OF A SPACE PRECURSOR ANALYSIS PROGRAM: A SATURN EXAMPLE

Joseph R. Fragola – SAIC/Rockville Centre, NY

Erin P. Collins – SAIC/New York, NY

ABSTRACT

There is general agreement that "near miss" or "close call" data is valuable to the space program and whenever NASA becomes convinced that such events have occurred, they have acted responsibly and quickly to address them. The problem lies in defining what constitutes a near miss in a system that is inherently very complex, such that 'abnormalities' are actually normal occurrences, and yet one that is relatively reliable because of the inherent strengths incorporated into the design, such as robustness, redundancy, and functional diversity. The question becomes: what would the consideration of failure precursors add to the insights to be drawn from history as it relates to forecasting future performance? This paper will use the example of the Saturn program to address the problems involved in forecasting the risk in complex, yet reasonably reliable programs and to indicate preliminary approaches for use in establishing a space industry precursor program.

Keywords: Anomalies, data, forecast, precursors, probability

INTRODUCTION

The value of investigating "near miss" or "close call" data is not new in general, nor is it new to the space program. There is general agreement that this type of information is valuable and whenever NASA becomes convinced that such events have occurred, they have acted responsibly and quickly to address them. The problem lies in defining what constitutes a near miss in a system that is inherently very complex, such that 'abnormalities' are actually normal occurrences, and yet one that is relatively reliable because of the inherent strengths incorporated into the design, such as robustness, redundancy, and functional diversity. Since these strengths act as barriers against the propagation of abnormalities to fully catastrophic failures, they can also act as barriers against the development of an understanding of the full implications of events, especially when their occurrence is signaled only individually, making underlying failure patterns difficult to identify. In such cases, the problem is not that there is not enough data (although this is an oft heard cry) but rather that there is too much data. True near misses can be obscured because the number of abnormalities observed prevents each and every one from being investigated thoroughly. In such an environment, the normal tendency is to categorize the occurrence of each abnormality without catastrophic or even mission-curtailing effect as benign, despite the fact that it may have demonstrated the erosion of some of the barriers that prevent its potential propagation. This inclination toward shifting the abnormal to the "normal" due to "escapes" has been a recurring problem in the space industry, as cited by Professor Richard Feynman [1]. However, the tendency is not unique to aerospace. Incidents leading up to the Three Mile Island accident in the commercial nuclear industry could be seen in the same light. To address this problem, the Nuclear Regulatory Commission (NRC), the regulatory agency for the commercial nuclear industry, established a program entitled "The Accident Sequence Precursor Program" (or ASP) with the specific intention of embedding individual events into a probabilistic framework in order to measure, in the probabilistic sense in terms of increased risk, just how close each event came to an accident and thereby, whether each should be considered as serious. This program has been addressed elsewhere [2-3] and the discussion will not be repeated here. However, given the success in the nuclear industry, a similar program was recommended for application to the aerospace industry and specifically to the space shuttle [4]. To date, though, no analogous effort has yet been implemented.

BACKGROUND

Following completion of the first comprehensive nuclear power plant risk assessments in the 1970s, the Nuclear Regulatory Commission (NRC) began a systematic analysis of reactor plant operating experience to identify precursors [2-3] or operational incidents that are important parts of reactor plant accident sequences. Precursors are identified in the NRC program using a methodology that permits a large number of incidents to be screened and potentially significant incidents to be quantified in a practical engineering effort. The effect of a precursor on elements of the risk model is assessed by reviewing the specifics of an incident against design and operating information, and translating the results of the review into revised failure probabilities for components included in the risk model. A precursor's significance is estimated by calculating a conditional probability of the risk end state using those revised probabilities, thereby providing an estimate of the measure of protection remaining given the incident, which is useful in ranking incident significance.

The success of this program in the nuclear industry over the past 20 years has inspired recommendations of a similar program for NASA in general and for the space shuttle program in particular [4]. Despite these initial efforts and meetings with the USNRC [5] at which NASA showed interest in the program, and recommendations from NASA internal reviews [6], no analogous effort has yet been forthcoming.

Why this is the case is a matter for speculation, but what has become apparent is that without a comprehensive program that systematically reviews post-flight information and evaluates observed anomalies in light of their potential for failure [7], NASA could be lulled into a "psychology of continued success" [8] posture that would deprive NASA decision-makers of the information necessary to make informed judgments concerning on-going shuttle flight safety.

LAUNCHER SUCCESS DATA

A recent review of historical launcher success rates [9] is presented in Fig. 1, which cites the "success rate" of a stable of current and historical launchers. Success rate is defined as the ratio of the number of successful launches over the number of total launches, but is a rather dubious measure of launcher reliability, especially when the number of launches is few. One obvious flaw is that the uncertainty in the estimate can be considerable because of the limited data set. However, a more subtle problem with the use of success rate is that the definition of "success" is mission dependent. This can be most clearly seen when the dual success criteria of mission completion and crew safe return are considered. For example, the Apollo 13 mission was clearly as much of a "success" in terms of the latter as it was a "failure" in terms of the former. This points out that the consideration of the desired end state in the definition of success or failure is critical to the understanding of the value of precursor information. Missions that are successful by one set of criteria might have very different implications under a different set of mission goals and associated end state definition.

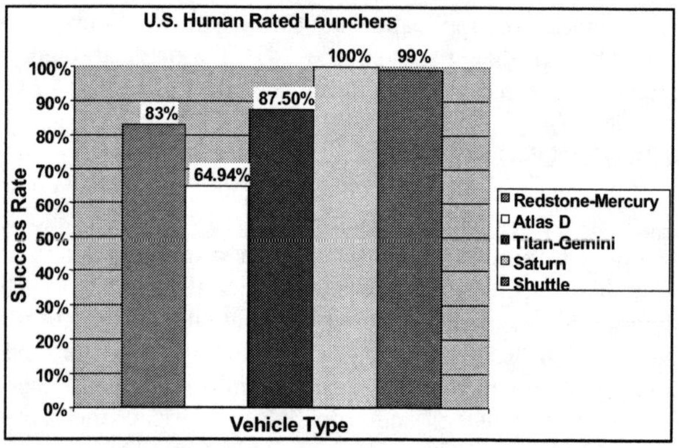

Figure 1. Saturn & Other Launcher “Success Rates” [9].

As an example, consider the Saturn launch system (the subject of a more detailed analysis, discussed later in this paper), which is shown in Fig. 1 as having a 100% success rate. From a crew safety success standpoint, this is certainly so, however from a mission success standpoint, the pattern is not so clear. In fact, it has been argued that one of the Saturn missions was a failure because if it had been a lunar mission, it would have resulted in a mission failure. [10]

Beyond this, it is clear from the review of post-flight launch data from the Saturn program [11], that even the successful missions had anomalies that could have "potentially" posed a problem to the success of the Saturn mission, but did not for one reason or another. These anomalies can therefore be described as "near misses" or "precursors to failure", since there are "virtual" failure states of the system that were not "occupied" on this particular mission or given this particular mission success definition. However, even though they are not failures per se, they certainly provide information to the decision-maker concerning the potential for failure on a future mission, or over a future set of missions, or with a different and more restrictive definition of mission success.

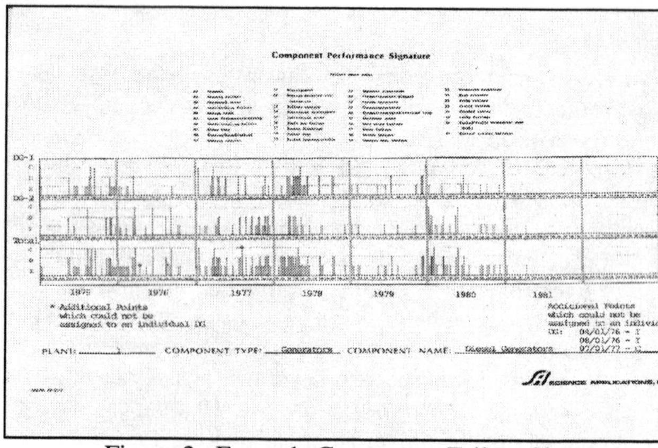

Figure 2. Example Component Failure Signature.

This potential for future failure has been referred to as a "potentiality" factor [12] and is used to measure the conditional probability of the risk end state (as defined in the analysis of interest) as compared to the condition observed in the data set. As such, it provides a measure of how close the event came to actual failure in terms of a fractional failure probability. In the case of a well-defined and implemented precursor program, [2-4] such as that of the US commercial nuclear industry, this fractional failure is evaluated systematically by embedding each failure into a pre-established risk structure. While an individual failure analysis approach has been recommended for the space shuttle program, other more heuristic and far simpler approaches have also been used in the past to provide useful information to decision-makers when failure data is scant. [12]

One classic case in the literature is the so-called "1/3 failure rule", used to estimate the most likely failure rate from a data set when zero failures have been observed [13] and another is the "Component Failure Signature" approach [14] shown in Fig. 2.

Both of these approaches will be applied to the analysis of the Saturn data in the sections that follow, as well as an additional approach based on recent research that will be recommended for consideration as an alternative. These analyses will then be compared to a bottom-up analysis of

Saturn risk developed by NASA just prior to the first manned lunar mission. [15]

SATURN PRECURSOR FORECAST EXAMPLE

In the past, standard reliability and risk approaches have sufficed to identify the dominant causes of failure in forensic analyses, and the dominant risk contributors for proactive risk investigations. These techniques are particularly applicable when individual or even common failure events of a similar type dominate the analysis. However, where failures are rare events or when even a single failure occurrence is intolerable, a standard approach is not a viable option. Alternative approaches are many and varied and are, in some cases, still in development, but their heritage is not new. They all involve a means of evoking subtle patterns from the data or from the physical phenomena themselves by "interrogating" the data until it yields its secrets.

This section will provide an example of how pre-cursor analysis and visual mapping might be used to discover problems in complex systems before they lead to failure. As a contrast to the historical launcher success rates presented previously in Fig. 1, an effort was undertaken to analyze the data further in an attempt to answer the question: "What would have been the risk of future launches if the Saturn launcher had continued in service?" The motivation was to try to show how the use of precursor data, properly accounted for and analyzed, can provide considerably better insight than scant failure data alone.

The first step in the analyses involved the development of a graphical presentation of catastrophic Saturn failures alone. Here the definition of a catastrophic failure implies the requirement to have exercised the escape system if the flight had been or was manned or, alternatively, the catastrophic failure (i.e. explosion or destruction) of the launch vehicle. Since none of the Saturn launches resulted either the requirement for escape or in explosion or destruction, the success ratio for the catastrophic failure case (defined as the number of successes over the total number of launches) is shown in Fig. 3 as 100%. This has been cited [9] to imply that the Saturn launch success rate is 100%.

Figure 3. Saturn Success Ratio by Launch Sequence - Catastrophic Failures.

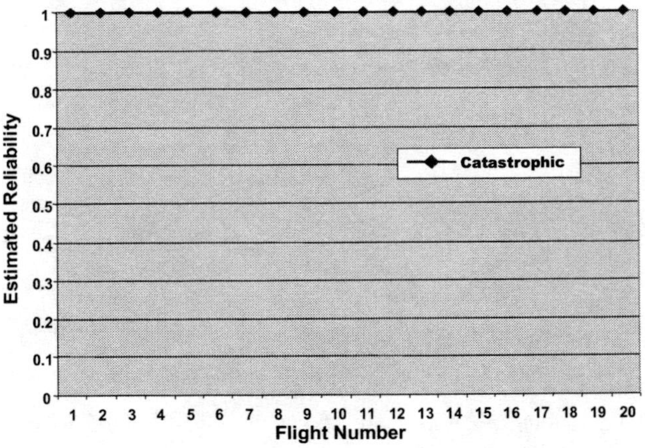

This certainly is true if the definition of success rate is considered to be launch success. However, this could send the wrong message to decision makers who might erroneously tie the success rate to a forecast of the Saturn reliability, because the reliability would instead be bounded by the statistics associated with the limited number of launches. The establishment of a statistical bound on a failure-free data set has been referred to as the "zero failure problem" and has been addressed elsewhere [14]. However uncomfortable the analyst might be with relying on these or other somewhat subjective bounds, there is little more that can be done if the data set is restricted to mission failures and successes. (Note: A bound developed according to the recommended procedure given in the reference is discussed further later and is shown in Fig. 7.)

The question then becomes: what would the consideration of failure precursors add to the insights to be drawn from the Saturn history as it relates to forecasting future Saturn performance? To address this question, the focus is shifted to the analysis of the post-flight engineering data.

Figure 4. Saturn Ratio of Catastrophic + Degraded Failures to Successes by Launch Sequence.

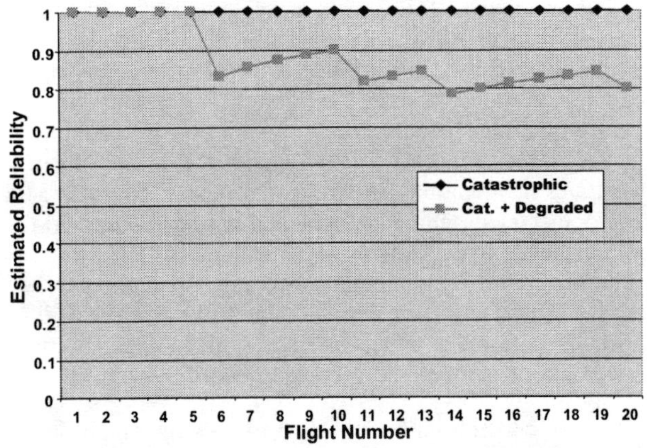

Fortunately, this data already exists in the records of the 45th Space Wing of the US Air Force, which have been analyzed previously [11]. In this prior study, data analysts were trained to review launch records to identify anomalous behavior recorded during launch and to categorize it according to its severity using a consistent set of criteria. In Fig. 4, the observed anomalous conditions categorized as "degraded failures" have been included. In this case, the degraded category applies to those observed anomalies that, while not resulting in mission failure, did degrade the performance of the system by eroding its functional margin.

From this chart, it can be seen that the picture is not nearly as rosy as it was when only the catastrophic (or mission failure-producing) anomalies were considered. As it turns out, the second Saturn V launch (Apollo 6), from the standpoint of a manned launch, would have necessitated an abort because of a third stage failure to restart. This failure, albeit benign, has been considered as a failure of the objectives for this mission [10], or a mission failure. In addition, there were several other failures along the way that, while they did not endanger mission success, constituted significant deviations from nominal Saturn performance and therefore, at the very least, eroded the robustness of the vehicle. This precursor to failure history would lead one to suspect that the Saturn was less than a "100%" vehicle from an engineering perspective, even without the statistical analysis.

Figure 5. Saturn Ratio of Catastrophic + Degraded + Incipient Failures to Successes by Launch Sequence.

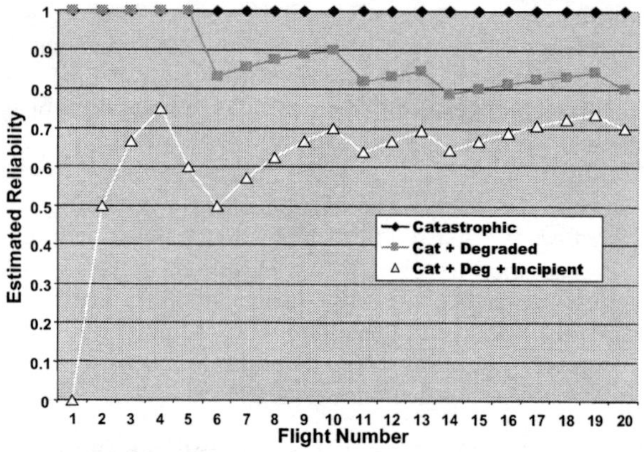

The category of incipient failures is added in Fig. 5. The definition of 'incipient' is an observed anomaly that did not degrade the vehicle performance, but might have done so under different conditions, such as if other things had failed or if the condition had persisted for a longer time. (The erosion of the O-rings in the shuttle booster field joints in missions prior to their failure during the 51-L mission would be an example of this type of condition). This incipient failure history demonstrates the traditional growth pattern that is characteristic of developmental systems in general and has been observed for other unmanned launch systems in particular. [16]

The 5th and 95th percentile uncertainty bounds are then added in Fig. 6 for the degraded and the incipient plus catastrophic failures; in other words, for all failures (because there were no catastrophic failures experienced).

Figure 6. Saturn Ratio of All Failures to Successes by Launch Sequence.

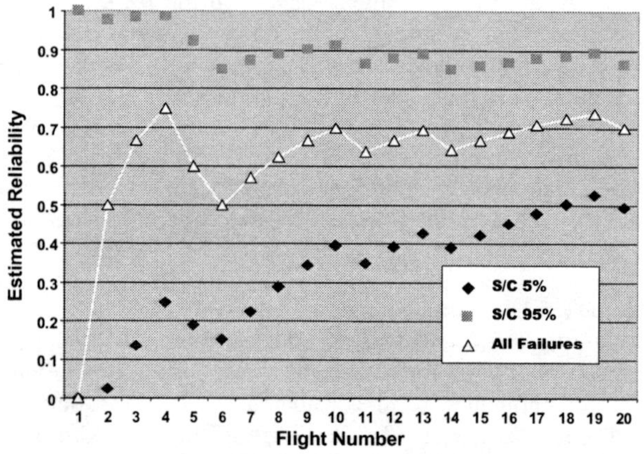

It is evident from this figure that even with 20 launches, there is considerable statistical uncertainty regarding what the forecasted Saturn reliability might be if it were measured by the total of all failure types. Of course, to use this combined failure case as a measure of Saturn reliability would be very conservative, at least from a crew safety and mission success perspective. For this reason, some other measures were investigated that would consider these failures probabilistically as precursors. The first precursor measure investigated was the 1/3 failure rule, which considers only catastrophic failures and the results of applying the rule itself.

Figure 7. Preliminary Precursor Estimation using One-Third Failure Rule.

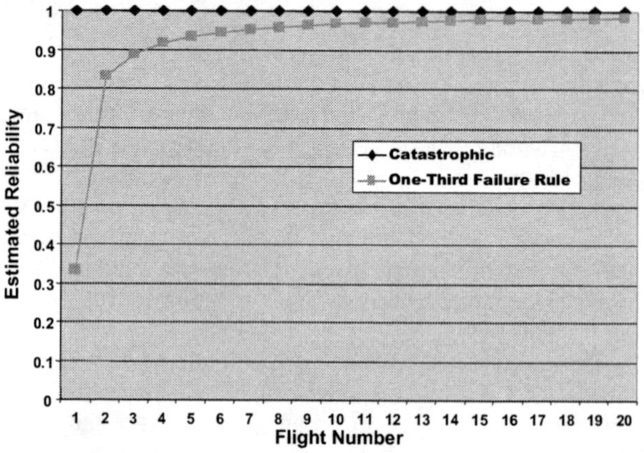

As shown in Fig. 7, the forecasted success ratio is developed at each launch assuming that 1/3 of a failure had occurred by that launch. Using this measure, the success ratio forecasted would be 1 in 60 after 20 launches (about the same as the shuttle's current success ratio).

While this one-third estimator sheds some insights, the question remains: can a better forecast be developed than that provided by the catastrophic failure data set alone by expanding the data set to include the precursor information? And if so, how might this be done?

For the purposes and limited scope of this paper, the focus is directed at the use of precursor information. To do this correctly, an infrastructure, similar to the nuclear ASP program [2-3] would have to be constructed. However, in this case, models would have to have been developed for each failure "measuring" how close each came to catastrophe. Such an effort was beyond the scope of the analysis included in this example. However, a crude measure can be obtained by weighting the failures according to their severity and developing a combined measure from the "expanded" data set.

Figure 8. Preliminary Precursor Estimation - Linear & Log Weighting

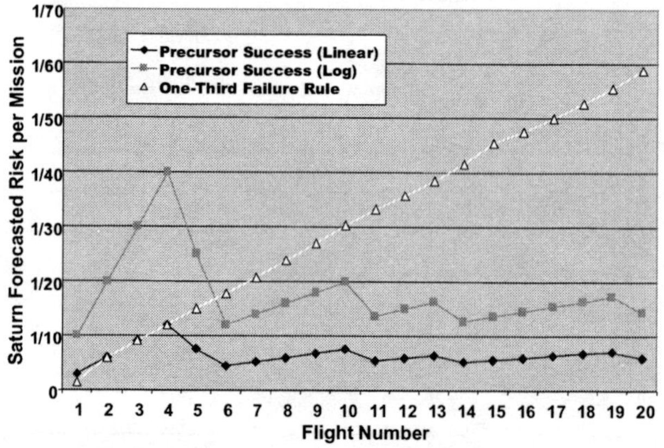

Two possible weightings are shown in Fig. 8, along with a comparison to the 1/3 failure results.

The first is a simple 3, 2, 1 linear weighting for Catastrophic, Degraded, and Incipient (C, D and I) failures, respectively, which was used for the development of the "component failure" signatures in the 1980's. The next is a log-weighting scheme, using weights of 10, 3, 1, respectively for the C, D, Is. Even these crude weights show the value of a precursor program for estimating the evolving risk. In particular, the log weighting indicates that the Saturn was a 1 in 14 vehicle, (0.93 reliability), as opposed to a 1 in 60 vehicle estimate from the 1/3 failure rule.

Figure 9. Preliminary Precursor Estimation - Log Weighting with Uncertainty

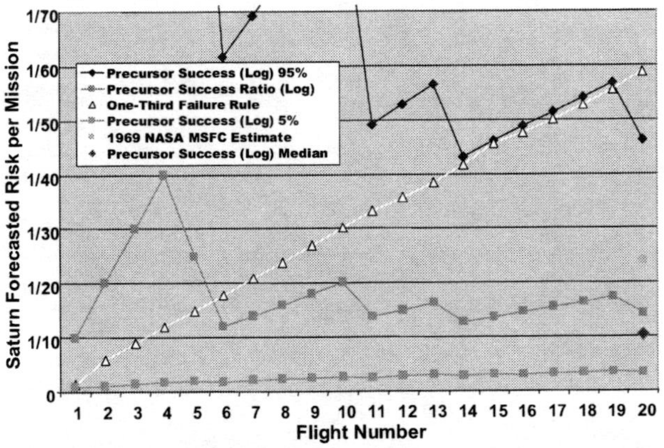

The best measure of forecasted risk is subjective, but some information of value to decision makers can be gleaned from even this crude analysis, namely:

1. The Saturn was not a 100% vehicle;
2. The 1 in 60 estimate obtained from the use of the one-third failure rule would have probably have been lower if a prior had been developed from heritage vehicles since the heritage was far worse than 1 in 60. (Note: The MSFC bottom up analysis performed in 1969 predicted a 1 in 20 vehicle);
3. The log-weighting scheme predicts a 1 in 14 number, which might be closer to the truth, close to the 1 in 20 MSFC number, and as Fig. 9 shows, indicates that the Saturn is stabilizing at that number;
4. When the uncertainties of the small statistics are considered, the Saturn's reliability performance is well within the heritage base range from 1 in 12 to 1 in 50 for the best of the best of launchers.

CONCLUSIONS

This analysis shows how the forecasting of launcher reliability is a complicated process, even for existing launchers, and how it can present a very inaccurate picture to the decision maker without the expansion of the data set to consider precursors.

It is the belief of the authors that in an actual precursor system, consideration would be given to the heritage of launchers both previous and related to the Saturn, as well as to the precursors in the Saturn data set. This heritage information would then be used to develop a prior distribution that would be combined with an expansion of the history to include the non-catastrophic events as precursors to actual Saturn catastrophic failure. A discussion of heritage-based surrogates is beyond the scope of this paper, but is provided elsewhere [16].

ACKNOWLEDGMENTS

The authors wish to thank Karl C. Knight of NASA MSFC for his review comments.

REFERENCES

[1] Gleick, James, *Genius: The Life and Science of Richard Feynman*, Pantheon Books, New York, 1992.

[2] Minarick, J.W. and Kukielka, C.A., *Precursors to Potential Severe Core Damage Accident Sequences: 1969-1979, A Status Report*, NUREG/CR-2497, Vols. 1 and 2, Oak Ridge National Laboratory, Oak Ridge, TN, 1982.

[3] *Precursors to Potential Severe Core Damage Accidents: 1985, A Status Report*, NUREG/CR-4674, Vol. 1 and Vol. 2, December 1986 and subsequent volumes for 1984, 1986-1998.

[4] Fragola, J.R., Space Shuttle Safety, Testimony before the Space and Aeronautics Subcommittee of the US House of Representatives, Congressional Record, 1995.

[5] Fragola, J.R., Minarick, J.W. and Maggio, G., "Systematic Application of Precursors in the Management of Space Shuttle Risk", Society for Risk Analysis SRA '96 Conference, December 1996.

[6] Fragola, J.R., Meeting Notes of Meeting with NASA management, NASA, Washington, DC, 22 October 1998.

[7] Chang, K., "Shuttle Investigators Have Theory About Hole in Wing Edge", New York Times, 2 April 2003.

[8] Schwartz, J. and Wald, M., "Shuttle Inquiry Considers Possible Role of Loose Panel", New York Times, 2 April 2003

[9] Rust, R., NASA Johnson Space Center, Safety Reliability and Quality Assurance Technology Division, Advanced Projects Group, presentation on "Historic and Current Launcher Success Rates", 29 May 2002.

[10] Isakowitz, S. J., *International Reference Guide to Space Launch Systems*, AIAA Press, Washington, DC, 1995.

[11] "NASA Space Flight Risk Data Collection/Analysis Project – Final Report", Dimensions International, Inc., Alexandria, VA and SAIC/New York, NY, prepared for NASA Headquarters - Code Q, 1994.

[12] Fragola, J.R. and Maggio, G., *Probabilistic Risk Assessment of the Space Shuttle*, SAICNY95-02-25, 28 February 1995.

[13] Welker, E.L. and Lipow, M., "Estimating the Exponential Failure Rate from Data with no Failure Events", Proceedings of the 1974 Reliability and Maintainability Symposium (RAMS), January 29 – 31, 1974 Los Angeles, CA, Institute of Electrical and Electronics Engineers (IEEE), New York, NY, pgs. 420-424.

[14] Component Failure Signature Report, SAI[C]/NY 83-11-9, November 1983.

[15] "Saturn V Reliability Analysis Model Summary", AS-507, MSFC Dwg. No. 10M308071969, NASA Marshall Space Flight Center (MSFC), Huntsville, AL, 23 June 1969.

[16] Fragola, J.R. and Perkins, D.R., "Propulsion Systems Reliability: A Historical Perspective", AIAA/SAE Joint Conference, Monterey, CA, July 10-13, 1989.

[17] Fragola, J.R., Putney, B., Go, S., and Mathias, D.L., "A Risk Evaluation Approach for Safety in Aerospace Preliminary Design", 2003 Reliability and Maintainability Symposium (RAMS), Tampa, FL, 27-30 January 2003.

FOREWORD

The current proceedings for the ASME Engineering and Technology Management (E&TM) Group, consisting of the Technology and Society (T&S) Division, the Safety Engineering and Risk Analysis Division (SERAD) and the Management Division, contain a collection of papers presented in sessions held under the sponsorship of E&TM Group at the 2003 ASME International Mechanical Engineering Congress (IMECE) November 15–21, 2003. The E&TM Group has offered a variety of sessions, shown in Table 1, for their division technical tracks and Engineering Business Management Track, and this section of the Proceedings contains a modest number of peer-reviewed papers, shown in Table 2, for the sessions sponsored by the Technology and Society Division.

We are privileged and honored to have distinguished speakers for our special events: Dr. Richard Hinder, an IEE Fellow and formerly an official of the Department of Trade and Industry, the United Kingdom, for Joint Technology and Society and Management Dinner and Session T&S-1 Management of Technology for Business Development; Dr. John A. Brighton, an ASME Fellow and Assistant Director for Engineering at National Science Foundation, for Henry Robinson Towne Lecture; and Dr. Michael Stamatelatos, Manager, Agency Risk Assessment Program, NASA Headquarters, for SERAD Dinner. Dr. Hinder speaks on "Changing Roles for the Engineer in the 21st Century — A European View," and Dr. Brighton, on "The Role of the National Science Foundation in Advancing the National Research and Education Agenda in Science and Engineering," respectively. Their URLs are shown in Table 1.

I owe a great deal to Dr. Winfred Phillips, ASME President (1998-99), Dr. Clifford Brubaker, Dean of Health and Rehabilitation Sciences at the University of Pittsburgh, and Mr. Marshall Graves of California Governor's Office, among other organizers and members of the T&S Program Committees shown in Table 3, for distinguished T&S sessions. We are blessed with distinguished speakers, Dr. John Watson of National Heart, Lung and Blood Institute and Dr. Gerson Rosenberg of the Pennsylvania State University for Session T&S-2, Dr. Michael Heinrich of National Institute of Child Health and Human Development, Dr. Donna Dean of National Institute of Biomedical and Bioengineering Research, Dr. Joel Myklebust of National Institute on Disability and Rehabilitation Research, and Dr. Rory Cooper, a bronze medalist at the 1998 Seoul Olympics for the 1500 m wheelchair competition and Professor at the University of Pittsburgh for Sessions T&S-5A and –6A, and Fred Allen, Editor for American Heritage of Invention and Technology, Dr. Thomas Hughes of the University of Pennsylvania, Ben Wu, Deputy Under Secretary for Technology at US Department of Commerce, and Dr. Heater Wicke, Senior Environmental Scientist at Office of Senator John McCain for Session T&S-8 among other speakers.

I also owe a great deal to members of the E&TM Group Operating Board and the division committees during the course of preparation for publication of this volume. Particular thanks go to Robert Simmons, E&TM Group Vice President, and Theodore Aanstoos, Engineering Business Management Track Manager, for their coordination work, Frank Spanitz, Jonathan Young, and Stephen Long, Program Representatives of the Technology and Society Division, SERAD, and Management Division, respectively, to Kenneth Horne, William Doerr, and Wolf Yeigh, IMECE 2003 Chairs for the three respective divisions, for their support during the course of organizing our E&TM Group sessions, and Flavio Franco, Chair of the ASME United Kingdom Section, for accepting my invitation to participate in a Technology and Society Division session. It is a pleasure for me to serve as the editor for the four consecutive E&TM Group symposium proceedings from IMECE 2000 to 2003, which were made possible through cooperation of those noted and many others.

Hiroshi Honda, Ph.D., P.E.
E&T M Group Representativ

Proceedings of IMECE'03
2003 ASME International Mechanical Engineering Congress
Washington, D.C., November 15–21, 2003

IMECE2003-43515

MANAGING A MAJOR CURRICULUM REFORM EFFORT IN A LARGE RESEARCH UNIVERSITY

Philip S. Schmidt, University of Texas at Austin **Joseph J. Beaman, University of Texas at Austin**

ABSTRACT

This paper describes a substantive undergraduate curriculum reform effort being undertaken over a five year period at the University of Texas at Austin, with special consideration to the management issues that are being addressed. The initiative, called Project PROCEED (for Project-Centered Education), grew out of a desire to strengthen our students' understanding of fundamental engineering theory by continuously tying it to tangible objects and systems, to strengthen our Department's connections with its industrial stakeholders by actively involving them in the development and delivery of curriculum content, and to provide our students with a broad range of team-based experiences which will better prepare them for growth and leadership in the corporate and professional world. The rationale for PROCEED and key program elements are described. Management issues, such as logistics, promoting faculty and student buy-in, financing, industrial interaction, and program evaluation are discussed.

INTRODUCTION

The Department of Mechanical Engineering at the University of Texas at Austin is one of the largest in the US, with a total enrollment of over 1000 undergraduates and 200 graduate students. In 2000-01, the Department awarded 170 BS, 78 MS and 24 PhD degrees; US News and World Report ratings for that year ranked ME at UT 10th nationally at the undergraduate level and 11th at the graduate level.

As is the case with departments nationwide, UTME faculty have been evaluating the strategic directions in which we need to move to stay in the top tier in the decades to come. One of our top priorities, along with excellence in research and graduate education, is to produce new graduates with exceptional preparation for further professional study and engineering practice. One result of this priority is a new undergraduate curriculum reform initiative called PROCEED, an acronym for "Project-Centered Education."

PROCEED was formally initiated in the fall of 2000. Overall curriculum goals were outlined and 13 pilot projects were initiated by the ME faculty. These projects cover the entire range of the curriculum, from the freshman introduction-to-ME course through the senior capstone design course. They include a number of reforms, including, among others, development of new labs which are closely integrated with core theory courses, using reverse-engineering of commercial products and systems to motivate analysis and computer modeling, development of an on-line student portfolio system to showcase project-centered work, and live-videoconferencing with corporate engineers to counsel with students on company-sponsored projects.

Implementing an ambitious curriculum reform program in a very large department with a strong research-oriented culture presents a number of management challenges. These include promoting faculty buy-in, building a base of industrial partnerships, financing of needed capital improvements and ongoing expenses, assessing educational effectiveness, dealing with faculty and student workload issues, balancing breadth versus depth through overall curriculum integration, and optimizing the involvement of practicing engineers as mentors and advisors.

We are now midway through PROCEED's 5 year implementation plan. This paper outlines the background, motivation, and strategy behind PROCEED, and discusses how we are dealing with the management issues described above.

RATIONALE FOR PROCEED

To put our rationale for PROCEED in perspective, it is helpful to briefly examine the history of engineering curriculum reform in the US over the past half-century. The launching of Sputnik by the Russians in 1958 and the ensuing space race of the '60s produced a huge demand for new technology, and with it a huge demand for more mathematically and scientifically-based engineering curricula. Engineering education traditionally had been heavily laboratory-oriented, but many lab courses went by the wayside to make room for more science and math. It is also worth noting that, up to this time, most young people came to engineering studies with a considerable amount of hands-on experience acquired by building and fixing engineered systems (cars, radios, appliances) and therefore a substantial body of conceptual understanding of how things work. This conceptual foundation provided a good framework on which to build a theoretical base of engineering science.

The emergence in the '70s of mainframe computers as engineering tools produced a demand for programming skills, and the introduction of programming and computing courses further forced lab studies out of the curriculum. The '80s brought the advent of personal computers and the rapid development of user-friendly application software. As increasingly powerful hardware and software tools became available, industry began demanding engineering graduates skilled in the use of these application tools, and showed less interest in students' ability to build their own tools from the ground up. The microprocessor also radically changed the way and extent to which pre-college students experience technology. Integrated electronics and the conversion of mechanical control to computer control effectively made function and form in everyday machines invisible; the conceptual database that engineering students formerly brought to the table declined substantially as "virtual" experience replaced hands-on experience and time spent "tinkering" was devoted instead to playing video games. The explosive growth of the Internet in the '90s brought instant access to detailed (but not necessarily accurate) information on any topic with little or no contemplation or reflection to provide broader perspective on its meaning. Taken as a whole, the trends described above have resulted in a phenomenon referred to by Frand [1] as "the information-age mindset", characterized by broad substitution of "virtual reality" for "reality", unquestioning acceptance of computer-based information, and a preference for "doing" as opposed to "knowing".

The traditional lecture/problem teaching approach that evolved in the latter half of the 20th Century has thus thrust today's engineering students into a highly theory-based learning environment with little intuitive understanding of the physical phenomena underlying the differential equations they are learning to solve. This represents an inversion of the Kolb Learning Cycle shown in Figure 1.

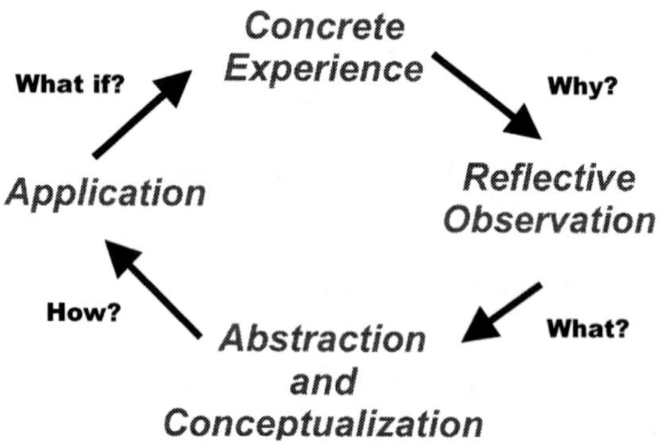

Figure 1: Kolb Learning Cycle

The Kolb Learning Cycle (Kolb[2]) describes a natural process in which individuals first acquire a "mental database" of experiences, more or less at random, by exposure to the real world. With repitition and time for reflection, connections between these experiences emerge and coalesce into concepts that lend themselves to formulation of abstract principles. These principles can then be applied to situations beyond those already present in the experiential database. Since, in the normal Kolb cycle, the abstract principles themselves are built on a base of actual experience, the individual has a natural basis for understanding when they can and cannot be reliably applied to new situations. On the other hand, if learning starts with abstract theory absent an experiential base, application of the theory becames largely a game of plugging arbitrary values into formulas that "work", not because they accurately represent reality, but because they conveniently match the variables in the formula.

The rationale for project-centered education, then, is to counter the information-age mindset by providing our students a real-world context on which to base their understanding of engineering science. In doing so, we hope to develop an appreciation for "knowing", rather than simply "doing", and a more solid basis for application of engineering theory to new situations.

HISTORY, TIMELINE AND ORGANIZATION

In the mid-'90s, substantial evidence began to emerge of the declining effectiveness of traditional learning methods in certain elements of our ME program, based in part on statistical information (e.g., low retention in the first two years) and in large measure on anecdotal results of course evaluations. This produced a general feeling of dissatisfaction on the part of many experienced faculty, and led to a series of informal brown-bag discussions in 1996 on what could be done to improve the situation. Out of these discussions came several curriculum and methodological changes which were implemented in the 1998-2000 catalog. These included:

- Consolidation of our separate freshman courses in engineering graphics and introduction to ME into a single course which centers instruction in graphics and concepts of engineering design around reverse-engineering of real devices.
- Converting our junior level course in machine elements, which was essentially a laundry list of topics in strength of materials and formulas for design of various components, into a project-centered experience, with theory and formulas being introduced as needed to reverse-engineer and extrapolate the design of real machine components.
- Converting our second thermodynamics course into a project-based thermal-fluid systems course in which theory topics are covered as needed to permit modeling of complex systems such as jet engines and cogeneration powerplants.

Based on positive results from these early efforts, a grant from Ford Motor Company was received in fall 2000 to broaden the scope of project-centered education to a department-wide effort. A second PROCEED grant from Applied Materials Corporation was received in 2001.The primary elements of this program include:

- Implementation of project-centered experiences in selected ME courses running throughout the curriculum from freshman through senior years ("just-in-time" theory).
- Redesigning and relocating "disconnected" lab courses to integrate them more closely with corresponding theory courses and reequipping related labs to provide a more flexible hands-on lab environment.
- Collaboration between faculty and practicing engineers from industry in the formulation and teaching of project-centered curriculum materials.
- Incorporation of computation, teamwork, and communication as common elements of courses to more accurately reflect the real environment of engineering practice.

The PROCEED initiative is being implemented over a five-year timeline, as illustrated in Figure 2. Phase 1 of PROCEED was officially inaugurated (i.e., the money was in the bank and the project was given a catchy name) in September of 2000. Over the next 6 months, a coordinating committee was established with broad representation across disciplines and ranks within the department. Members of the committee convened planning sessions of each of the various disciplinary areas to provide input on the extent to which curriculum in

that area might be improved through project-centered activities and what approach might be most effective for specific courses.

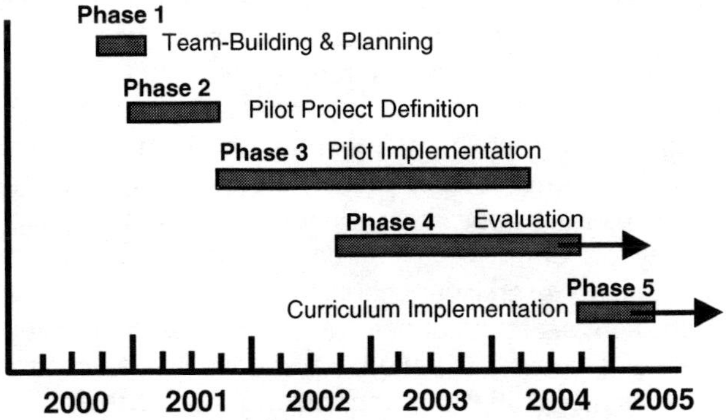

Figure 2: PROCEED Implementation Plan

This grass-roots planning period was critical to the future success of PROCEED, not only for developing specifics of the program plan, but for building enthusiasm, across the board buy-in from faculty, and team spirit. In a major research university like UT Austin, where faculty are hard-pressed to fulfill their research missions, broad commitment to a time-consuming undergraduate curriculum reform effort was a major achievement. The planning period culminated in a full-day workshop in January 2001 which was attended by about two-thirds of the department's faculty. At this workshop, a set of desired curricular outcomes, similar but not identical to ABET's "a-through-k" outcomes, was established and an intial list of pilot projects was laid out.

Phase 2 was carried out during the spring and summer of 2001; thirteen pilot projects were defined in detail, actively involving about 20 faculty members. Budgets were established, initial corporate contacts were made, and supporting materials were prepared to permit pilot implementation in experimental sections of the respective courses in the 2001 fall semester.

Pilot implementation began with the 2001-02 academic year, and was continued in 2002-03. Also during this period, an extensive evaluation process was initiated, both to provide formative assessment for ongoing improvement of individual courses and summative assessment of the effectiveness of the PROCEED program as a whole. This evaluation process is being integrated into PROCEED as a permanent element, and the assessment methods that are being developed will in the future be applied to both project-centered and conventionally-taught courses in the department. Most of the curriculum reforms now being piloted on a limited enrollment basis will be expanded during the 2003-04 academic year, and full-scale implementation is planned to begin in 2004 with publication of the next biennial catalog.

PROCEED PILOT PROJECTS

The PROCEED projects currently being implemented on a pilot basis cover the curricular spectrum from the first introductory ME course through the senior capstone design course. We should emphasize that the term "project-centered" is not meant to imply that an entire course is necessarily built around a single all-encompassing application. Projects may, in fact, take a variety of forms, ranging from short lab or computer simulation exercises coupled to a conventional homework set to major system design/analysis problems extending over a period of weeks. The common denominator is that the application must be real, actual hardware and data should be used when possible, and elements of teamwork and communication should be incorporated. Faculty and student contact with a practicing engineer familiar with the problem is desirable, and in the case of industrially-sponsored projects, essential.

Initiatives which focus on a specific course or group of courses in a given area include the following (note several projects may be grouped under one heading).

Integration of CAD/FEA modules and rapid prototyping in freshman Introduction to Engineering Design and Graphics: Faculty in our Graphics area have developed modules to teach students how to couple a 3D CAD model to a widely used finite-element package (COSMOS) for stress analysis and also to a simple rapid-prototyping machine which can produce visual prototypes (Barr et al[3]). Assembly modeling and kinematic simulation modules are under development. Student projects include reverse engineering of industrially-supplied automotive components and robotic assemblies for handling of semiconductor wafers during processing.

Integration of industrial case studies in Engineering Economics and Engineering Statistics and Probability: Students in our junior-level engineering statistics and probability course are using a powerful statistical analysis package to evaluate manufacturing quality-control data supplied by an industrial partner. The purpose of the analyses is to pinpoint the origins of component failure.

A semester-long project centering around the design and production of a car to meet a particular market requirement has been utilized in our Engineering Economics course as the unifying theme to address such topics as capital and production cost analysis, time value of money, present worth, useful life, etc. As a spin-off of these upper-division initiatives, faculty have also produced short teaching modules that can be used in freshman courses to acquaint students with the relationship of economics and statistics to important engineering problems.

Integration of labs and theory in core Materials Science and Engineering courses: Faculty in the MS&E area have created a series of "reference modules" on various topics covered in our sophomore and junior-level core courses in basic materials science and materials processing. Modules include lecture presentations and examples, demo materials and video clips focusing on real parts and processes developed in coordination with industrial partners. Appropriate problem sets and lab exercises built around these applications have been developed.

In the junior-level materials processing laboratory course, a project is introduced during the first several weeks focusing on parts provided by an industrial partner. The project provides a frame of reference for a series of modularized laboratory experiments carried out throughout the semester that provide the students with experience working in teams as well as hands-on time using materials processing equipment.

A scanning electron microscope and an xray diffractometer system have been acquired as the centerpieces of a new student materials-characterization laboratory. Use of these instruments in the two core courses mentioned above has been initiated and they are also being used in materials-related projects in our senior capstone design course and in undergraduate elective courses in the MS&E area.

Integration of labs and projects in core Thermal-Fluid Systems courses:

As mentioned earlier, one of our early initiatives in project-centered education was the introduction of a junior-level thermal-fluid systems course to replace the traditional second thermo course. Most projects undertaken during the past several years have focused on computer modeling of complex systems, such as gas turbine/cogeneration powerplants, jet engines and associated aircraft performance, and heating/cooling systems for semiconductor processing tools (Schmidt et al[4]). One unique project entailed reverse engineering of a domestic refrigerator. Two identical refrigerators were purchased, one of which was maintained in running condition and instrumented so that students could acquire actual performance data, while the other was "sacrificed" to allow for observation and measurement of insulation, heat transfer surfaces, and compressor parameters. These data were used to model the system and predict power requirements, which could be compared with measured data and published specifications.

Two automobile-related projects have been implemented in our fundamental heat transfer class, one on cooling of an under-the-hood circuit board and the other on design of a radiator. These projects served as unifying themes for basic studies of conduction, convection and radiative heat transfer and were used in parallel with conventional textbook problems in the respective topic areas.

Two new laboratories in fluid mechanics and heat transfer, respectively, are being implemented in parallel with the corresponding fundamentals courses in these areas. Extensive redesign and renovation of the TFS labs is being undertaken to convert them from large single-purpose apparatus used mainly in the demonstration mode to flexible small-scale apparatus which can be economically replicated to foster more hands-on experience. A number of simple "kitchen experiment kits" are being designed and fabricated to be used either as in-class demonstrations or take-home experiments.

Introduction of computer simulation projects in Kinematics, Dynamics and Controls courses: Faculty in the Dynamics and Controls area are implementing use of Working Model, a dynamic simulation software package, in our sophomore-level kinematics and dynamics course. Students use Working Model to perform parametric analyses related to textbook problem assignments and at the end of the course build a dynamic simulation model of an off-road bicycle. New project-centered modules are being developed to simulate automobile engine components.

We are also using Matlab and Sim-20, a Bond-graph modeling software package, in our junior-level dynamic systems and controls course to model noise, vibration and harshness (NVH) in automobiles.

Reverse engineering, redesign, and prototyping in Machine Elements: Our junior-level course in machine elements has been reoriented to teach machine element design through product emulation. Students reverse engineer an existing product, formulate and model design changes, and then recreate a functional prototype using LEGO Mindstorm kits (Campbell[5]).

Expansion and upgrading of Student Reverse Engineering and Prototyping Laboratories: In years past, our senior-level design methodology and capstone design projects courses were largely taught "on paper". Students had little access to hands-on facilities to tear down and measure characteristics of existing mechanical systems or to build functional prototypes of their own designs. We are now equipping two student labs with a variety of tools and instruments to facilitate reverse-engineering studies, as well as wood, plastic, and metal-working equipment for prototyping. The availability of this lab has added a significant dimension to our design courses, as well as upper-division electives and independent projects courses.

In addition to the course-centered initiatives described above, several cross-curriculum projects have been undertaken which support the overall PROCEED effort.

Videoconferencing to support student and faculty interaction with engineers in industry: The costs of travel and lost time can be a major obstacle to frequent interaction between faculty, students, and industrial sponsors, a key element of the PROCEED program. To help overcome this obstacle, a freestanding videoconferencing system has been installed in a large conference room equipped with both ethernet and ISDN communication lines to permit dialup videoconferencing with virtually any industrial or academic facility in the world. The system is portable and can easily be moved to other rooms in the building which have the necessary communication infrastructure. A distance learning classroom with a capacity of 35 students is also equipped for videoconferencing, and plans are in the works for similarly equipping a 100 seat lecture room. This will give us videoconference capability ranging from small project groups to large classes.

Development of a web-based student portfolio system: Media professionals in our Faculty Innovation Center, working with faculty and students, have developed a web-based electronic portfolio system which permits students to display the work they have done throughout their project-centered academic careers (Campbell and Moore[6]). The system consists of a preformatted template with a very user-friendly interface which allows students to easily create and update portfolios; it also includes provisions to allow student control of access to portfolios. These portfolios will provide a valuable added resource beyond transcripts and resumes for students to use in presenting their capabilities to prospective employers and graduate schools.

Web-based learning modules on Engineering Ethics, Professional Responsibility, and Technical Communication: Faculty in our Technical Communication and Legal/Ethical areas have collaborated to produce a series of web-based modules that can be used on an as-needed basis for a variety of project-centered courses throughout the curriculum (Hart and Moore[7].) Modules have been prepared on global and social responsibility, research ethics, scholastic dishonesty and plagiarism, working in groups, and writing executive summaries. Additional modules are currently in development.

Evaluation and assessment of project-centered learning: Assessment is a core element of PROCEED, both for real-time improvement of ongoing courses and long term evaluation of curricular design and teaching methodologies. We have engaged a full-time specialist in this area, with education and experience both as a practicing mechanical engineer and as an educational designer/evaluator, who is developing a variety of assessment tools for project-centered courses. These tools are being refined during the pilot phases of PROCEED and plans are to implement them on a department-wide basis during the 2003-04 academic year.

MANAGEMENT CHALLENGES IN IMPLEMENTATION

PROCEED represents significant departures from traditional engineering education practices, and as such there are many issues to consider and challenges to overcome from a management perspective. These include the logistics of handling a very large student population, financing both the transitional and steady-state costs, motivating faculty and student buy-in, assessing program efficiency and efficacy, and interacting with corporate partners.

Logistics of large-scale implementation: As pointed out in the opening section, the ME Department at UT Austin is one of the largest in the U.S., with over 1,000 undergraduate students. The problems of handling such a large number of students are many: scheduling and staffing of multiple sections, providing and maintaining adequate lab equipment, and supporting and housing large numbers of TAs, to name a few. We are now implementing PROCEED on a mainstream basis. While it is too early to claim success in meeting all these challenges, we can comment on how we are approaching them.

Most important, we have designed PROCEED as a five-year effort and not tried to make it happen overnight. Typically, we have introduced a new lab or major curriculum modification in an existing course on a pilot basis for at least a year prior to mainstreaming it. We have first tried out the project-centered approach in a single section, in some cases given faculty reduced teaching load, and provided extra TA assistance to allow them to work out kinks and prepare new course materials. As new lab equipment is proven out, we have replicated it to accommodate an expanded student load.

Space has been a major challenge: lab space, classroom space, student workspace, and TA office space. It seems axiomatic that there is never enough space to meet all of our research, teaching, and administrative requirements, and there is intense competition for it. We are approaching this problem both by reclaiming underutilized space, and by expansion of our existing space. As the need for more lab and student workspace to support the project-centered teaching mode became clearer, we did a careful inventory of all space in our building and worked wth faculty to consolidate or relocate some activities to make room for new ones. We were also fortunate to obtain the support of one of our corporate sponsors to renovate and expand a large centrally located area of our building as a student meeting/work area; this beatiful new facility will greatlly enhance the overall student academic environment, and will be especially valuable as a place where student teams can meet to work on project-centered activities.

Financial support: Meeting the development costs and operating expenses of PROCEED has required a major fundraising effort, both external and internal, and planning for a long-term continuous income stream. The enabling events for implementation of PROCEED were the major center-of-excellence grants in 2000 and 2001 from Ford Motor Company and Applied Materials Corporation. These grants have been matched in part with academic development funds from the College of Engineering, equipment and renovation funds from the University, and smaller grants from private donors and other corporations. It is impossible to overstate the value of the corporate grants, which have enabled us to provide both the expanded facilities and equipment and the special faculty/staff support needed to initiate major curricular changes. For ongoing support, we are assessing new instructional fees that will help cover maintenance, upgrading of lab equipment, and staff to ensure continuity of the project-centered program.

Faculty buy-in: An important consideration in PROCEED is faculty workload and incentives to participate in a major undergraduate reform effort, particularly given the demands of a research-oriented university. Project-centered education is inherently labor-intensive for both faculty and students. The key to faculty buy-in has been to build PROCEED as a grass-roots effort, rather than trying to impose it from above, and to build in incentives for faculty to participate in the program. The initial planning for PROCEED emerged from faculty brown-bags and workshops, as described earlier, and this continues through the activities of the various discipline-specific groups in the department and the PROCEED Coordinating Committee.

We have implemented a system of industry-sponsored summer PROCEED Fellowships which provide part-time salary support for a faculty member and a grad student to pilot new projects to be used during the coming academic year. This support has been a significant incentive for faculty involvement. We have also encouraged faculty to publish articles about PROCEED initiatives in the engineering education literature and have provided travel grants to ASEE, ASME, and similar conferences to present these papers. Because PROCEED is a department-wide effort, our Productivity Committee and Budget Council (which passes on promotion and tenure decisions) have become more sensitized to the importance of educational publications as well as traditional research publications in assessing faculty performance. This is especially important in a research university where publications are a major metric of faculty performance. Beyond these direct incentives, many faculty have been motivated simply by the belief that PROCEED is a better way to educate engineers and by the positive student feedback they have received in pilot courses.

Student buy-in: Project-centered courses are typically more-time consuming for students than conventional lecture-problem courses. Indeed, it can be argued that one of the central motivations for challenging students with open-ended problems is to break the "information-age mindset" that assumes that a "correct answer" to every problem can be obtained in a matter of minutes. In the PROCEED environment, resolving mistakes in consultation with teammates (rather than counting on the instructor to do it) can be among the most valuable learning experiences, since it demands contemplation, reflection, and collaboration. In PROCEED courses, we have placed a high priority on providing good TA support, and course evaluations have confirmed that students view the opportunity for in-depth consultation with the TA as a very important element in the overall learning process.

We have also found that an important factor in student buy-in is clearly articulating why we are using this approach, how it contributes to desired learning outcomes, and how they must approach learning differently in order to successfully achieve these outcomes. Most students initially feel uncomfortable with the more open-ended nature of project-centered assignments and the more subjective approach to grading that it requires. Part of our investment in PROCEED has been to support a professional study of student attitudes and expectations in project-centered courses (Jones[8]), and this study is providing valuable insight into how to promote a positive attitude toward the demands of project-centered education.

Another major factor in building student buy-in is to insure that project-centered courses are both efficient and effective. Some of our initial PROCEED projects were implemented in full sections right from the start, without having been thoroughly worked out in advance. Invariably, unanticipated delays occurred which caused a lot of student (and faculty) frustration. We learned that it is much wiser to "pretest" large projects (especially if they are industrially-sponsored) with a small group (2 or 3 students who have already completed the targeted course) in a summer session to develop an initial roadmap for the project. This is especially important if experimental data is needed during the course of the project.

Finally, we have learned that not all projects need to be large and time-consuming or open-ended. Small motivating projects, if they have engineering significance and effectively motivate the need to learn and apply engineering science and theory, can be quite effective with less time commitment by the students and faculty.

Assessment: The bottom-line question for PROCEED is "Does project-centered education work better than traditional lecture/problem methods in engineering?" There is a substantial body of research literature in the domains of K-12 science education and medical education to support the thesis that project-based learning substantially improves long-term retention and "deep understanding" (Barron et al[9], Blumenfeld et al[10], and Williams[11]). Relatively little hard data are available specific to project-based engineering education. While it is reasonable to assume that similar benefits might accrue, this has yet to be proven conclusively.

In order to develop a quantitative metric of the efficacy of PROCEED, we are piloting a modified instrument based on the QQI (Quantity, Quality, and Improvement) assessment instrument developed at VMI (Addington and Johnson[12]). This instrument measures quantity and quality of learning opportunities, and students' self-reported learning gains relative to the desired outcomes specified by the instructor (and consistent with EC 2000 criteria). The modified QQI instrument (Jones[13]) was piloted during Summer and Fall 2002, and implemented in selected PROCEED courses in the Spring and Summer of 2003. Based on our early data, students in PROCEED courses show high self-reported improvements in such areas as application of fundamentals to complex problems, ability to apply computers in an engineering context, teamwork, time management, and project planning skills, and ability to communicate technical information. In general, project centered courses are rated as more time-consuming than conventional courses, but the time spent is rated as high quality. Overall course satisfaction ratings have typically been above average. Obviously, this is a rich field for further research and we plan to integrate continuous in-depth assessment as a fundamental element of PROCEED.

Industrial interaction: A significant effort has been devoted to learning how to most effectively interact with our industrial partners. This is not just a matter of fundraising; one of the fundamental principles of PROCEED is direct involvement of industrial experts in the teaching process. This includes collaboration with faculty in formulating relevant and exciting projects, and direct interaction with students in the course of project execution. The primary benefits which attract our industrial sponsors to the program include early access to potential new hires and early in-depth exposure to industry-specific problems for the entire ME undergraduate population. Regular communication with technical staff in our sponsors' corporations is critical to establishing and maintaining this fundamental element of PROCEED. To facilitate this communication, we have made a significant investment in videoconferencing facilities and infrastructure; we now have the ability to hold live teleconferences ranging from a conference room to a classroom. We have also found that it is imperative to have an enthusiastic "champion" for PROCEED within the company to identify and coordinate appropriate technical staff contacts, and to assure that the champion's involvement in PROCEED is properly blessed and supported by top management.

Other considerations: Every project-centered course entails a tradeoff between time spent on broad coverage of a greater number of topics and time spent on application of fewer topics to realistic engineering problems. There is no simple formula that applies across-the-board; it is clear that some courses demand a more traditional topic-centered focus, while others, particularly those at the more advanced level, offer more flexibility to substitute depth for breadth. As we implement and evaluate PROCEED, we hope to refine our ability to effectively balance these competing objectives.

For large projects, teams are often used effectively in PROCEED, but organizing effective teams, teaching teamwork skills, and allocating credit for team-based work are significant challenges. Fortunately, the literature offers some guidance in this area (e.g. Jensen and Wood[14]). As more faculty in our Department become engaged in this mode of teaching, we are working together to formulate best practices on how to implement team-based learning.

CONCLUSION

PROCEED is a work in progress which we believe has the potential to substantially enhance the learning experience, technical competence, and marketability of our graduates. The knowledge, skills, ethics and attitudes which we are trying to instill are prerequisites for leadership in the professional world. We hope that this paper provides some insight into our motivations for undertaking the PROCEED program and the methods we are using to implement it in a large research-oriented university. As we progress in this process, we look forward to reporting both our successes and our failures as our contribution to the continuing quest for the best in engineering education.

ACKNOWLEDGEMENTS

The authors wish to express their appreciation to Ford Motor Company and Applied Materials Corporation for their financial and in-kind support of the PROCEED Program and to the many members of the faculty and staff in the Department of Mechanical Engineering at UT Austin who are committing their time and energy to making the program a success.

REFERENCES

1. Frand, J. (2000): "The Information-Age Mindset: Changes in Students and Implications for Higher Education", Educause Review, September/October 2000.

2. Kolb, D. (1984): **Experiential Learning: Experience As the Source of Learning and Development**, Prentice-Hall, Englewood Cliffs, NJ.

3. Barr, R., P.Schmidt, T.Krueger, and C.Twu (2000): "Introduction to Engineering Through an Integrated Reverse Engineering and Design Graphics Project", J. Engineering Education, 89 (4).

4. Schmidt, P.S., J.Jones, G.Vliet, and T. Jones (2003): "A Project-Centered Approach to Teaching of Thermal-Fluid Systems Analysis and Design", Proceedings of the 2003 ASEE Annual Conference, Paper 2003-460.

5. Campbell, M.I. (2002): "Teaching Mechanical Design through Product Emulation", Proceedings of the 2002 ASEE Annual Conference, Session 2366.

6. Campbell, M. I. and C. Moore (2003): "Web-Based Engineering Portfolio System", Proceedings of the 1999 ASEE Annual Conference, Session 1520.

7. Hart, H. and O.C. Moore (2002): "Developing Instructional Modules on Engineering Ethics", Proceedings of the 2002 ASEE Annual Conference, Paper 2002-2123.

8. Jones, T.L., (2003): "Conveying Instructor Expectations in a Project-Centered Course", Proceedings of the 2003 ASEE Annual Conference, Paper 2003-619.

9. Barron, B.J.S., D.L. Schwartz, N.J. Vye, A.Moore, A.Petrosino, L.Zech, and J.D. Bransford (1998): "Doing with Understanding: Lessons from Research on Problem- and Project-Based Learning", J.Learning Sciences, 7 (3&4).

10. Blumenfeld, P.C., E.Soloway, R.Marx, J.S.Krajcik, M.Guzdial, and A. Palincsar (1991):"Motivating Project-Based Learning: Sustaining the Doing, Supporting the Learning", Educational Psychologist, 26 (3&4).

11. Williams, S.M. (1992): "Putting Case-Based Instruction into Context: Examples from Legal and Medical Education", J. Learning Sciences, 2 (4).

12. Addington, J.S. and R.A. Johnson (1999): "Incorporating the Design and Use of Surveys with Other Engineering Assessment Methods Under Criteria 2000 Guidelines", Proceedings of the 1999 ASEE Annual Conference, Session 1332.

13. Jones, T.L. (2003): "Self-Reported Instrument for Measuring Student Learning Outcomes", Proceedings of the 2003 ASEE Annual Conference, Paper 2003-920.

14. Jensen, D. and K. Wood (2000): "Incorporating Learning Styles to Enhance Mechanical Engineering Curricula by Restructuring Courses, Increasing Hands-On Activities, and Improving Team Dynamics", Proceedings of the ASME International Congress, November 2000. (ASME Curriculum Innovation Award Paper, April 2001).

Proceedings of IMECE'03
2003 ASME International Mechanical Engineering Congress
Washington, D.C., November 15–21, 2003

IMECE2003-43533

THE TREASURE IN TECHNICAL INFORMATION: A RESEARCH PROJECT FOR ALL DISCIPLINES

Bruce A. Harding, Purdue University

ABSTRACT

Engineering practitioners often find themselves required to locate technical and non-technical information from a variety of sources. They may generate quotes, research equipment specifications, determine market shares, delineate a proposed manufacturing process, find company profiles, retrieve demographics data, certify compliance with international or national standards, or research any one of countless other pieces of information. This may involve books, CDs, periodicals, newspapers, corresponding via email, fax, letters or the worldwide web, across several languages.

Few classroom projects address practitioner's needs for broad information research skills. This paper details strategies for a unique student research project leading to useful experience applicable to engineering and engineering technology practitioners in any discipline, in any locality. The project could be instituted in virtually any class at any level and run in any language. It is readily adaptable to changing technology, any textbook, and has now been in use since 1982. As implemented at Purdue University, it has been cited by multiple ABET re-accreditation teams for innovation and as an excellent example of continuously improved instruction. Over the years it has grown to become one of the more noteworthy experiences cited in both student exit surveys and in graduate surveys.

IN THE BEGINNING

The impetus for this project began in the early 1980s as a way to teach students to use one of their required classroom texts quickly and efficiently. The book, *Machinery's Handbook*, is an expansive 2500 page industrial tome on standards, fasteners, engineering materials, mechanics, machining, quality assurance, manufacturing processes, CNC and just about everything mechanical. The class, *Production Design & Specifications* (PD&S), enrolls about 200 second-semester freshman students per year. It is a core class and the first of several classes requiring *Machinery's*. In this class, students are expected to become skillful in the book's use. Unfortunately the book is not practical for assigned readings, since it is mostly charts, tables and other practitioner information.

During the same timeframe, a popular TV show followed the lives of Ivy League law school students learning law and lessons in life from venerable actor John Houseman as the cranky, but wise, law professor. In one episode, Houseman, telling the class that he was trying to bolster the students' knowledge of the library references they would use as practicing attorneys, assigned a weekend to answer a set of 100 obscure, very detailed questions spanning all reaches of the law. Because of the impossibly short timeframe, the class nearly rebelled until they realized that through teamwork they could divide the questions among the class and complete the assignment. They did, of course, and afterwards discovered that teamwork was one of the real lessons in the assignment.

That approach sparked the genesis of a project about science, technology and engineering practice plus other topics, as a way to encourage students to become more deeply familiar with *Machinery's*, their other texts and various references. It has since been expanded to encompass nearly every technical discipline and now heavily leverages resources of the worldwide web. In completing the project, it was envisioned that along the way students would learn to apply a good dose of creativity in finding sources and discover the value of teamwork as well. From the beginning, the intent was for the answers to become secondary to the process and then only as affirmation of the experience of the search. The project was dubbed, "the Treasure Hunt," because of the 'treasured' knowledge to be gained from the project.

Altogether, over four thousand students have experienced this project during the 20 intervening years.

THE MECHANICS OF RUNNING THE HUNT

GENERATION OF QUESTIONS. Questions are generated from a list, which has grown from an initial 20 to over 1200 questions.

Currently questions are stored in an MS Excel database. The database includes columns labeled question, answer, source and standard if applicable. A Visual Basic macro is used to randomly select questions for each student, resetting the random number generator after each question. Statistically, duplications occur but have diminished as the database has grown.

ORIENTATION. Students receive an introductory presentation on the rationale, hunt rules, timetable and organization. They also review the generic bibliography (see Fig. 2) and a sample of Part 1 of the report (see Fig. 1). Also discussed are advantages and disadvantages of working alone or optionally forming teams of two to four persons.

Once oriented and prior to distribution of the actual questions, teams are formed. Students working as individuals are responsible for 10 of the 12 questions. Teams of two are responsible for any 20 of the 24 questions received. Teams of three are responsible for any 30 of 36; teams of four, any 40 of 48 questions. Because of the random selection of the questions, team members may have a few duplicate questions among the lists received. These duplicates may used.

After distribution of questions, the reference librarians meet with the classes to discuss the use of the library, general types of resources available locally and online, and strategies for information research which might aid them in their hunt.

TIMETABLE. The Treasure Hunt is assigned near the end of the semester, typically on a Friday with a week from the following Monday being the due date for the written report. This gives them two weekends and five full weekdays to complete the assignment. Between the assignment and due date, formal class meetings are cancelled but instructors remain available as consultants.

Students are encouraged to begin by identifying any questions with answers that must be sought via letters or fax. They are advised to address these immediately so there is sufficient time for a reply. Students are encouraged to use letters rather than phone calls or faxes. Because replies could be delayed beyond the report deadline, a secondary deadline seven days later allows inclusion of any source information received too late to accompany the written report.

REPORTING RESULTS

The report format requires a bound and word-processed two-section document. Teams must submit a monolithic document representing their combined work.

SECTION ONE – QUESTIONS/ANSWERS. Section one lists questions, answers and sources. It may be as little as one page depending on the extent of the answers. Resources must be referenced properly. If the answer requires a drawing or sketch, the artwork must be computer drawn or scanned and pasted into the section. If a team report, the questions must be numbered with both the report question number plus the name and question number from whomever's list the question originated. See the question numbering scheme in Fig. 1.

SECTION TWO – DOCUMENTATION. This section is reserved for supporting documentation authenticating answers and sources in section one. If the documentation is from a printed source, a Xerox of the page, plus one of the cover of the publication is required for each question. If the source is online or a CD ROM, an original printout of the pertinent page is required. Facsimiles may be original or copies. The documentation must be bound and annotated in the same order as questions in section one.

Occasionally documentation is received in forms other than paper. For instance, one question asked the model numbers and capabilities of a series of forklifts manufactured by Caterpillar. Company reps often send promotional videotapes rather than printed tech sheets, thus students answer this question by viewing the tape and including the tape as source documentation.

Students may use any source for their answers with the understanding that verbal sources or otherwise unsubstantiated sources only receive the one-half credit for correct answers. The remaining credit is awarded for verifying documentation.

Sample Treasure Hunt Report - PART 1 (by team Smith & Jones)

No. Actual

1. (Smith Question 1) *[Your text as source]*
   - Question: What is the free diameter of a 6 mm ANSI Metric Reduced Cross Section Retaining Ring (external)?
   - [10 pts.] Answer: 4.70 mm.
   - [5 pts.] Standard: ANSI B27.7-1977
   - [5 pts.] Source: R. E. Green (Ed), Machinery's Handbook, 25th Ed, Ind. Press, p. 1594.
2. (Smith Question 2) *[A standard as source]*
   - Question: What angle is the 'V" of a surface finish mark drawn to?
   - [10 pts.] Answer: 60°.
   - [5 pts.] Standard: ANSI Y14.36M - 1996
   - [5 pts.] Source: ASME, Surface Texture Symbols, Y14.36M-1996, ASME, p. 2.
3. (Jones Question 1) *[A book as source]*
   - Question: Who was the first woman head of a western hemisphere country?
   - [10 pts.] Answer: Isabel Perón.
   - Standard: N/A
   - [10 pts.] Source: The Good Housekeeping Woman's Almanac, p. 250, Appendix A, 1977.
4. (Jones Question 4) *[A text as a source]*
   - Question: What are the limits on the height dimension for a 610 Woodruff Key?
   - [10 pts.] Answer: 0.547" max., 0.537" min.
   - [5 pts.] Standard: ANSI B17.2 - 1967.
   - [5 pts.] Source: Geisecke, et al, Technical Drawing, 11nd Ed., Prentice Hall, p. 642.
5. (Smith Question 5) *[A call for answer with suporting fax or product literature coming later]*
   - Question: What types of tools does Thaxton,® Inc. make?
   - [10 pts.] Answer: Tools for pressure testing pipes, tubing & vessels.
   - Standard: N/A
   - [0 pts. until rec'd, then 10] Source: (IN MAIL or TO BE FAXed next week) Product literature coming from Product Mgr., Thaxton, Inc., 25 Leanburg Rd., Mars, Pa. 16046.
6. (Smith Question 9) *[A periodical as source]*
   - Question: What is the smallest diameter abrasive cord sold by E. C. Mitchell Co.?
   - [10 pts.] Answer: 0.012" dia.
   - Standard: N/A
   - [10 pts.] Source: Industrial Equipment News, Vol. 57, No. 5, p. 20.
7. (Jones Question 10) *[A web page as source with printout and FAX attached for timely pricing]*
   - Question: For a yet unidentified manufacturing process, you have been asked to spec out a source for 99% pure platinum (Pt) wire in Ø1mm X 100 m, to use as a catalyst. Identify cost and vendor (name address, phone, fax, URL, etc.) and any safety precautions.
   - [10 pts.] Answer: As of yesterday's date: $1612 per spool (Ø1mm X 100m). Kirby Risk Supply, 1815 Sagamore Parkway, Laf, 47905 (317) 448-4567, FAX (317) 448-4568, http://www.kirbyrisk.com/html/krhome.htm. No safety precautions.
   - Standard: N/A
   - [10 pts.] Source: K-R Online catalog. FAX quote on price (would be attached).
8. (and so on ...)
   - Question: ......
   - Answer: ......
   - Standard: ......
   - Source: ......

Figure 1. Sample report format for Part 1, showing various sources, plus point values assigned to answers, sources and standards if applicable. Each question is worth 20 points total.

LATE DOCUMENTATION. Initially students have about nine days to complete their hunt; however, sometimes company documentation is not received by the due date. In such cases they must indicate this in their report (see sample question #5 in Fig. 1). Credit is not received for the source — only the answer — even if correct. Students then have up to seven days after the graded report is returned to submit any missing documentation. Once submitted and verified, full credit for the source information is granted and the grade changed. While this may sound complex, it effectively provides over three weeks from project assignment for total documentation submission.

GRADING

As stated before, the overriding goal of this project is the process of research and teamwork, not the answers. By completing the project, students experience the types of references available to them, both locally and worldwide. In so doing they sometimes must exhibit highly creative solutions when locating sources not in the campus or public library.

To reinforce this, grades are based on two parts. One-half credit is awarded for the correct answer, regardless of how it was obtained — no documentation required. The other half of the credit is earned from providing documentation to confirm the answer. If a standard is applicable, the source credit is split again between the documentation and the identification of the standard. Thus a less than fully documented answer, albeit correct with documentation, would still yield only partial credit if the student failed to reference an applicable standard. See the point value breakout in Fig 1.

TEAMWORK. For the Treasure Hunt students may form teams of two to four members or work alone. Once formed, teams may not be dissolved, regardless of who does the work. Each team member receives an equal portion of the total grade. The students are counseled on the advantages and disadvantages of teaming and receive questions after teams have been formed.

They are told that advantages to teams are: a) they may divide the labor, perhaps based on interest or knowledge or contacts, b) they may discard two questions per team member, regardless of whose list the question originated, and c) that because of the random generation of each person's list, there may be question duplication among the members' lists and they may exploit the anomaly by answering the duplicates. Students are also informed that a disadvantage could be that because they must trust each member of the team to do their portion of the work, all team members might not do their share of the work, yet receive the same grade.

WRONG ANSWERS MAY BE ACCEPTABLE. Information changes and standards are revised; some material is updated, some not. Opinions change and authors of competing books on the same topic may disagree. As such, any database of questions can become out-of-date.

Nevertheless, this does not seriously affect the intent of Treasure Hunt because the overriding goal is the process of searching, not the answers per se. Where answers disagree or information has been updated, full credit is still given for answers and/or sources differing from the answers database, as long as the included documentation supports the answer. Differing answers (and sources), when encountered, are recorded to maintain currency. In this way the project becomes self-updating and allows flexibility when a library collection is not maintained with up-to-date material. Credit is not given for an incorrect answer from an otherwise correct source. This occasionally happens when a student identifies the correct table or passage but misreads the question, or is careless in recording the answer from the source.

ORIENTING SUPPORT STAFF

Regardless of other resources used, the campus library may initially receive the brunt of Hunter demands. Librarians may feel overwhelmed when 100+ students descend en masse on the library reference desk. Thus, if only out of self-defense, librarians will most likely be eager to help in orienting students to efficient library use. This is especially important with freshmen who may never have visited the campus library.

Therefore, campus librarians need to be involved in planning the timing and scope of a hunt. Often the librarians suggest resources that the instructor had not considered; however, no matter how extensive campus facilities may be, not all resources are available locally. In those cases the question set could be modified to suit the resources available. Regardless, this project was never conceived to be solely a local library project. Among the resources discussed, especially with worldwide web access, it would be a disservice to students if this project were limited only to what was available in a campus library.

Alloy digests	*Infoworld* Magazine
ANSI standards	*Intech* Magazine
ASM standards	Intros to ASME codes & stds
ASTM standards	*Machinery's Handbook*
Bar code books	Macintosh magazines
Bearing journals	*Manufacturing Engineering* mag.
Building estimator's handbooks	Materials handbooks
CAD/CAM books	*Materials Handling Product News*
Compressed air & gas hdbks	Mathematics texts
Computer dictionaries	*Mechanical Engineer* magazine
Datamation magazine	Metals handbooks
Dict. of electronics terms	*NASA Tech Briefs*
Diesel & gas turbines hdbks	National Electrical Code
Dir. of US private comps.	*Newsweek* magazine
Electrical motors handbooks	Nozzle user's manuals
Engine handbooks	OSHA Regulations
Engineering graphics texts	Patent & trademark pubs
Glass engineering handbooks	PC magazines
Graphic arts handbooks	Product catalogs
Hazardous materials pubs	Safety supply catalogs
Histories of invention & tech.	Seal design handbooks
Human development handbooks	Semiconductor guides
Human factor handbooks	Solid state books
ID News magazine	State & local newspapers
Ind. Equipment News mag.	Structural design data handbooks

Figure 2. Generic bibliography provided to students and library staff.

BUILDING A DATABASE OF QUESTIONS

Questions can be about anything remotely affecting the practice of engineering or engineering technology, but most practically should begin with sources close at hand, then expanded outward.

A FOUR-WAY TEST FOR VIABILITY. The actual selection of questions is subjected to an informal test of viability.

1. Can the question be answered and documented? *(not a, 'how many stars are in the sky?' type question)*
2. Is the answer publicly available? *(not proprietary, confidential or based on references with limited access)*
3. Is the question in some way related to science, engineering or technology? *(things, people, facts, places, processes, specifications, costs, etc.)*
4. Does the process of the search require students to delve into useful references or require them to make beneficial contacts? *(books, periodicals, vendors, the web, etc.)*

TEXTBOOKS. In this case, the database began with questions from required textbooks for PD&S and other classes in the discipline. It was expanded with questions designed to reach students who do not thoroughly read their texts and/or references. This was accomplished by identifying obscure portions which require careful reading to discern answers. Footnotes became a great source because students often skip them, especially in handbooks. Questions were built on key words not listed in indices or tables of content. These, too, required research and forced students to become familiar with the details of their books.

REFERENCES AND HANDBOOKS. Mark's *Standard Handbook for Mechanical Engineers*, *Machinery's Handbook* and electronics manufacturer's data handbooks all became significant sources. *Thomas Register* and *Comprehensive Dictionary of Electrical Engineering* are others. Most libraries stock these and similar references. Some are available via the worldwide web.

STANDARDS. Standards are excellent references and represent the original source for much of a text's or handbook's technical information. Most commercial products and engineering processes are standardized or use some form of standardization. Yet few practitioners seem to realize that every engineering field uses applicable national or international standards.

The importance of standards when contrasted with practitioner knowledge of standards, can be characterized by the following quote from a U.S. congressional white paper.

> *"Standards generally go unnoticed. They are mostly quiet, unseen forces, such as specifications, regulations and protocols, that ensure that things work properly, interactively, and responsibly. How standards come about is a mystery to most people should they even ponder the question."*
>
> *Global Standards: Building Blocks for the Future*
>
> OTA Congressional Report, March 1992

ANSI, the American National Standards Institute, is the United States standardizing body. It organizes and promotes American national standards throughout the country. Standards are actually written by professional societies such as ASME, IEEE and dozens of others. Even ASEE, the American Society for Engineering Education, has been involved in writing engineering standards. ANSI in turn is a member body of ISO (the International Organization for Standardization). As a member body, ANSI distributes English language versions of most ISO standards.

Many engineering libraries have ANSI, IEEE, and ASTM standards as well as OSHA codes and national building codes. If not, local industries may serve as sources. More and more libraries are beginning to house ISO standards as well.

PRODUCT CATALOGS AND EQUIPMENT / MATERIAL SPECIFICATIONS. Vendors eagerly furnish literature and catalogs on their products. Questions based on vendor information could be based on capabilities, size, weight, color, options, shipping specifications, cost, etc. Related questions could request the shipping weight of an XYZ oscilloscope, or a current price quote for a certain type, purity and configuration of exotic material. Much of this type of information is available on the web.

It should be noted that occasionally vendors won't talk to students, but most are helpful because they recognize that students will be their customers in a few years. If questions are written about products and/or services from vendors who cater primarily to practitioners in one's discipline, students may establish a lasting relationship with these contacts. More than a few students have parlayed jobs after graduation out of initial contacts made searching for Treasure Hunt questions.

DICTIONARIES OF DISCIPLINE-SPECIFIC TERMINOLOGY. Technical subjects present many specialized terms. Questions on terms could be gleaned from the class subject, the processes associated with the subject, the end use or the vendors and raw materials of the subject. For source material, most libraries have technical dictionaries and handbooks which define terms. The internet can be another valuable source of definitions.

DICTIONARIES OF ACRONYMS. Students often do not realize that dictionaries of acronyms exist, yet technical fields are fraught with acronyms for materials, processes and organizations. Acronyms are an excellent source of questions because every discipline has them. Another attraction to acronym-based questions is that acronyms sometimes have several definitions. With the Treasure Hunt goal being the documentation of the answers, alternate definitions become acceptable if documented. Again, most libraries have dictionaries of acronyms.

PATENTS AND TRADEMARKS. Because patents and trademarks are now available on CD ROM, some libraries are becoming patent and trademark repositories. These CDs, plus what information can be obtained via web sites, make patent and trademark information great sources for questions. Who invented what, when or what was the patent number, etc., could be asked. Trademarks are of interest because of the enlightenment in learning who actually owns those familiar names and processes and how ownership passes from company to company.

Trademark lore about computer companies forms the basis for some questions. One involves Apple Computer and the company name 'Apple.' It has been related that Apple Computer almost lost use of the name in a trademark dispute with Apple Record, at the time the record label for the *Beatles*. Apparently after much negotiation, Apple Computer was allowed to retain the name as long as the company did not become involved in the music industry. However as computer industry historians may remember, Apple eventually began to tout itself as the hub of a digital lifestyle, including *music*. This raised the ire of Apple Records, thinking that Apple Computer had reneged on their agreement. Eventually the issue was resolved with an undisclosed settlement between the two Apples, but the tale remains a legend among computer professionals.

Other questions could spring from related topics such as the process of obtaining patents and trademarks, the costs, etc. Many libraries have at least some trademark reference material. Two web sources for patent information are www.*womplex.patents.ibm.com* and *www.patents.uspto.gov*.

QUOTATIONS. Technical quotes can be difficult for students to find unless they are timeless classics from Ben Franklin, Henry Ford or another prominent individual. The trick is to find technology-oriented people whose quotes become classics. In the early years of the Treasure Hunt, contemporary quotes from among others, Bill Gates and Steve Jobs, were used. The quotes were widely circulated at the time (or so it was thought) but few students were ever able to document them. Eventually they were dropped from the database. Few quotes questions have survived in the current version of the Treasure Hunt.

COMPANY NAME / HISTORIES / PROFILES / ADDRESS / PHONE. This type of question may ultimately prove to be extremely useful to students as they approach graduation. If questions are chosen with possible employers in mind, students will learn the resources needed to identify company contacts and research background information as they start job hunting. Some very interesting questions can be based on companies known for popular cultural icons to which students readily relate. For instance, the Treasure Hunt has questions on Nintendo, motorcycles, music CDs, cars and computers. Related questions could be queries on market shares for specific years or market projections, or possibly comparisons between what a company projected five years ago contrasted with its present circumstances. These questions can be derived from product literature, company web sites, references such as *Thomas Register* and countless others.

'OFF-THE-WALL' QUESTIONS. This type of question and its apparent absurdity, energizes a sometimes dull set of questions. Off-the-wall questions can almost seem like riddles but have a basis in science; for example, questions include, "What makes your knuckles crack and why must you wait to crack them again?" (a fluid dynamics problem) and "Why do clocks run clockwise?" (mimics the shadow movement on a sundial in the Northern Hemisphere).

CALCULATED ANSWERS. Often calculated, or interpolated, data from charts and tables can form the basis for useful questions. This could be as simple as reading the value on a stress-strain curve to interpolating the average height of a 183-month old child on a pediatric growth chart. In these cases the source may be a set of derived calculations, a text, or a graph published for medical practitioners.

SPECIFICATIONS AND COMPOSITION OF MATERIALS. Material use, selection and processing are major components in engineering and technology curricula. Because this project began as a way to propel students deeply into a handbook, many of the questions deal with uses and composition of engineering materials, especially metals and polymers. Two favorite sources for questions include the *Materials Data Handbook* and the *Plastics Encyclopedia.* These questions could be derived from references relating to nearly any engineering discipline.

INVENTIONS AND INVENTORS. Unless the invention or its creator is extremely obscure, questions about inventions and inventors could become simply encyclopedic ventures. Nevertheless questions requiring research into the lives of inventors or other historical figures yield great possibilities. Histories of science and technology abound in print or on the web and a few such questions make a good mix with other more technical questions. Students sometimes enjoy this because their research often answers the 'why?' of a particular concept.

CURRENT EVENTS AND PROFESSIONAL PERIODICALS. Current, or at least recent events, especially as they relate to engineering fields can make useful questions because they often form the basis for case studies. Sources can be newspapers, magazines or the web. This type of query especially lends itself to local customization where minimal other resources may be available.

HISTORICAL PUBLICATIONS ON FACTS, DATES AND PERSONS. A number of publications exist which discuss the historical roots of discipline-specific technologies. Often these publications feature lesser known facts general encyclopedias would not cover. These sources can represent an introduction to the rich history of a discipline, its technology, people, and occurrences a professor's generation takes for granted, but about which traditional-age students may know nothing. With the internet, one would be hard pressed to write an imponderable question about a public person, technology or engineering event.

CAMPUS FACTS AND HAPPENINGS. The greatest way to generate a graduate who keeps in touch after graduation and from the university's point of view, later contributes to the university, is to produce a graduate who has a sense of belonging. What better way than to introduce students to their university's rich history? Questions could be as simple as identifying who was also in the class with a famous alum or identifying who was the editor of the student newspaper when that peace rally occurred in the 60s, or even the current address of a school roommate, or listing the contents of that time capsule buried in the engineering mall. All colleges and universities keep yearbooks and often archive student newspapers. Local city papers are often kept on microform and the internet certainly could be utilized to locate addresses by name anywhere in the country.

It could be argued that these types of questions have little direct educational value, but the resources are available proximally and add a local flavor to the Hunt. The answer and the process of locating it may be remembered long after graduation.

OTHER SOURCES FOR QUESTIONS

COLLEAGUES. Interestingly, questions initially supplied from colleagues do not always require answers and/or the sources to be viable Hunt questions.

Colleagues proposing questions are asked to generate questions and references about something they want students to learn, or at least learn how to locate. They are also encouraged to suggest questions even for which they may not know the precise answer or source. If a question can be written from these suggestions and probable resources are available, it is included, with or without an answer and/or a source.

This unorthodox practice works because hunters must document the answer for full credit. As such, a Treasure Hunt instructor will quickly know if the answer is correct — the documentation supports the answer. If documentation is not supplied, an educated guess can be made as to if the answer is correct or at worst, the question could be thrown out and credit given for the effort.

When questions from a colleague are received without answers and/or sources, the colleague is cautioned that it may take time before an answer is obtained due to two factors. First, because each student's questions are generated randomly, this may result in some questions rarely being selected by the random generator, and second, even if the question is assigned, students may never actually choose to answer it (See Mechanics of Running the Hunt). In the current database, several questions have yet to be chosen and/or answered correctly, even after over forty semesters of use. For example, one of these questions, which was contributed by an engineering materials colleague is: "Which elements spark 'easily' when held to a grinding wheel?".

STUDENTS. Just as fraternities and sororities store memorabilia for future members, some students relish contributing to future students' learning. Often current students, having just experienced this project, will go to great lengths to generate additional questions and even stumpers for future treasure hunters. The advantage/disadvantage of student generated questions is that they may often be based on personal interests and hobbies, such as cars, motorcycles, tools, computers, boats, etc. This is good because they may be of popular interest to students, but they may be too specific or may go out of date quickly. Regardless, the practice is beneficial because their contributions give them a sense of ownership in the process. Graduating seniors in another class are asked to remember the project after graduation and consider contributing to the Treasure Hunt when they run into that 'special' question as they begin professional practice. Usually 5-10 questions are received in this manner each year from graduates.

ALUMNI AND INDUSTRIAL PRACTITIONERS. Alumni and industrial colleagues are excellent sources for questions that are focused, often gritty and always on-the-job related. They remember the types of questions they had to answer as new practitioners. Alumni, after entering the 'real world,' quickly recognize the value of the Treasure Hunt and are often pleased to help. A secondary benefit may be to give alumni, and non-alumni alike, a sense of connection or even pride in providing questions to challenge future graduates. Needless to say, involving alumni can generate excellent industry relationships and be a way to maintain contact with students well after graduation.

CONTINUOUS IMPROVEMENT

The need for continuous improvement has been widely recognized for many years among industrial practitioners worldwide. However only within the last decade have accreditation bodies required the concept to be applied to engineering and technology academic programs. Still, continuous improvement occurs by necessity rather than just to meet accreditation requirements. This project exemplifies that necessity and became obvious well before this author had ever heard of Kaizen (the popular industrial term of Japanese origin, for continuous improvement of working practices that underlies total quality management). The need persists to this day.

Each semester after the Treasure Hunt project concludes, the instructors and others involved discuss any problems and improvements which might be made. Unanswered questions are revisited to see if wording needs revision. Alternate answers and sources are documented in the master database. Among others, these efforts resulted in the following opportunities (and solutions) for continuous improvement.

IMPROVEMENT RESULTS. One of the first opportunities came about at the insistence of the library staff and resulted in the specific requirement for Xerox copies of documentation. Even though it was obviously destructive, a few students turned in documentation as actual pages cut from books and periodicals, much to the horror of the library staff and professors alike.

Another example occurred during the first semester when the whole list of questions was given to all students in the hope that they would recognize the need to form teams to answer all questions (as the idyllic law students had done).

The students did cooperate as desired. They also passed on the full set of answers to students in the next semester's classes, resulting in a new database of questions having to be written.

The improvement became one of growing the questions database sufficiently so that each student would have a different set.

While the goal was to have students answer all questions assigned to them, some questions proved to be much more time-consuming than others. The improvement which suggested itself was to give each student a subset from which they could choose not to answer a few of the questions. It was determined that the project would still achieve its goals if the students received twelve questions and answered any ten.

Another opportunity became apparent with the need for a unique set of questions for each student. The solution was to develop a randomization routine in the selection of questions. This was accomplished by writing an Excel macro that copied a student's name from a grade roster, then randomly copied twelve rows of question/answer/source and pasted all into a blank worksheet. The macro then looped back to select another student name, another twelve random questions and so on. With this scheme, an individualized grading key could be printed with each student name, plus their randomly-chosen set of questions, answers and sources.

A further refinement was the need to generate a sufficiently large database of questions so that each person received a different set of questions with minimal duplication. This resulted in a goal of 3000 questions (250 students per year x 12 questions). The database is still growing, but has not met that goal, with about 1200 questions currently in the database. Around 50 questions are added each semester with approximately ten culled.

Sometimes improvement was necessary in the form of explanation on what not to do. For instance, one question caused consistent problems because it appeared to require a phone call to Japan. The question asked the tolerance on the diameter of a popular brand of 0.5 mm mechanical pencil lead. As it turned out, that lead is distributed worldwide, but made only in Japan. Rather than search industry standards or contact pencil manufacturers listed in *Thomas Register* and elsewhere, a few students each semester would try to call the company in Japan. While those who did call Japan often found the answer, their phone bills became problematic. The solution was to reiterate that all questions could be answered from multiple sources.

Another need for improvement became evident in two problems relating to the library staff being given a complete set of questions and answers. First, early on when students were having difficulty, the librarians would tend to find the answer for the student. As word got around, students stopped looking themselves and expected the librarian to do the research. The improvement took the form of a 'mini' training program for the librarians which stressed helping students locate resources, but not divulging the actual answers. This also reinforced the significance of the process of inquiry, not the answers. Second, once the librarians stopped doing the research for students, another more serious problem arose. Students discovered that the librarians had a copy of the questions/answers/sources, and security became a major concern. The improvement was removing the database from the library and issuing a generic bibliography to both the librarians and the students. See Fig. 2.

A further improvement was needed to encourage students to use actual ANSI, ASTM and ISO standards when possible. To do so, a number of questions were written specifically so they could be answered only by referring the actual standards. This practice obviously depends on having access to actual standards. Still, many of the questions could be answered from general references, even though a standard was the original source. It was felt that students needed to be aware of this resource, even when the answer came from a text or handbook.

To induce the use of actual standards, an additional requirement was added for notation of an applicable standard. The requirement stated that if a standard was the original source, it had to be specified regardless of where the answer was actually obtained. It was not necessary to actually visit the standard, but just note it as applicable.

This was felt to be important because a few questions do not initially appear to be based on standards and periodically students lose points because they do not research the question sufficiently. For example, one question asks the botanical name of a particular tree. Some students gamble that there is no ANSI standard for tree names and don't search for an applicable one. In this case, an ANSI standard classifying botanical names of trees does exist and must be referenced to receive full credit.

CONCLUSIONS

The Treasure Hunt project described in this paper presents a unique opportunity to give students experiences not readily accomplishable in traditional class projects. It can be customized to work in virtually any discipline, with any text and any level of instruction. It can include aspects crossing national and international borders and can be implemented in any language. The project demonstrates both the need for, and the benefits of, continuously improved instruction. Because questions can be gleaned from almost any source or location, it is readily adaptable to the smallest campus or one with few technical library resources. It yields concrete experiences which will serve students both in school and well after they evolve into engineering and engineering technology practitioners.

Proceedings of IMECE'03
2003 ASME International Mechanical Engineering Congress
Washington, D.C., November 15–21, 2003

IMECE2003-43648

CASE-BASED PROBLEM SOLVING EXERCISES FOR ENGINEERING EDUCATION

Venkat Allada
Engineering Management Department
University of Missouri-Rolla, Rolla, USA

Vivas Jose
Manufacturing Engineering Program
University of Missouri-Rolla, Rolla, USA

ABSTRACT

This paper describes the development of a case-based problem solving exercise that emphasizes the managerial and the engineering aspects of a real life project. Using the real life example of a quality initiative project that was completed at a laundry detergent plant of a international company, a series of modular exercises have been developed on topics such as organizational structure, process capability, six sigma, and Taguchi methods. The exercise modules have been developed by referring to the Bloom's six levels of cognitive learning process. This paper provides some of the highlights of the exercise development process that may be useful to educators interested in developing their own case study-based exercises or adapting the available ones to suit their needs.

INTRODUCTION

Engineering is an applied science field in which problem solving techniques play an important role and hands on experience is a desirable skill. There still exist gaps in the engineering curriculum between theory and practice -- a gap that can be partly filled through the deployment of case based problem solving exercises in the classroom. Traditionally, the case study method of learning is widely used in the business education curriculum. In contrast, engineering education relies on focused problem solving learning techniques and the use of case study based learning is quite limited. We believe that a learning paradigm that blends the "soft science" business approach and "hard science" engineering approach will be immensely useful to the engineering graduates. A case is a factual description of events that actually happened at some point in the past. A case is designed to elicit discussion and analysis of a particular situation so that students can develop an ability to determine appropriate criteria by which alternative solution sets can be measured [1].

This paper focuses on the development of one such case-based problem solving exercise relating to quality engineering that involves process capability improvement in laundry detergents spray drying operations of a multi-national company. The exercise has the following objectives:

1. Develop "plug and learn" exercise modules that can be used in a variety of manufacturing and quality related courses or to educate industry professionals. Alternatively, the instructors can plug the appropriate modules into various manufacturing and quality engineering courses and create a rich learning experience for the students in a relatively short period of time.
2. Provide students with opportunities to discuss and learn various "soft issues" such as managerial decision making, organizational structure, project management, and people issues.
3. Provide students opportunities to discuss and learn some of the quality engineering concepts and tools such as process mapping, process capability indices, Taguchi method, and six-sigma.
4. Provide a detailed case study along with teaching notes and student notes on the relevant business and engineering tools that can be applied for case analysis purposes.

Apart from describing the specifics of the exercise, we will provide our ideas on various issues relating to the actual development of the exercise itself. These insights can serve as a guide to other engineering educators who may be involved in developing engineering case studies.

CASE-BASED PROBLEM SOLVING EXERCISES APPROACH

Engineering education places heavy emphasis on problem solving. The hard part, however, is identification of the right problem before we go about solving the problem. Identification of the right problem in the real world is not very simple. This may need not just engineering skills but also many other "soft skills" such as communication skills, and business skills. The case based problem solving exercises approach proposed in this paper seeks to integrate numerical problems solving skills in the context of an actual case. It is hoped that the students do not just experience the "plug and solve" number crunching aspects of engineering tools but also gain a deeper understanding of the engineering tools in the context of a real life project. Some of the advantages provided by case based problem solving exercises approach include the following:

- A practiced-oriented learning style in which the engineering tools and methodologies are understood and applied in the context of a real life case.
- Ability to incorporate various levels of learning (discussed in the next section) so that educational process is optimized.
- Enhancement of the learning environment. It has been observed that compared to lecture-based courses, students found the experiential learning methods more enjoyable, informative and less work-load oppressive [2].
- Reduce the gap between academic work and practical solving skills. Case-based problem solving exercises approach reduces the gap between theoretical and practical training. These type of cases let students to engage in real engineering problems.
- Incorporate basic concepts into a student body of knowledge. The approach explained in this paper allows students to grasp basic concepts on methodologies that he/she might not be aware of, but are important to know. The basic concepts taught through a case (and their relationship with real engineering problems) hopefully may encourage students to seek further in-depth knowledge.

BLOOM'S TAXONOMY

In 1956, Benjamin Bloom conducted research in the field of education to capture the intellectual behavior important in the learning process. The levels of learning identified on the cognitive domain are listed below [3]:

- Knowledge: This refers to those behaviors that emphasize remembering (either by recognition or recall) of specifics and ideas, terminology, and phenomena.
- Comprehension: This refers to those objectives and behaviors that represent an understanding of the material without necessarily relating it to other material.
- Application: This refers to the ability to use previously learned material in situations that are either new or contain new elements.
- Analysis: This refers to the breaking up of material into its constituent parts and detection of the relationship of the parts and of the way they are organized.
- Synthesis: This is defined as the putting the elements together so as to form a whole.
- Evaluation: This involves the judgment about the value of methods, and solutions.

We used Bloom's taxonomy for the development of the case presented in this paper. The Bloom's taxonomy served an important purpose of "dividing" the case exercise into modules that address the various levels of learning. The description of various modules and their relationship with the Bloom's taxonomy will be discussed later in the paper.

CASE BACKGROUND

The case describes a project involving the improvement of process capability indicators in a laundry detergent manufacturing plant. It describes the case of a company involved in standardizing quality related procedures in all its subsidiaries worldwide. The standard set of policies for the company included various issues such as technology transfer, process capability, documentation, change control, utilities, recall of products, and processing wastewater. The subsidiaries received the standards and it was responsibility of each vice-president-operations (of a subsidiary) to implement those policies within approximately one year. Teams of international auditors (employees of parent company in different subsidiaries and experts the relevant fields) were formed to audit the implementation of the quality standards. This case relates to the work that was pursued by the vice-president of one such subsidiary for process capability standards. The process capability standard of the company states that each manufacturing facility has to have appropriate statistical measures to ensure that the processes are capable of consistently produce products that conform to all specifications and then implement procedures to monitor and control process variation during ongoing manufacturing. It is also expected that each facility would define the key process attributes, measure their variability and implement process improvement projects based on statistical data.

CASE MATERIAL ORGANIZATION

The case has been organized into two major categories:

1. Case study
2. Instructor teaching notes.

As discussed earlier, the case study provides a description of the project involving the improvement of process capability indicators in a laundry detergent manufacturing plant. The teaching notes are intended to help the instructor to guide the class discussion and cover one or more learning modules. The teaching notes for the case have been divided into four modules as listed below:

- Module 1: General discussion about the project to help students understand how the project was approached, what are the lessons learned, and what could have been done differently?
- Module 2: The objective of this module is to make students understand the meaning of process capability indicators (Cp and Cpk), and gain proficiency in the calculation of these indices. Also, discuss the methods data collection that has to be put in place to determine the process capability indicators.
- Module 3: This module aims to expose the students to Six Sigma concepts through the application of the DMAIC model (Define, Measure, Analyze, Improve and Control) to the project described in the narrative.
- Module 4: This module introduces the basic Taguchi method concepts for quality improvement by using the project described in the case study.

The details of the case study narrative and the teaching notes are depicted in Table 1.

PLUG-AND-LEARN APPROACH

The current reality is that industry perceives a short fall in the way graduating students are prepared to enter the workforce. A study by the Society of Manufacturing Engineers (SME) involving industry managers from six different types of industry revealed that a vast majority of managers believed that freshly graduated engineers were not prepared for "...real world engineering applications" [4]. Gaps in competences were identified in a wide array of areas including communication skills and teamwork; application of scientific first-principles to specific real-life engineering problems; deep and detailed understanding of the principles, tools, and processes of modern manufacturing; appreciation for business and economic factors in engineering; skills for managing change and continuous learning [5]. While this study was specific to the discipline of manufacturing engineering it is arguable that the same critique is applicable, with different terminology, to all fields of engineering [4].

The work discussed in this study uses a focused case study as a platform to construct four learning modules. These four modules facilitate the "plug-and-learn" teaching paradigm. The instructors can plug the appropriate modules into various manufacturing and quality engineering courses and create a rich learning experience for the students in a relatively short period of time. The case study also provides supplement material on the basic theoretical frame or tools such as Taguchi method, and six-sigma that could be used by the students. The instructor is expected to encourage students to look for more in-depth material that is not covered by the supplementary material.

IMPORTANCE OF THE LEARNING LEVELS

When developing the case-based problem solving exercise modules, the objective of the developers must be to relate the modules to the levels of learning such as the Bloom's taxonomy. Some of the key points while writing and developing a case are listed below:

- When writing the case, provide as many relevant details as possible. Good information, especially in an engineering based case, enhances the learning possibilities for the students.
- Integrate the numerical problem solving exercises with the case in hand. The students should be able to see a clear value (to solve the case) to their number crunching exercises.

CASE STUDY	TEACHING NOTES
• Company background. • Implementation of quality policies among subsidiaries worldwide. • Description of the detergent manufacturing process. • Steps done to implement the process capability standard. • Sources of process variability found on the detergent's spray-drying process. • Specific actions taken to reduce or eliminate sources of variability. • Conclusions	• **Module 1:** Discussion points on: - Approach taken to execute the project. - Importance and minimum requirements for a quality system. - Organizational issues. - Process attributes selection - Importance of soft / technical skills - Process control measures. • **Module 2:** - Process capability calculations - Evaluate the results obtained. • **Module 3:** - Formulation the project using the DMAIC model. - Create an organizational chart to identify project team members and their areas of responsibility for Six Sigma implementation. • **Module 4:** - Quality loss function calculations. - Justify potential investment in process capability improvement. - Construct a P-diagram

Table 1. Topics covered by the case study and teaching notes

- Based on the discipline of the case study, think of engineering tools and methodologies that are appropriate and relevant for the students. In this particular case study, the process capability improvement project at the company was done using common sense and empirical approaches. The authors believed that six sigma and Taguchi methods could have been valid tools for such project, and hence, it was decided to include these tools as part of the learning modules.
- Write appropriate questions for the different levels of learning. Bloom's taxonomy has different verbs for each level and is quite appropriate for technical topics. Table 2 shows how the questions prepared for the case based problem solving exercise described in this paper fit into each of the levels of learning.
- Think about usefulness of the case study for life-long learning. The knowledge of an average engineer becomes obsolete in two to five years [6]. The learning needed can furnished from many sources including short courses, seminars, and professional conferences and courses [5]. The "plug-and-learn" features of the modules make then suitable for life-long learning.

Levels of learning (cognitive)	How the case-based problem solving exercise satisfies those levels?
Knowledge	• Summarize the case study. • Recall of situations narrated in the case study.
Comprehension	• Interpret the approach taken by the team that executed the project described in the case. • Explain discrepancies in the way the project was conducted. • Provide thoughts about alternative approaches for the project
Application	• Calculate the capability index, quality loss. • Prepare of a cost/benefit analysis for supporting the project.
Analysis	• Analysis of the improvement actions taken.
Synthesis	• Categorization and formulation of the project in terms of the DMAIC model (Six Sigma). • Formulation of a P-diagram
Evaluation	• Evaluation of Cp and Cpk data obtained, quality loss and the cost/benefit analysis. • Assessment of the importance of using Six Sigma and Taguchi methods in real-life projects.

Table 2. Mapping case-based problem solving exercises into the levels of learning

ASSESSMENT PLANS

One of the best known and comprehensive approaches to assessment of training was developed by Kirkpatrick [7]. The approach comprises four discrete and progressive levels of evaluation: student reactions, learning achieved, transfer of changed behavior to the workplace, and results achieved in the workplace. We plan to test our case using the Kirkpatrick model as a reference. The evaluation would be provided to students, faculty and people in industry. The information expected to gather from such broad audience is listed below:

- Students. The focus will be to test the case based problem solving exercises approach on qualitative indicators such as stimulating, relevance, material clarity, and overall learning.
- Faculty: The idea would be to explore how other faculty members feel about the modules introduced in the teaching notes and the overall applicability of this method in their courses.
- Industry: The objective is to gather industry people opinions to test the applicability of case-based problem solving exercises (especially the plug-and-learn approach) in life-long-learning. Qualitative indicators such as potential application of the material learned and importance of introducing new concepts and methodologies will be used in the evaluation forms.

CONCLUSIONS

The gap between theory and practice can be bridged to a certain extent through deployment of case-based problem solving exercises in the classroom. Traditionally, the case study method of learning is widely used in the business education curriculum. In contrast, engineering education relies on focused problem solving learning techniques and the use of case study based learning is quite limited. We believe that case-based problem solving exercises learning paradigm that blends the "soft science" business approach and "hard science" engineering approach will enhance the educational learning process for students and life-long learners.

REFERENCES

[1] Naumes, W., and Naumes, M., 1999, "The art and craft of case writing", SAGE publications, Thousand Oaks, CA, p. 10.

[2] Lightsey, R., "Engineering management training: comparing experiential vs lecture training

methods", Acquisition Review Quaterly: Winter 2000.

[3] Crawley, E., 2001, "The CDIO syllabus, a statement of goal for undergraduate engineering education", CDIO Report #1, Massachusetts Institute of Technology, Boston, MA. http://web.mit.edu/aeroastro/www/cdio/cdiodocuments/CDIO.pdf

[4] Manufacturing education plan phase I: Industry identifies gaps among newly hired engineering graduates, Manufacturing for the 21st century, Vol IV., Society of Manufacturing Engineers: 1997

[5] Wells, D., and Lagenfeld, G., "Life long learning for manufacturing engineers", 2nd International Conference on Manufacturing Education, Society of Manufacturing Engineers, San Diego, CA: 1998.

[6] Motorola Corporation. "1994 Annual Report", Schaumburg, IL: 1995.

[7] Kirkpatrick, D.L., "Evaluating training programs: evidence versus proof", Training and development, 31, 9-12: 1977.

Proceedings of IMECE'03
2003 ASME International Mechanical Engineering Congress
Washington, D.C., November 15–21, 2003

IMECE2003-55237

INDUSTRY FOCUSED COURSES UTILIZING COLLABORATIVE LEARNING APPROACH

Hiroshi Honda*
E-Mail Address: Hondah9876@aol.com Tel/Fax No.: 81 + 47 453 6144
Institute for the International Education of Students (IES) Tokyo
Contact Address: 201, Tsudanuma 5-3-24, Narashino City, Chiba 275-0016

Shuichi Fukuda
Tokyo Metropolitan Institute of Technology (TMIT)
6-6 Asahigaoka, Hino City, Tokyo 191-0065

ABSTRACT

Industry focused, full semester Courses A "Industry and Information Systems" and B "Overview of the Market and Introduction to Marketing" are lectured by the lead author at Tokyo Metropolitan Institute of Technology (TMIT). In these courses, overview of the global and Japanese economies and trends of each industry sector are introduced by the lecturer. Students are assigned to write a report on a specific industry sector, a specific market and other related topics of his or her interest, and opportunities for presentation and discussion on their draft reports with their classmates and the lecturer are given toward the end of the semester. Apart from these courses, a full semester Course C "Evolution of Industry, Business and Engineering Practice in the Global Economy," utilizing collaborative learning approach, was proposed by the lead author in 1998, and was partially taught at American and Japanese universities, utilizing the Proceedings of the ASME Engineering & Technology Management Group Symposia at IMECE 2000 –2002.[1)-3)]

The surveys were conducted concerning the student's interest of industry sectors, lecture topics, and international topics both at the beginning and at the end of the semester. In addition, surveys to sense students' new discovery from lectures for all course modules and those to sense useful knowledge in their career planning and in general were conducted at the end of all classes. The results of surveys show that students found the discussion opportunities valuable to gain broader perspectives and insights into their own subject of study when writing their reports, by learning what their classmates are studying and from classmates' comments. This paper also discusses what motivates students in learning, what influences students' general, specific, and career interests, and how students' interests may change from the beginning to the end of the semester for which these courses are given.

INTRODUCTION

It has long been said that Japanese college engineering education lacks teaching of industry focused subjects, due to relatively less personnel exchanges between industry and academia, as compared with the United States and other industrialized western nations. The lead author had experiences of working at research laboratories, engineering headquarters, corporate planning department, business planning department and marketing department in heavy industry, and at positions with government related organizations such as the Institute of Energy Economics, Japan (IEEJ), the Japan Microgravity Center (JAMIC), and Japan Space Utilization Promotion Center (JSUP), where he did collaborative work with the U. S., Asian, European, Russian, Canadian, Australian, and international organizations. These experiences were incorporated into his lectures for the aforementioned courses to stimulate students' interest in industrial and international activities and to motivate them to write a report on a topic of their own interest.

The discussion of paper provides opportunities for marketing the subject courses at universities and other institutions, and for arranging the course contents, based on the students' needs and interests. Course EC320 "The Contemporary Economy of Japan" to be lectured for American students at IES Tokyo is a good example of such arrangement. [4)]

COURSE OUTLINES AND STUDENTS' REACTON TO EACH COURSE MODULE

Course A "Industry and Information Systems"

Table 1 shows contents of Course A, and ranks and degrees of interest in each course module of students surveyed at TMIT. It can be seen from Table 1 that, at the end of the course,

* Visiting Research Professor and Lecturer at TMIT

students at TMIT showed higher interests in Course Modules 10, 11, 14, 15 and 4, i.e. "Trends of Amusement and Service Industries, and Public Organizations, and Role of Information Systems," "Trends of Automobile, Information and Communication, and Software Industries, and Role of Information Systems," "Presentation of Students' Reports, Discussion and Guidance to Refine Reports," "Discussion to Refine Students' Reports and Instruction on Abstract, Report Format and Writing," and "Globalization of Various Flows and Future Information Systems," respectively, while at the beginning the course, students showed lower interests in Course Modules 14 and 15. This is due to the fact that students found presentation and discussion of their draft reports were interesting and helpful in value-adding and refining their reports, as can be seen from Table2. As students learned more from a series of lectures, the ranks of interest changed from the beginning to the end of the course, as shown in Table 1. However, the students' high level of interests in Course Modules 10, 11 and 4 remains unchanged throughout the semester, as their major is production, information and systems engineering.

Table 1 Results of Survey on Course A "Industry and Information Systems" for Junior and Senior Students of Department of Production, Information and Systems Engineering (DPISE) at Tokyo Metropolitan Institute of Technology
(Degree of Interest 5: Very Interested 4: Interested 3: Neutral 2: Not much Interested 1: No Interest)

Course Title and 15 Course Modules		**TMIT DPISE Junior and Senior Students**			
		A. Response at the Beginning of the Course (October, 2002)		**B. Response at the End of the Course (January 28, 2003)**	
ID No.	**Industry and Information Systems**	Interest Rank	Interest Average	Interest Rank	Interest Average
1	Introduction: Survey for Students and Course Work Orientation	12	3.00	*7*	3.58
2	Global Economy, and Trends and New Paradigms for Industry and Technology	*9*	3.50	*8*	3.54
3	National Economy, and Trends and Issues Associated with Internationalization and Globalization of Industry	②	4.50	*9*	3.31
4	Globalization of Various Flows and Future Information Systems	⑤	4.00	⑤	3.77
5	Trends of Industry and Role of Information Systems in General	⑤	4.00	13	3.17
6	Trends of Agribusiness, Food, Life Science, Pharmaceutical, and Medical Industries, and Role of Information Systems	⑤	4.00	*9*	3.31
7	Trends of Natural Resources, Energy, Electronics, and Computer Industries, and Role of Information Systems	*9*	3.50	*9*	3.31
8	Trends of Materials, Engineering, Ocean and Frontier Development Industry, and Role of Information Systems	*9*	3.50	14	3.08
9	Trends of Machinery and Aerospace Industries, and Role of Information Systems	⑤	4.00	15	2.83
10	Trends of Amusement and Service Industries, and Public Organizations, and Role of Information Systems	②	4.50	①	4.17
11	Trends of Automobile, Information and Communication and Software Industries, and Role of Information Systems	①	5.00	②	3.92
12	Trends of Sports, Home Equipment and Systems Industries, and Role of Information Systems	②	4.50	*6*	3.75
13	Trends of Other Industry Sectors, and Role of Information Systems	12	3.00	12	3.18
14	Presentation of Students' Reports, Discussion and Guidance to Refine Reports	12	3.00	③	3.82
15	Discussion to Refine Students' Reports and Instruction on Abstract and Report Format and Writing	15	2.00	④	3.80
Average for Average Interest of Each Course Module		(Note)	**3.73**	(Note)	**3.50**
Average for Each Individual's Average Interest		2 student	**3.92**	13 student	**3.52**
Grand Interest Average for All Interest Points		responses	**3.88**	responses	**3.50**

Table 2 Comments of Students at Tokyo Metropolitan Institute of Technology after Completion of Course A "Industry and Information Systems" at the End of the Semester

I could really understand that this course was intended for senior students. I could have done more detailed research, if this course were given in the first semester of a senior year.
"Globalization" related themes were lectured from a variety of perspectives with tightly packed contents.
Presentation and discussion on draft reports were helpful to improve them. This was a kind of approach I could not find in other lectures.
I missed supplementary classes as I went to U. K. The final report meant me something, although I did not do it perfectly.
I wrote a report on women's leadership, as I was interested in an opposite sex.
As I already had some knowledge of information and communication, I could learn and understand new knowledge effectively and efficiently.
It took a lot of effort to do homework, which turned out to be very good for me. I would assess this lecture very highly.
Thank you very much for your classes given for this semester.

Table 3 Questionnaire Given at Each Class

1.	List items of new discovery for you in the order of magnitude of impact.
2.	List items of helpful but common knowledge which you have learned.
3	List items of knowledge which you find helpful in your career planning.
4	Any other comments

Table 3 shows the questionnaire given to the students at the end of each class, and Table 4 shows a summary of students' responses concerning **"1 New Discovery"**, **"2 Helpful but Common Knowledge"**, **"3 Knowledge Helpful in Career Planning"**, and **"4 Other Comments."** It can be seen from Table 4 that a variety of topics interest students and that some **new discovery** items for one student are also categorized as **helpful but common knowledge** and/or **knowledge helpful in career planning** for another student. During the course of writing reports, students gain further insight into the knowledge learned, and utilize them in creating their own concept.

Table 4 Summary of Reponses to Questionnaire Given at Each Class for Course A "Industry and Information Systems"

1.	**Items of New Discovery** (Course Module Numbers are shown below) **Note: Original responses were given in the Japanese language.** 1. None 2. Trends and contrast of serious Japanese economy and growing Chinese and other Asian economies, Future of the Japanese economy and industries, Difference between GDP and GNP, Similarity between Japanese and German Economies, The world economic growth rate, 3. American autoworkers' crushing Japanese cars in the 1980's, GDP gaps, Types of globalization and internationalization of companies, Bad assets, 4. Huge number of foreigners coming to Japan, Advanced nations' huge investment in developing nations, Influences of currency exchange rates, 5. Wide areas of IT application, Popularity of CATV in the U. S., Existence of numerous sea bottom cables, 6. High technology greenhouse, Automated cucumber packing system, TMIT-Stanford Project, Huge amount of kitchen and other garbage per day, Telemedicine system, Hospital information system, Network systems for manufacturers, Application of biotechnology, 7. Various new energy such as methane hydrate, hydrogen energy, and photovoltaic energy, Difficulty of CO_2 emission reduction in advanced nations and role of governments, Dispersion and networking of electric power sources, Increasing demand of electronics technology for medical and welfare areas, Various robots, Greater demand for note-type P. C. over desktop P. C., 8. Application of fine ceramics, Plant engineering, Planning, design and construction of floating structures, Functional multilayer glasses, Consulting engineering, Remotely operated vehicles, Offshore and polar development, Process of steel production, Space development, Quality aspects of textile production, Super techno-liner, Growth of materials industry of great demand, 9. U. S. leads Aerospace Industry. Russia leads space flight duration. Japanese electronics technology was used for Gulf War weapons, Trends of aerospace, machinery and apparel industry, Value adding with a combination of various machinery, Advanced management systems used by successful retail companies, Students can make satellites. 10. Internet lecture on organic food, Online governmental authorization and other service systems, Difference in the meaning of Inter- and Intra-, LAN=Local Area Network, Telemedicine system, 11. Japanese automobile plants in the U. S. and other countries, Global alliances of automobile companies, Digitalization of TV sets, Toyota is Number One in the total share value. 12. Advancement of computer analysis has contributed to higher quality of sporting goods. **(Continued to Page 4)**

2.	**Items of Helpful but Common Knowledge** **Table 4 (Continued from Page 3)** 1. None 2. Strength of the English language, History of Japanese industry, Slump of Japanese economy, Worldwide GDP, 3. Various world products Japan lacks, Restructuring due to globalization, Unemployment rate, 4. Commodities of international trade, Interdependence of nations, Close economical relationship between Australia and Japan, 10% of U. S.-born citizens are from foreign couples. Globalization of money flow, 5. Highly functional cell phones and its wide applications, High popularity of broadband communication systems in Korea, ADSL, 6. Clone cows for enhanced milk production, Application of electronics to agribusiness, Kitchen garbage recycling system, Necessity of technology for highly aged society, Corporate management of large farms, Interaction between agriculture and industry, Organic farm products, 7. Current state of energy resources and means to avail them, ASME Paper "Oil, Gas and Engineering" by Keith Thayer and Hiroshi Honda, Research on energy conservation and new energy, Advancement and increased production of IC and PC, Factory Automation, Various robots, Digitalization, 8. Engineering qualification, Demand drop of cement, Engineering qualification, Aluminum car ferry, Recycling of paper, Negative effect of Yen's appreciation on Japanese industry, Various technologies applied to shipbuilding, Offshore engineering and floating structures, Growing demand for PET bottles and aluminum cans, and decreasing demand for tin cans, Deep sea development, Increasing production of nonferrous metal products for IT industry, Value-adding is a key to sustainable materials industry in Japan. 9. Heat island phenomena in large cities, Success of Special store retailer of Private Label Apparel (SPA), Japan's high level export of industrial robots, Dependence of everyday life on satellites, Trends of apparel industry and China, Issues of space pollution, I want CD-RW PC. 10. Intranet, Expansion of amusement/service business into overseas market with foreign finance/investment, Necessity for internationalization of Sumo wrestling. 11. Advancement of sporting goods, Global alliance of automobile companies, Impact of information & communication industries on other industries, Advertisement agent should be deleted from TV program production. Cell phone use has greatly increased from 1994 to 1998. 12. Trends of China and hotel industry, Introduction of DVD, BS, and CS to communication industry, 13. Threatening by China, Hotel industry, How difficult it is to construct a systems diagram! 14. A report on women's leadership was interesting.
3.	**Items of Knowledge Helpful in Career Planning** 1. None 2. Trends of Japanese industry, A temporary drop of GDP of Asian countries, Importance of the English language, 3. Management plan of automobile companies, Types of globalization and internationalization of Japanese industry, Investment necessary for growing areas, Interdisciplinary nature of technology and academic subjects, Importance of international view, Trends of giving chances to the young in this age of economic slump, 4. Globalization of various flows, Contents of ODA, 5. Impact of IT on communication business, 6. Transplantation of organs from mini-pig to human being, Production and distribution network system, Global networking of medical laboratories, Multidisciplinary collaboration of researchers, High-tech farms, Automation of medical treatment, 7. Trends of energy and other resources, Trends of new energy, Trends of electronics and consumer products, Utilization of new energy, Role of information systems, Networking of electronic equipment, Various service systems, Dispersion model of new energy and electric power systems. 8. Instruction on systems construction, Increasing demand for photo-fiber cables, Arctic and Antarctic development, Role of Japanese engineering in the global market, Cultural management issues, trends of textile and other materials industries, Role of new manufacturing and measurement technology in materials industry, New steel, 9. Role of space development and aerospace industry, Need for raising efficiency of photovoltaic systems, 10. Expansion of Internet business, Regional intranet systems, Advanced IT has contributed to international exchanges in sports and other areas, and Asians and Japanese became recognized through them. 11. Global trends/locations/alliances of automobile industry, IT industry has a future. Satellite industry needs to be expanded. 12. Distribution of T-shirts, Advancement of IT in media sector will facilitate development of new control systems.

(Continued to Page 5)

4	**Other Comments** **Table 4 (Continued from Pages 3 and 4)** 1. None 2. Need more explanations on English documents. Digital divide makes globalization very difficult. 3. Many reading materials sometimes confuse me. 4. Want to know more about truth of finiteness of energy resources, Company activities, and international trends of grouping of companies. 5. Want to know the reason why internet popularity is high in Korea, their national policy and people's nature. My part time job relates to vegetable treatment and I am therefore interested in agricultural systems. 6. Looking forward to the future of hydrogen energy. Concerned about restructuring of companies and introduction of robots to reduce labor. 7. None 8. Underground development should be promoted. 9. I found the theme of my report was deeper and broader in contents than I thought. 10. American football helmets are still now under development for jaw protection. 11. None 12. Themes of reports of other students are bigger than mine. A report theme on sports is interesting. Other students' presentations were very good, and I am concerned about mine.	

Table 5 shows a list of report themes chosen by the students for Course A. These may reflect their career interests, interests in prospective graduate research, and other personal interests, depending on each individual.

Course B "Overview of the Market and Introduction to Marketing"

Table 6 shows students' comments primarily on Course B at the end of Courses A and B. It should be noted that all the students who took Course B also took Course A. It can be seen from Table 6 that students appreciate real stories and are eager to learn about the real world. Table 7 shows a list of report themes chosen by the students for Course B. Table 8 shows contents of Course B, and ranks and degrees of interest in each course module of students surveyed at TMIT. It can be seen from Table 8 that, at the end of the course, TMIT students showed higher interests in Course Modules 9, 3, 2, 6, and 10, i.e. "What is Marketing?," "Overview of the Japanese Market from International Perspective, " "Global Economy and International Market," "Service Market (including Distribution and Media Market), Technology Market, Human Resource Market, and

Table 5 Themes of Reports of TMIT Students for Course A "Industry and Information Systems"

No.	Themes of Reports
1	Hotel Industry and Information Systems
2	Information Systems for Electric Power Industry
3	Study on Locations of New Skiing Grounds and Facilities
4	Financial Service Utilizing Internet
5	Discussion on Service System
6	Women's Leadership
7	Clothes Order and Sales System
8	Use of Information System to Explore New Markets
9	Freight Systems for Chinese Market
10	A Proxy System for Professional Sport Players
11	Medical, Healthcare and Welfare System for Rural Area
12	Air Ticket Reservation System
13	T-Shirt Net System

Table 6 Comments of Students at Tokyo Metropolitan Institute of Technology on Course B after Completion of Classes for Courses A and B at the End of the Semester

I became very interested in the term "Global" from the contents of your lecture.
I had been interested in management, and I could deepen my interest by taking this course.
The data on the real markets were very helpful for me.
The contents of classes were very good for me. I especially enjoyed the lecturer's talk on American life.
It took me a lot of time to read and understand a lot of English materials the lecturer gave us.
Each course module had separate topics, which were very good for me in learning things.
Your lecture was full of very real stories, which I could hear with much interest, unlike the other lectures.

Table 7 Themes of Reports of TMIT Students for Course B "Overview of the Market and Introduction to Marketing"

No.	Themes of Reports
1	Trends of Market in Hotel Industry
2	Trends of Electric Power Market and Its Marketing
3	Trends of Steel Market
4	New Service Proposed for Horse Race
5	Discussion on Service System
6	Marketing for Air Ticket Reservation System
7	Use of Information System to Explore New Markets

Educational Market," and "Historical Overview of Marketing and Its Theory." Also, at the beginning of the course, the students showed similar trends of having higher interests in Course Modules 6, 3, 9, 2, and 4. The far right column shows the results for all the students who responded to the questionnaire. It can be concluded from Table 8 that the students are highly interested in marketing and its history, the Japanese market, the international market, marketing and information technology, and the markets of service, technology, human resources and education.

Table 9 shows a summary of students' responses concerning **"1 New Discovery", "2 Helpful but Common Knowledge", "3 Knowledge Helpful in Career Planning",** and **"4 Other Comments,"** in response to the questionnaire given to TMIT students at the end of each class, shown in Table 3. A variety of topics interest students and some new discovery items for one student is also categorized as helpful but common knowledge and/or knowledge helpful in career planning for another student. Students gain further insight into the knowledge learned and utilize them during the course of writing their reports on themes of their interest, shown in Table 7.

Table 8 Results of Survey on Course B "Overview of the Market and Introduction to Marketing" for Junior and Senior Students of Department of Production, Information and Systems Engineering (DPISE) at Tokyo Metropolitan Institute of Technology
(Degree of Interest 5: Very Interested 4: Interested 3: Neutral 2: Not much Interested 1: No Interest)

Course Title and 15 Course Modules		**TMIT DPISE Junior and Senior Students**					
		A. Response at the Beginning of the Course (October, 2002)		B. Response at the End of the Course (January 28, 2003)		B. plus A. (excluding those responded for B)	
ID No.	**Overview of the Market and Introduction to Marketing**	Interest Rank	Interest Average	Interest Rank	Interest Average	Interest Rank	Interest Average
1	Introduction: Survey for Students and Course Work Orientation	13	3.17	12	3.33	12	3.27
2	Global Economy and International Market	④	3.91	③	3.91	③	4.00
3	Overview of the Japanese Market from International Perspective	②	4.09	②	4.09	②	4.07
4	Capital Market, Financial and Currency Market, and Real Estate Market	④	3.91	*7*	3.55	***8***	3.53
5	Commodity, Energy and Futures Market	***8***	3.55	15	3.18	12	3.27
6	Service Market (including Distribution and Media Market), Technology Market, Human Resource Market, and Educational Market	①	4.18	③	3.91	④	3.93
7	Trends of Market for Selected Industry (1)	***10***	3.27	***8***	3.40	***10***	3.36
8	Trends of Market for Selected Industry (2)	***10***	3.27	***8***	3.40	***10***	3.36
9	What is Marketing?	③	4.00	①	4.18	①	4.13
10	Historical Overview of Marketing and Its Theory	12	3.18	⑤	3.64	***6***	3.60
11	Marketing at Organizations	***6***	3.64	***8***	3.40	*7*	3.57
12	Marketing and the Socio-Economic Environment	***9***	3.36	11	3.36	***9***	3.47
13	Marketing and Information Technology	***6***	3.64	***6***	3.60	⑤	3.71
14	Presentation of Students' Reports, Discussion and Guidance to Refine Them	14	2.86	12	3.33	14	3.09
15	Discussion to Refine Students' Reports and Instruction on Abstract and Report Formats	14	2.86	14	3.25	15	2.73
Average for Average Interest of Each Course Module		(Note)	**3.53**	(Note)	**3.57**	(Note) 11 responses	**3.54**
Average for Each Individual's Average Inerest		11 Student	**3.60**	11 Student	**3.67**	from B, and 4	**3.66**
Grand Interest Average for All Interest Points		Responses	**3.57**	Responses	**3.66**	responses from A	**3.52**

Table 9 Summary of Reponses to Questionnaire Given at Each Class for Course B "Overview of the Market and Introduction to Marketing"

1.	**Items of New Discovery** (Course Module Numbers are shown below) **Note: Original responses were given in the Japanese language.** 1. Can't we study, focusing on "Pachinko" Industry? 2. Electric power in Indonesia and necessary power to operate Shinkansen (bullet) trains, Various economic forecasts after the September 11 event, Big gap between the rich and the poor in Brazil, Formation of EU type union in Asia, Less-than-expected investment in Africa, Low growth rate of Japanese economy, Low shares of high technology products in trade, 3. Almost everything comes from the other part of the world, often from unexpected places. Does the legal constraint weaken Japanese medicine and biotechnology? CIS, Lecturer's overseas experiences, Japan's huge trade amount with East Asia, Dependence on coal for power generation in U. S. and Japan, Great amount of import of electrical machinery from Asia, Dependence of oil on Middle East, 30% of Japan's export amount is for U. S., Japan's strength in materials but weakness in biotechnology and software, Import amount of service to Japan is twice its export, 4. National military power influences strength of the currency. Stock market of each nation, Currency of each nation and EU, TOPIX, Shares prices are determined like votes for beauty contests, Influence of NASDAQ on Japanese stock market, Four factors that determine the rate of national currencies, Share price and public interest rate, Current Nikkei average is one fifth those during the bubble economy in 1991, The value of shares does not always reflect performance of companies. 5. Derivatives, Correlation between Population, GDP and Energy Consumption, Scheme of futures, Impact of wars on commodity prices, Futures of gold, Linkage between oil and gas prices, and their demand and supply, Reliance on oil and coal, Forecast of energy prices, Japan premium prices of oil, Increasing consumption of fossil fuels, 6. Some service sectors are not growing. Gamble industries such as pachinko, horse races, and bicycle races are much bigger than I thought. National characteristics and nationality play a big role in business. Electronic commerce is growing. Growth of industries depends also on the popularity of systems from time to time. Sales of the domestic and overseas travel services are about the same, contrary to my expectation. 7. Role of manufacturing industry in Japan's overall industries, Japan's broadband fare is relatively low. Impressive growth of Japan's information and communication industries. Sweden is No. 1 in Internet popularity, while Japan is No. 16. I thought Japan's investment in information areas was bigger that by the U. S. contrary to the fact. I thought Automobile Industry of Japan stopped growing due to trade frictions with the U. S., but it has grown to the present. Japan's Internet popularity using portable digital assistance is No. 1 in the world. Japan's Internet fare is about the same as that in France. 8. Global standing and alliances of Japanese steel manufacturers, Global sales share and standing of Japanese inorganic chemical industry, Growth of Chinese four-wheel vehicle market, Environmental issues will become most important in materials industry in the future. China is trying to catch up with other nations in automobile industry. Will the market become that big in 2020, as forecasted? Russia's space flight duration is No. 1 and occupies about half that of the world total. More than half of imitation products to Japan are from P. R. China, Taiwan and Korea. Production life cycles become shorter in every industry sector. Some steel industries have entered into electronics, energy and the environment business. 9. What is marketing? (Market trend, Marketing activity, Marketing Mix 4P (Price, Product, Place, Promotion)), Market making, Satisfying customer's desire, 10. Lecture on marketing began in 1908 in the U. S., while Japan lagged behind. Marketing has advanced through interaction between the government and industry. Change of 4P by the marketing environment, 11. Research and introduction of marketing approaches have benefited customers. This is due to increased competition between service providers. In contrast, I hear that some Pachinko shops do behind-the-scene collaboration to protect themselves. The term "Fixed price" was changed to "Maker's Preferred Retail Price" due to a change in law. I thought the rank of Desante in sporting good sales was low because their shopping area was smaller than their competitors, Mizuno and Assics, but their rank was higher than I thought according to Nikke's article in 1998. I did not know that there is a law concerning price and advertisement. It was a discovery that there are various laws to protect consumers. Governmental organizations do various marketing researches. Both sampling food at stores and marketing research utilizing IT are part of marketing activities.
	Items of Helpful but Common Knowledge 1. It is important to spend meaningful extra years at universities, when needed to gain credits for graduation. 2. English economic terms, Worldwide distribution and flow of money, Structural slump of Japanese economy, Influence of the September 11 event, American economic recovery from the 1980's to the 90's, Impact of American economy on other economies, High economic growth rate of China, the other East Asia, and Latin America, 3. Almost all commodities exported from Japan are in a similar category. Close relationship between trade and politics, Japan's ranking of industrial competitiveness, Great import amount of food from CIS and coffee from Latin America to Japan, Most westerners do not eat octopus. Australians are less industrious than Japanese in general. **(Continued to Page 8)**

<table>
<tr><td></td><td>Table 9 (Continued from Page 7)
Japan's reliance of energy on overseas resources, Decrease of service import due to declines in tourists from 1999 to 2000, Most goods exported from Japan are manufactured from imported raw materials,
4. Mechanism of changes in share prices and currency conversion rates, and great influences by the super power, GDP share, Deflation in Japan, Taxation systems in Asia,
5. Difference in energy mix between Asia and the world, Contents of futures, Changes of energy consumption caused by wars, etc., Currency exchange rate and sales of energy, Variation of energy prices, Electronics commerce, Production management utilizing network, Dominant consumption of fossil energy in Asia,
6. Services for plants and business offices included those I did not expect. Japan's surplus of technology trade against the U. S. Ranking of universities made by industry, Trends of service industries for individuals and for plants and business offices are influenced by the social trends such as aging of the society. Service industries, except public sector services, are growing. Information service sector is steadily growing. Images of professionals differ from one country to another.
7. Japan's automobile industry is still growing, if not as highly as that in the past. The only (and the highest) growing industries in Japan are found in the information and communication sector and its related sectors. There is still some more room for Internet sector to grow, considering its current popularity. Industries related to automobile industry will grow if the automobile industry itself grows. Domestic demand in Japan for automobiles is decreasing.
8. Japan's manufacturers must make mid- to high level products to survive. Japan must produce high grade products. Digital equipment production will grow. Artificial intelligence software will become most important, even though it may not be needed as much as it is produced. Ex-work production of textile industry of Japan is decreasing.
9. None,
10. Impact of wars spreads in various manners. The origin of what is focused in Japan can be found elsewhere in the world. BCG Matrix, Establishment of Amazon.com in 1994, Korean War stimulated Japanese economy in 1950.
11. Corporate brand, M&A and management integration are subject to various laws. Sony is still No. 1 in corporate ranking in Japan. News about integration of Sumitomo Chemical and Mitsui Chemical (which was postponed later), Hong Kong's news of stopping public sector's development of housing to vitalize private sector development of housing, Not only service, but naming, branding and corporate images have much influence on marketing.</td></tr>
<tr><td>2.</td><td></td></tr>
<tr><td>3.</td><td>Items of Knowledge Helpful in Career Planning
1. None
2. Impressive economic growth of Asia, Share of East Asia in the world trade, Low GDP growth rate of Japan,
3. Japan's strong areas and weak areas, Forecast of biotechnology and clone markets in 2020, Japan's total energy demand and energy mix, Cost factors facilitate import of goods even if Japan can make them.
4. Company's performance and share prices, Impressive Asian GDP growth rate, Trends of financial and currency markets, A good chance to buy real estates because of low interest rate, Difficulties in stock investment,
5. Futures, Chinese resources, China can be the U. S. in the future. Dominant consumption of coal in China, Huge coal resources in the world,
6. Systems engineers need to be proficient in English. Twenty years of engineering experience differs from twenty repetitions of one year experience. Need to communicate in consideration of characters of local people and local culture. Trends of amusement industry reflect those of media industry. Ranks and images of universities determined and held by industry,
7. Japan leads some, if not all, sectors in information and communication industry. Differences in investment in information and communication, stemming from difference in nature of American and Japanese people,
8. The world map of Japan's and other nations' industries, Production of fine ceramics, photo-fiber cables and their parts, silicon wafer, and new glass is growing. Japan's cost of protecting intellectual property right is high. Reasons for Japan's low patenting ratio, The reason for recent lower supply of products in Japan is shifts of manufacturing bases to overseas countries of lower labor costs.
9. Excursion talk,
10. Practical aspects are closely taken into consideration in current marketing. IT is extensively applied to marketing. Some industrial maps of the U. S., Characteristics of product life cycles are linked with the nature of commodities.
11. Marketing approaches are used in almost all areas. I went to a newly opened UNIQLO, and was surprised at its popularity by seeing a number of cars in front of their parking lot. Eight functions of marketing, Translation of Japanese into English is difficult for me. Service will make difference in types of customers we may get, even if we sell similar products at similar prices.
(Continued to Page 9)</td></tr>
</table>

	Other Comments Table 9 (Continued from Pages 7 and 8)
4.	1. Good to know that there are growing industry sectors even in the midst of Japan's economic slump. 2. Want to learn common knowledge of economics and other useful information. 3. Today's lecture was easy to understand, as there was a lot of numerical data on the market. 4. I found it difficult to understand English materials. Please speak more slowly. 5. Net casino is a decent business. 6. None. 7. Is Volvo a Swedish manufacturer? I want to write a report on sports industries, but do not know how, yet. 8. Your lecture on how to write reports and papers (is needed). How can I get materials on Chinese market? 9. Activities of individual companies were interesting. 10. After all, global standards are American standards. 11. Play Stations 1 and 2, as well as cell phones, correspond to "the long term maturing products by product improvement," glasses and watches correspond to "the long term maturing products," while Nike's Air Max may correspond to "the temporary trendsetting products." 12. Our class has become better, as we have more questions and answers, as compared with the beginning.

Course C "Evolution of Industry, Business and Engineering Practice in the Global Economy"

Table 10 shows contents of Course C, and ranks and degrees of interests of students, surveyed at TMIT and the Pennsylvania State University, in each course module. It can be seen that DPISE students at TMIT show higher interests in Course Modules 5, 7, 3, 11, 12 and 2, i.e. working in Japan: an overview of the society (as Compared with the U. S. and international counterparts) and an insider's view of employment cases for American and international engineers and scientists, [5] internationalization and globalization of industries and various organizations, the historical view and outlook for engineering, technology and business: (3) space utilization and development and (4) IT industry, and the global a economy.

The Penn State "Entrepreneurship as a Minor Program" students show higher interests in Course Modules 2, 3, 4, and 9, i.e. the global economy, internationalization and globalization of industries and various organizations, international framework and international organizations for global coordination, and historical view and outlook for engineering, technology and business: (1) manufacturing industry. The Penn State students also have mid-level interests in Course Modules 5, 6, 7, 10 14 and 15, i.e. Japan related themes,[5] energy industry issues, and presentations and discussion of their reports. The differences between the two universities may stem from differences in their majors, their personal interests and the social environment.

CONCLUDING REMARKS

TMIT undergraduate students who took Courses A and B faced serious situations in securing their jobs in this age of a difficult national economy, and thus had a keen interest in their job markets. Some had to spend extra years at TMIT due to a shortage in credits, and had sufficient time to contemplate and plan their future careers. The aims of these courses were to educe students' interest in various topics related to industry, and stimulate the students to design a new information system and to conduct marketing studies in the areas of their own interest, with a combination of their passion for pursuit of study in those areas and the lecturer's instruction, based on his career experiences. The results of survey and students' report show that these aims are fulfilled to various degrees for the students who took these courses.

The authors would like to express their gratitude to Associate Dean Robert Pangborn and Professor Elizabeth Kisenwether of College of Engineering at the Pennsylvania State University for arranging for a seminar in which the lead author could conduct a survey for students of the Engineering Entrepreneurship Minor Program (also popularly referred to as "Entrepreneurship as a Minor Program"), shown in Table 10.

REFERENCES

1) Honda, H., ed., Proceedings of Engineering & Technology Management Group Symposium on Industry, Technology, Society and History, International Mechanical Engineering Congress and Exposition (IMECE) 2002, ASME, (November 2002).
2) Honda, H., ed., Proceedings of Engineering & Technology Management Group Symposium on Partnering to Succeed: Keys to Managing Technology Development, Risks and Globalization, IMECE 2001, ASME, (November 2001).
3) Honda, H., ed., Proceedings of Engineering & Technology Management Group Symposium on Successfully Managing the Risk and Development of Your Business and Technology, IMECE 2000, ASME, (November 2000).
4) Honda, H., Syllabus for EC320 "The Contemporary Economy of Japan," IES Tokyo (August 2003).
5) Honda, H., editor/author, Working in Japan: An Insider's Guide for Engineers and Scientists, ASME Press (July 2000).

Table 10 Results of Survey on Course C "Evolution of Industry, Business and Engineering Practice in the Global Economy" for Junior and Senior Students of Department of Production, Information and Systems Engineering (DPISE) at Tokyo Metropolitan Institute of Technology and Undergraduate Students of Entrepreneurship As A Minor Program (EAMP) at The Pennsylvania State University

(**Degree of Interest** 5: Very Interested 4: Interested 3: Neutral 2: Not much Interested 1: No Interest)

Course Title and 15 Course Modules		**DPISE Students at Tokyo Metropolitan Institute of Technology**						**Penn State Univ.**	
		A. Response at the Beginning of the Course (Oct, 2002)		B. Response at the End of the Course (January 28, 2003)		B. plus A. (excluding those responded for B)		Five EAMP Students (November 11, 2002)	
ID No.	**Evolution of Industry, Business and Engineering Practice in the Global Economy**	**Interest Rank**	**Interest Average**	**Interest Rank**	**Interest Average**	**Interest Rank**	**Interest Average**	**Interest Rank**	**Interest Average**
1	Introduction: Survey for Students and Course Work Orientation	14	3.20	***10***	3.25	13	3.20	14	2.8
2	Global Economy	③	3.69	⑤	3.50	***6***	3.59	①	**4.4**
3	Internationalization and Globalization of Industries and Various Organizations	⑤	3.44	②	3.75	③	3.65	②	**4.2**
4	International Framework and International Organizations for Global Coordination	13	3.25	14	2.91	15	3.06	③	3.8
5	Working in Japan: An Overview of the Society [5)] (As Compared with the U. S. and International Counterparts)	③	3.69	①	3.92	①	3.94	⑤	3.6
6	Political Environment of Japan and Its Exposure to Global Politics	15	3.13	***10***	3.25	***10***	3.35	⑤	3.6
7	Working in Japan: An Insider's View of Employment Cases for American and International Engineers and Scientists [5)]	⑤	3.44	③	3.64	②	3.69	⑤	3.6
8	Working in Japan: An Insider's View of Employment Cases for American and International Business Professionals	⑤	3.44	*7*	3.42	*7*	3.47	***11***	3.4
9	Historical View and Outlook for Engineering, Technology and Business: (1) Manufacturing Industry	⑤	3.44	15	2.75	14	3.12	③	3.8
10	Historical View and Outlook for Engineering, Technology and Business: (2) Energy Industry	11	3.31	13	3.17	12	3.24	⑤	3.6
11	Historical View and Outlook for Engineering, Technology and Business: (3) Space Utilization and Development	②	3.75	④	3.58	③	3.65	***11***	3.4
12	Historical View and Outlook for Engineering, Technology and Business: (4) IT Industry	①	3.88	***9***	3.36	⑤	3.63	***11***	3.4
13	Historical View and Outlook for Engineering, Technology and Business: (5) Bioengineering and Materials Industry	11	3.31	12	3.18	***8***	3.38	14	2.8
14	Workshop (1): Presentation of Reports on Individual Students' Choice, Discussion, and Guidance	***10***	3.33	***6***	3.45	***9***	3.36	⑤	3.6
15	Workshop (2): Presentation of Reports on Individual Students' Choice, Discussion, Guidance and Summary	***9***	3.42	***8***	3.40	11	3.31	⑤	3.6
Average for Average Interest of Each Course Module		(Note)	**3.45**	(Note)	**3.37**	(Note)	**3.44**	(Note)	3.57
Average for Each Individual's Average Interest		16	**3.5**	12	**3.40**	12 from B	**3.47**	5	**3.57**
Grand Interest Average for All Interest Points		replies	**3.5**	replies	**3.4**	5 from A	**3.5**	replies	**3.6**

APPENDIX: DISCUSSION ON FRUIT OF INDUSTRY FOCUSED COURSES UTILIZING COLLABORATIVE LEARNING APPROACH

Various factors, such as the subjects taught by the lecturer, students' personal interests, their interests in career planning, and approaches learned from other courses, have helped and motivated students in writing reports of the themes, shown in Tables 5 and 7. Some cases and discussions for Courses A, B and C are introduced below.

Course A "Industry and Information Systems"

One student, who had much interest in services and management, described the current state of information systems employed by western hotels such as Hilton, Hyatt, and Four Seasons, and leading Japanese hotels such as Imperial Hotel and Kagaya, a traditional inn (ryokan) in a hot spring area, and pointed out positive and negative impacts of IT systems, referring to kinds of services customers would most value. He warned an excessive emphasis on cost reduction by the use of IT systems may spoil a true service spirit customers may want.

Another student investigated human flows from urban areas to existing, various skiing grounds, their total skiing distances, numbers of visitors, capacities of facilities, etc., in Japan, and used correlation analyses to obtain ideal conditions in planning a new skiing ground. Others proposed "T-net System," (where designs of T-Shirts are solicited, and orders can be placed for the designs, and those received feasible numbers of orders only are produced and sold), and information systems for electric power industry, including photovoltaic systems for houses, with descriptions of necessity and safety measures for nuclear power.

Those cases clearly show that combinations of the aforementioned factors motivate students to write a report on the topics of their interest and/or propose new information systems among others, and most of the students appeared to have enjoyed during the course of doing their work. In addition, a final examination of ninety minutes included construction of a proposal on new coaching systems for a ball game of their interest, and students revealed various ideas on this subject, based on their experiences, as some of them were members of tennis, baseball, volleyball, and basketball teams.

Course B "Overview of the Market and Introduction to Marketing"

One student wrote about hotel markets in Japan, and their four major services, i.e., accommodations, restaurants, services for parties and wedding ceremonies. His report included state of the aforementioned and other hotel groups, marketing research for new hotels, yield management, and niche markets for senior people and those divorced. Another student reported on Chinese markets in general and prospects of express delivery services in People's Republic of China. A few others reported on deregulation of electric power market, electric power fare systems and a proposal on a new product, a new service for horse race customers, and a proposal on marketing of new products in steel industry.

It appeared that students' interests in career planning were the major driving force in writing these reports. Many of the ideas shown in the reports were primitive, but some were new ideas. It is considered that the process of investigating the subjects and compiling them in a report contribute to the students' future career and other planning. It should be noted that a survey on the students' interests in industry sectors are conducted at the beginning of Courses A and B so that they can focus on and/or pay attention to the subjects of their study for an entire semester.

Course C "Evolution of Industry, Business and Engineering Practice in the Global Economy"

Reasons for the differences in interest ranks and interest averages between the TMIT and Penn State students may be summarized as follows:

1) As the Japanese economy had been heavily dependent on conventional manufacturing industry, the national government began to focus on new areas such as service, IT and aerospace industries. Consequently, the TMIT students showed higher interests in space utilization and development, and IT industry, while the five Penn State EAMP students showed higher interests in manufacturing and energy industries, which are common but have wider skirts currently, as compared with the fields such as space utilization and development.

2) The Penn State EAMP students showed interests in Hebrew, Chinese, and Japanese languages among others, while TMIT students' primary focus was on improvement of the English language. As English is the major language in the international community, the Penn State students are considered to take an advantage and thus have higher interests in being engaged in international framework and organizations, while TMIT students showed higher interests in how other nationals may see the Japanese society, people, and job markets.

3) Another survey was conducted for over 200 mechanical engineering students at Doshisha University, a U. S. influenced university in Kyoto, and trends of their interests in industry sectors were closer to the Penn State EAMP students than the TMIT students. Therefore, differences in majors are also considered to be one of the main factors in the shown differences in their interests.

Proceedings of IMECE'03
2003 ASME International Mechanical Engineering Congress
Washington, D.C., November 15–21, 2003

IMECE2003-43767

THE IMPACT OF STEAM INNOVATIONS ON SHIP DESIGN - AN ABBREVIATED HISTORY OF MARINE ENGINEERING

CDR Matthew A. Carr, USNR, PhD, PE
United States Naval Academy
Annapolis, Maryland

ABSTRACT

The adaptation of steam engines for marine propulsion caused a dramatic shift in naval and commercial ship design during the 19th Century. The transition from sail to steam hastened the demise of several classes of ships and altered shipping routes from the trade winds to great circle routing. The conduct of naval warfare was always influenced by the limits of available propulsion technology. Throughout maritime history, innovative naval commanders sought ways to overrun, outmaneuver, and outlast their opponents. Coincident developments in armaments and armor, facilitated by this "new" propulsion technology, rendered the world's sailing navies largely obsolete within a relatively brief period of the 19th Century. This presentation highlights the major technological advances in steam propulsion from the early combination of low-speed single-acting reciprocating engines driving paddle wheels through high-speed turbines and reduction gears driving multiple-blade variable-pitch propellers; and, boilers heated by hand-fed wood and coal through nuclear fission.

INTRODUCTION

While the conduct of naval warfare has always been influenced by the available technology, the world's navies have also always been constrained by the treasuries that produced, operated, and maintained them. Other, enormous variables in "modernization" were the senior leaders' comfort with proven, older (and sometimes clearly obsolete) technology and the fact that new technology usually had disadvantages, as well as advantages, that made the optimum choices less clear. Significant developments in gun and projectile technology during the 19th Century forced naval shipbuilders to improve the methods and materials of ship construction and ultimately to replace the centuries-old standard of oak hulls with iron, and later, steel. The interjection of steam propulsion into the cycle of armament versus armor innovation facilitated the development of warships driven solely by machines. However, unless urgent military necessity dictated otherwise, budgetary constraints and the fact that steam technology had commercial applications generally convinced the senior naval leadership to encourage private individuals, operating under business-related profit motives, to conduct the necessary research and development of new propulsion technologies.

In contrast to the cost minimization philosophy of the navies, steam-powered merchant ships and passenger liners developed under a significantly different motivation – maximizing profit. The low power and fuel efficiency of early propulsion systems produced slow speed and short ranges, as a result steam-powered ships first found their profitable niche in the river, harbor, and coastal environments, where winds, currents, and tides slowed commerce conducted via sailing vessels. The market demand to move goods and passengers between seaports and inland areas in a timelier manner led to steamships and ferries especially suited to the river, bay, and lake environments. Steam-powered tugboats were developed to provide maneuverability to the merchant ships and warships in the harbor and estuary environments. Moreover, as the economics of coastal and riverine trade, especially where competition played a role, provided the incentive to produce faster steamers with more cargo capacity and lower operating costs, steam propulsion systems improved to the point that they could augment, and eventually replace, sails.

OVERVIEW OF SHIP DESIGN CONSIDERATIONS

Merchant ships are designed for cost-efficiency of moving cargo (and in the days before commercial airliners, people). Cargo type and capacity in a given hull shape, time between ports (speed and distance), fuel costs and potential cargo volume taken up by fuel, minimizing manning requirements and labor costs are all factors that influenced and continue to influence the design of merchant ships. Clearly, the capitalization represented by existing ships and their availability and convertibility plays into the decisions for new ship construction. Governmental subsidies and wartime conversion/mobilization requirements also influenced merchant ship designs. Over the last few centuries, insurance costs also weighed in on the economics of ship construction and trade decisions. Taxes, and historically, tribute payments, are included in the costs of doing business. Historically, many merchant ships also included weapons for self-defense against pirates, privateers, and smaller warships of enemy navies. Restrictive laws intended to protect a nation's merchant marine or punish other nations are yet another variable in the situation. Clearly, the cost-benefit decisions of shipping companies in deciding whether and what to build and how to use ships in any trade have included a complex set of considerations.

Conversely, warships operate under a significantly different set of requirements than commercial ships. Warships typically carry more crew than merchant ships, both to operate the weapons and to allow for damage control and casualties in combat. While some countries possess a navy only for coastal defense and the form of those warships is adapted to that operating environment, the major navies of the world design their ships to operate at great distances from homeport for long periods and are therefore exposed to more hazardous sea conditions. Throughout history, whether the mission involved protecting commercial and diplomatic interests, conducting scientific research, blockading ports, or just "showing the flag," on-station time typically carried a higher priority than transit time. These realities encouraged a cost-minimization approach to naval forces that often resulted in periods of relative neglect or dismantlement during peacetime. It generally took perceived or actual vulnerability to a potential adversary to force large expenditures on new warships, weapons, and crew, especially in the larger standing navies.

Regardless of whether the application was naval or commercial, the core technological areas during the transition from sails to steam involved steam cycle components, propulsors, and hull materials. It is noteworthy that in the evolution of marine engineering, while the "firsts" may be interesting and illustrate the cleverness, perseverance, or risk tolerance of their inventors, the adoption of the demonstrated technology for general application is more significant. As will be discussed below, many of the "firsts" did not work out as anticipated and were abandoned or converted to more conventional configurations.

NAVIGATION CONSIDERATIONS

Sailing vessels, by their very nature, exploit the desirable feature of being propelled by a "free resource" - the wind. However, wind is not always available in the velocity or direction needed and sailing ships are limited in their ability to sail "into the wind." Sailing into the wind was nearly impossible for the great square-rigged sailing ships and especially for the naval "ships of the line" with the extra wind resistance caused by additional structure above the waterline. Most of the square-rigged ships could not carry enough momentum to tack the bow through the wind and had to gibe or, as known in those days, "wear ship." This explains the sailor's historical concern for sailing off a "lee shore," where the wind direction was nearly perpendicular toward the coast. Fore-and-aft rigged ships, such as schooners could sail better into the wind, but the practical limit for even modern racing sailboats is around 30 to 40 degrees on either side of the bow. Because of the relative inability of the "square-riggers" to "sail to windward," the typical method of navigating was to sail north or south fairly close to the coasts, taking advantage of the land-affected winds and coastal currents until reaching the "trade winds" that allowed east or west transits across the oceans. For a typical Atlantic crossing, vessels sailed south to the Canary Islands and then west to the Caribbean. The return trip was east via trade winds to the Azores. In addition, seasonal relatively stable high-pressure weather centers created circulatory wind patterns that were exploited by seamen.

Knowing the ship's location is an essential part of voyaging. Celestial navigation started with using the angle from the horizon to the star Polaris, or to the sun at its local apparent noon, to determine the ship's latitude. Determining longitude was possible when more-accurate chronometers were invented in association with detailed knowledge of additional celestial body locations. By the time that steam engines were being installed on ships, navigators could determine their ships position with reasonable accuracy.

SAILING MERCHANT SHIP OVERVIEW

Merchant ships developed into specialized forms depending upon their routes, cargoes, and competition. At the time when the early marine steam engine experiments were taking place, maritime trade was roughly divided into transoceanic, coastal, and river/lake/canal. At the end of the American Revolution, the ocean-going ships tended to be large, full-rigged ships that had good cargo capacity but were very slow (speed of a typical East Indiaman was around 4 to 6 knots). Departures were frequently held up by weather and dependent upon if there was enough cargo on board to make the trip worthwhile. In the late 18th Century, packet ships developed for regular scheduled service between ports. Packets carried mail, passengers, and general cargo, but their primary operating characteristic was that they sailed on a schedule whether they were full or not. Packets developed better sailing characteristics as competition between shipping companies rewarded speed, but tended not to be overly large ships.

After the American Revolution, American shipping companies explored trade in Europe, Africa, the Mediterranean, South America, the West Indies and the Far East. Having recently lost the protection of the Royal Navy and not having a very capable navy made the American merchant marine vulnerable to piracy and to seizure associated with violating the embargo or blockade of warring nations. While these factors influenced the United States to develop a standing navy, American merchant ships were developed that sacrificed cargo capacity for speed.

At first, Yankee traders used the faster, but smaller schooners that were successful smugglers during the American Revolution. Schooners were adapted for general cargo and increased in size over the years. They were built up through the 1920s, seeing their last commercial use as fishing vessels with many converted to diesel power.

The Baltimore clipper was another design that reached its peak as a blockade-runner during the War of 1812. Following this war, speed was given up for capacity with more full-bodied ships. The economic conditions around 1830 rewarded shippers for getting goods and passengers to their destination before competitors with the lure of large profits and premiums. In addition, the gold rushes to California in 1848 and Australia in 1850 helped to emphasize faster ships. The Baltimore clipper and packet ship designs evolved into "clipper ships" around 1840. The generic name "clipper ship" was applied to three hull types: the extreme clipper, the clipper, and the medium or half-clipper. The extreme clippers had the highest speed but the lowest cargo capacity. Speeds above 20 knots were common. Naval architects and ship builders, such as William H. Webb (Founder of the Webb Institute), Donald McKay, and John W. Griffiths were famous for their designs. In the decade of the 1850s, 395 clipper ships were built. Clipper ships faded out in the 1870s. [1]

Another class of long-haul sailing merchant ships that are worthy of note was the "Cape Horners," also known as "Down Easters." These large square-rigged ships were used from about 1870 through the First World War and typically carried bulk cargoes such as grain, nitrate, and sugar.

Typical sailing ships for the coast and river routes continued to be schooners. These smaller ships were better suited to shallower water and, with their ability to sail better to windward than the square-rigged ships, were more maneuverable in the variable coastal wind conditions. In river and canal areas, where towpaths could be accommodated, shallow-draft barges pulled by animals, or poled downstream by humans were the vessels best suited to the environment.

As related in the balance of this article, large merchant sailing ships were mostly replaced by steam-powered ships. Those that lasted well into the 20th Century may also have been replaced by diesel or gasoline engine-driven ships. Some sailing vessels were converted to gasoline or diesel drive and their sailing rigs cut down. However, sailing merchant vessels have not been altogether eliminated. Niche routes, such as the copra trade in the Pacific, continue to be served by sailing vessels even today.

RECIPROCATING STEAM PRIMER

At this point, some basic definitions associated with reciprocating steam engines are probably useful. The simplest reciprocating steam system consists of a boiler connected to a piston and cylinder engine via a valve configuration that, at first, admits steam to the cylinder to push the piston from one extreme position to the other. A valve then vents the cylinder to allow the piston to move back to its initial position. Condensing refers to a condenser being installed that receives the steam from the cylinder after its final expansion. Condensers fall into two main categories – jet (in which water is sprayed into the entering steam) and surface (where cooling water flows around the condenser's shell or through tubes). A "condensing engine" is generally a closed-cycle system in which the condenser operates at a vacuum and the condensate produced is pumped back into the boiler. "Single-acting" refers to steam pushing a piston in one direction only. A "double-acting" engine alternately admits steam to one side of the piston and then the other, thereby causing a work-producing stroke in each direction of the piston's travel. Single-expansion exhausts the steam from the cylinder to either the atmosphere or a condenser. The concept of using cylinders in series is referred to as compounding. In a double-expansion engine, the steam from the first cylinder exhausts to a second cylinder for another expansion at a lower pressure. Triple-expansion, double-acting, condensing engines were common by the time that steam turbines were introduced.

EARLY MARINE PROPULSION DEVELOPMENTS

Prior to 1800, almost every marine propulsion device known today had been described in theory, patented, or attempted. The history of this technology shows that great inventions are never and great discoveries are seldom, the work of any one mind. Hydraulic jets using steam pumps, internal paddles, and exploding gases were suggested. Various propeller configurations, some similar to modern designs, had been tried. Duck-feet, oscillating umbrella cones, vibrating boards, and mechanized paddles had all been tried or proposed. Under the paddle wheel category, bow-wheel, stern-wheel, side-wheel, submerged-wheel, center-wheel (boxed and between catamaran-type hulls) had been tried. Feathering and adjustable draft paddles had been proposed. The power for these propulsors was from windmills, waterpower, animal power, and various steam engine configurations. [2, 3] The enabling technology that facilitated the final solution of the screw propeller was rotational speed. Early reciprocating steam engines produced low rotational speeds that were better matched to paddlewheels. Propellers required increased shaft speed to perform effectively.

A Roman army under Claudius Caudex was reportedly transported to Sicily in boats with paddlewheels driven by oxen.[4] Blasco de Garay, a Spanish naval officer demonstrated a steam-powered ship using paddlewheels in 1543, of which no details survive. De Garay's ship is thus considered the first steam-powered ship.[5] David Ramseye (1630), Thomas Grant (1632) and Edward Ford (1640) all patented concepts for moving ships against wind and tide.

[1] Taggart, R., 1983, Evolution of the Vessels Engaged in the Waterborne Commerce of the United States, US Govt Printing Office, pp. 14 ff.

[2] Thurston, R.H., 1879, A History of the Growth of the Steam Engine, C. Kegan Paul, London, p. 2.

[3] Taggart, p. 25.

[4] Thurston, p 223.

[5] Thurston, p. 12.

Not much is known about their specific designs, but they imply steam engines. In 1736, Jonathan Hulls was awarded an English patent for a steam-powered tugboat using a Newcomen engine and a ratchet-like drive to a stern-wheel.[6] The mathematician, Daniel Bernouilli, was awarded a prize for his 1752 essay on ship propulsion using screws.[7]

In 1784, James Rumsey built an 80-foot steamboat using hydraulic jet propulsion. He demonstrated this on the Potomac River in 1787 before a large group, including George Washington. Rumsey went to England and built a larger vessel working of the same principle, the *Columbia Maid* in 1792, but died before it was finished. In 1785, John Fitch had completed a model steamboat using a chain of buckets as the driver. He later modified this to a vertical paddle design in which six paddles were mounted on each side and mechanically connected so that half of the paddles were in the water on the power stroke while the others were retracted on the reset stroke. Fitch also initiated steamboat passenger service on the Delaware River, in 1790 and experimented with screw propulsion.[8] In 1790, Captain Samuel Morely ran a steamboat with a stern paddlewheel from Hartford to New York, averaging five miles per hour. He demonstrated another steamboat in Philadelphia a few years later.[9]

In 1789, John C. Stevens, a colonel in the American Army during the Revolutionary War, was unsuccessful in his attempt to obtain the exclusive right to navigate the waters of New York by steam-powered vessels. He thereafter advocated the development of the U.S. Patent system to protect the rights of inventors. The first U.S. patents for steam engines were granted in 1791 and included patents issued to Stevens, Rumsey, and Fitch.[10] Stevens and his sons, John and Robert L. continued work in steam transportation, canals, and railroads. The Stevens Institute of Technology was founded by this family on the original estate in 1870.

THE AGE OF SAILS AND STEAM

The grand view of the conversion of ships from sail to steam started with adapting wooden hulls to the new propulsion system but retaining sailing capability. Various forms of propulsor were tried and the paddlewheel won out until a screw propeller was developed that could handle low shaft speeds of the early reciprocating steam engines. Iron was adapted to ship construction, both for additional strength to allow for larger ships and for armor for warships. The introduction of iron went through a period of composite hulls, where interior structural elements were shifted and later iron hull plating was used, as well. Armor also went through a composite stage with ships known as ironclads. Improvements to steam-engine power density and efficiency, accompanied by establishment of a refueling infrastructure, allowed abandoning the sailing rig altogether. Sailing merchants were relegated to niche routes and cargoes and as the economy of the larger and faster steamships won out.

In 1803, Robert Fulton and Chancellor Livingston achieved the rights sought by Stevens in 1789. They had to demonstrate a viable steamboat by 1807. Fulton demonstrated his *Clermont* in time to secure the rights for twenty years. The 1803 act and subsequent renewal froze out Oliver Evans, who had constructed a steam-powered amphibious barge, the *Oruktor Amphibolis* in 1805, and John and Robert L Stevens' *Phoenix* of the same year. The fact that Stevens could not use the *Phoenix* in New York waters prompted him to move it to the Delaware River, thereby making the first open ocean transit of a steamship.[11]

In 1814, in response to the vulnerability of New York to attack by the Royal Navy during the War of 1812, Robert Fulton designed the first steam-driven warship. The *Demologos*, which after the death of its famous designer in 1815 was renamed the *Fulton*, was driven by a single-cylinder steam engine. The steam engine drove a paddle wheel, which was mounted between two hulls. The *Fulton*, although steam powered, still maintained sails for reserve power. She demonstrated 5 ½ knots in trials in 1815. With the war over and Robert Fulton having died, the *Fulton* became a station ship until destroyed by fire in 1829.[12]

The combinations of the Hudson River and Long Island Sound and the Mississippi River and its tributaries were perfect locations for the development of steam navigation. Due to wind and current, steamboats mostly replaced small coastal schooners, animal-drawn barges, and drift barges that were poled downstream. Between 1811 and 1831, 402 steamboats were built for use on the Mississippi, and 220 were in use as of 1830.[13] In that same year, there were 86 steamers on the Hudson and Long Island Sound.[14] Engine and boiler designs were improved and speeds increased, especially under the motivation of competition between lines, once the exclusive rights to New York waters awarded to Robert Fulton expired.

Early steam propulsion systems pumped water from the sea to the boiler and to their jet condensers. Saltwater used in these applications produced chemical scale and corrosion that limited the heat transfer and weakened the metals in the propulsion system. Steam leaks and boiler explosions were common events in early steam propulsion systems and the oftentimes severe consequences of these explosions influenced the establishment of the ASME and its Boiler and Pressure Vessel Code. The limits of metallurgy in the saltwater environment forced lower steam pressures in the range of 15 to 25 psi. In contrast, merchant vessels in the freshwater Great Lakes were fitted with propulsion systems that operated with 200-psi steam by mid-century. The surface condenser, originally invented by James Watt, with another patent by Hall in 1834, was tried in several seagoing applications in about 1835, but was

---

[6] Thurston, pp 225 f.

[7] Thurston, p. 228.

[8] Thurston, pp 239 ff.

[9] Taggart, pp. 21 ff.

[10] Thurston, p. 246.

[11] Taggart, pp 26 f.

[12] Tomblin, B.B., 1988, From Sail to Steam: The Development of Steam Technology in the United States Navy, 1838-1865, (PhD Dissertation) Rutgers – UMI, pp. 10 f.

[13] *New England* magazine, 1831, Vol I, Issue #6,; Cornell University Making of America Project, p. 546.

[14] Thurston, p. 283.

abandoned due to clogging by the grease used to lubricate the pistons. It was not until about 1860 that surface condensers came into common use.[15] Surface condensers allowed greater condenser vacuum with an accompanying fuel efficiency increase. Surface condensers also facilitated the use of freshwater makeup. The softer water reduced boiler scaling, but accelerated corrosion. This led to the development of chemical boiler corrosion controls and additional metallurgical improvements.[16]

Early steam propulsion systems drove paddle wheels. The paddle wheels were usually mounted on the sides of deep-draft ocean-going vessels and at the stern of shallow-draft river vessels. From a naval perspective, paddle wheels increased drag while under sail, were susceptible to battle damage, and took up valuable area for cannons. The first Royal Navy steamship was the HMS *Congo* (1815). The first steam warship was the HMS *Comet* (1822). By 1830, the Royal Navy had eight steamships, only two of which carried guns.

The second U.S. Navy steamship was the USS *Fulton II*, commissioned in 1838, and commanded by CAPT Matthew C. Perry, at that time a leading expert on steam propulsion in the U.S. Navy. The *Fulton II* was capable of 12 knots, but was laid up in 1842. Despite the early success of the *Fulton II*, by 1839, the U.S Navy only had two steam vessels, the other being a small tug named *Engineer*. The Navy started building steam warships in 1839. Notable ships include the USS *Princeton* (1842) was the first warship to use screw propellers and the USS *Pensacola* (1859) had the first surface condenser. HMS Rattler (1843) was the first warship to use a two-bladed propeller.[17]

The American Civil War saw a huge increase in the size of the U.S. Navy. At the war's outset, the Navy consisted of 90 warships of which only 45 vessels were in commission. Of the commissioned vessels, there were 28 steamships of various sizes. By the war's end, the U.S. Navy had 671 commissioned vessels.[18]

The post-Civil War period saw a return to the emphasis on far-ranging deployments for diplomatic presence. Most of the steam vessels procured by the Navy during the Civil War were better suited for river and coastal environments and were disposed. In 1867, Benjamin Isherwood, the first Chief Engineer of the Navy, designed the *Wampanoag*. This revolutionary warship was the first completely steam-driven ship. The *Wampanoag* used four super-heated boilers and a four-bladed propeller. However, the *Wampanoag* was deemed unfit by the U.S. Navy, and was transformed into an inferior sailing vessel, which simply relied upon steam as a auxiliary source of power. In the politics of the post-Civil War Navy leadership there was considerable tension between the traditional seagoing line officers and the new engineers - one side seeing warships as ships and the other considering them as machines. This played out through the next several decades, with technological advancements generally forced on those who saw too much power given to the technical bureaus of the Navy (Ships and Ordnance), but a general "anti-engineer" bent in the highest levels of Navy leadership.[19]

IRON HULLS AND ARMAMENT VS ARMOR

At this point, it is probably useful to look at the influence of iron and steel on ship construction. The maximum size of sailing vessels was constrained by the wood materials used in their construction. Iron started to make inroads into ship construction with iron rods and bolts replacing wooden treenails and iron straps added for diagonal bracing. In 1839, William Watson patented the idea of "composite ships" with iron transverse framing (ribs and deck frames) and wood planking fastened with rivets or nuts-and-bolts. Iron was later adapted to the longitudinal framing (keel, stem, sternpost, and stringers) in a patent issued to John Jordan in 1849.[20]

The next step was to adopt iron plating for the hull. While iron hulls solved the problem of shipworms boring holes into the hull, corrosion and fouling became issues. Antifouling paints did not exist for many years after iron hulls were adapted. It was common for extensive marine growth to cause severe speed reductions. Ship captains managed this by taking their ships up a river to reach freshwater where the seawater marine growth would die and flake off. Alternately fouling forced dry-docking to restore the hull. Copper sheathing was not an option due to galvanic corrosion issues. Copper sheathing below the waterline was adopted by the Royal Navy for antifouling and to protect against shipworms after 1761. Copper hull sheathing soon thereafter became the general standard for ships. Cathodic reactions between copper and iron were studied after several incidents, including the 1780 loss of over 900 on board the HMS *Royal George*. The *Royal George* sank after iron nails holding copper sheathing contributed to accelerated decay of the hull planking and catastrophic failure of the hull.[21]

The evolution from wood to steel ships produced lighter and stronger ships. The development of the machine trades was a facilitator for steam engines and shipbuilding. A pioneer of the machine tool industry, John Wilkinson was an "ironmaster" who invented an improved boring machine for cannons that also had application in creating precision cylinders for steam engines. He earned a virtual monopoly for cylinders for Boulton & Watt engines. As a side note, engrossed by the capabilities of iron, Wilkinson also built the first iron boat in 1787.[22]

Cast iron was discarded in favor of wrought iron. This was followed by another shift in hull material, to mild steel, in the 1870's, which produced a weight savings of about 15% in the hull beyond wrought iron and was likewise incorporated into engine construction.[23]

---

[15] Taggart, pp. 41 f.

[16] Rippon, P.M., 1988, Evolution of Engineering in the Royal Navy, Spellmount Ltd., Kent, UK, p. 60.

[17] Taggart, p. 36.

[18] Tomblin, pp. 1ff.

[19] McBride, W. M., 2000, Technological Change and the United States Navy, 1865 – 1945, Johns Hopkins University Press.

[20] Taggart, R., p. 17.

[21] http://www.rhosybolbach.freeserve.co.uk/Coppersheathing.htm, 07/15/03.

[22] http://www.gsn.uk.com/wilkinson.html, 07/03/03.

[23] Taggart, p. 19.

The development of naval guns forced the navies to adopt armor and to shift away from vulnerable paddlewheels to propeller drive, even before propellers were very effective. At the end of the Napoleonic wars in 1815, naval guns were cast-iron, smoothbore, solid-shot firing muzzleloaders with an extreme range of about 2500 yards, but more typically employed at 300 yards or less. Over the next few decades, developments included bigger and stronger guns, explosive shells instead of cannon balls, rifled barrels, breech loading and revolving gun mounts. Explosive shells had been discounted as an anti-ship weapon in favor of the heavy, but short-ranged carronade firing solid shot. In 1824, Henri Paixhains, a French artillery officer demonstrated a shell-firing gun against a ship that caught the attention of the navies. The "Paixhan" gun was adopted as a specialty weapon in several navies by the late 1830s, but it took the devastation of the Turkish Fleet by the Russian fleet using shellfire at Sinope in 1853 (Crimean War 1853-1856) to convince the navies that the shell was the anti-ship weapon of choice.[24] Another innovation employed in the Crimean War was armored batteries. Both the British and French built armored batteries for the Crimean War, but only three saw action: the French *Tonnante*, *Lave*, and *Davastation*. These ironclad ships had 4-inch total thickness of iron plates with 14-inches of wood backing. Their propulsion system was only strong enough to maneuver during battle, but not to transit. These three ships were used to subdue Russian forts at Kinburn in 1855. The French and Royal Navies, as well as the US Navy took notice of the ineffectiveness of the Russian shot and shell against these armored ships.

The SS *Great Britain*, launched in 1843, was first screw propelled, ocean going, wrought iron ship. The first iron-hulled naval vessel was the USS *Michigan*, commissioned in 1844 for service on Lake Erie.[25] The first ironclad warship in a European navy was *La Gloire*, launched in 1859 by the French. The Royal Navy reacted by building the HMS *Warrior*, which was launched in 1860. Both of these ships are classified as ironclads, in that they had wood behind the iron plates as a composite armor. *La Gloire's* hull below the waterline was wood, whereas *Warrior's* was iron.[26]

At the start of the American Civil War, the Union Navy conducted a competition for ironclads. They selected two designs to build, which became the USS *Galena* and the USS *New Ironsides*. Interestingly, the USS *Monitor* was not a choice because its designer, John Ericsson, was distrusted by the Navy board since an experimental gun explosion on the USS *Princeton* in 1844. The gun was designed by Ericsson, but not built by him. The Navy refused to pay him and the resultant controversy left Ericsson disqualified from future Navy business until the urgency of the Civil War caused reconsideration of Ericsson's designs.

The USS *Merrimack*, built in 1854-1856 was the first of a class of six 40-gun frigates, displacing about 3200 tons, and that were steam-powered and screw-propeller-driven. The propulsion plant consisted of two 72-inch diameter cylinders producing 869 horsepower at 40 rpm. Two brass water tube boilers consumed 2880 pounds of coal per hour, drove a 13-ton, 17-foot diameter twin-bladed propeller, and produced 6 knots of speed.[27] The problematic propulsion system was one of the reasons that the *Merrimack* was in the Gosport Navy Yard (now known as the Norfolk Naval Shipyard) when the Yard was taken over by the Confederates in 1861. Although scuttled and set on fire, she was raised and converted into the ironclad, CSS *Virginia*. The *Virginia* retained the *Merrimack's* original engines and had an armored casemate of 4 inches of iron plates over 22 inches of wood.[28]

John Ericsson's *Monitor*, completed in 1862, was the first all-steam all-iron ship. Unable to make long transits unassisted and marginally stable in rough seas, the *Monitor* had many innovative features including watertight bulkheads, forced draft blowers producing a pressurized engine room, the rotating gun turret, and a host of steam-powered auxiliaries.[29] The story of how the Monitor was funded and John Ericsson restored to the good graces of the U.S. Navy is interesting, but beyond the scope of this article.

THE AGE OF STEAM

By the late 1880's, steam power became the prime mover for ships. No longer did ships need to rely upon steam and sail power. The class of ships known as "monitors" represented the first all-steam warships. The U.S. Navy started a post-Civil War renaissance with the construction of the *Atlanta, Boston, Chicago* and *Dolphin*, authorized in 1883. Steam power had taken over the passenger liner business and was making inroads into the long-haul bulk freight commerce.

Arthur Woolf patented double expansion compound engines in 1803 and 1804. These were not generally adapted to seagoing ships due to the low pressures necessary when using salty feedwater. After the surface condensers were improved and the compound engines tried in merchant ships *Valparaiso* and *Inca*, the Royal Navy adopted compound engines after 1871. This was followed by triple expansion, designed by Kirk (1880). The U.S. Navy adopted compound engines after the Royal Navy and then even tried quadruple expansion engines in the battleships *New York, Texas* and *Oklahoma* in 1912.[30]

By 1890, over 100 patents associated with marine steam turbines had been issued. Steam turbines were commercialized by the electric power industry. A steam turbine operates most efficiently at very high speeds, while a ship propeller operates most efficiently at much lower speeds. In 1894, Charles Parsons built the *Turbinia* with steam turbine drive. Over the next several years of experimentation and improvement, the *Turbinia* demonstrated 34.5 knots at the Royal Navy review of 1897 using three high-speed propellers on three shafts. The shaft

[24] Potter, E.B., and Nimitz, C.W., 1960, Sea Power, Prentice-Hall, Englewood Cliffs, NJ, pp. 240 f.

[25] http://www.hazegray.org/danfs/steamers/michig1.htm, 07/15/03.

[26] Rippon, pp. 54 f.

[27] Quarstein, J.V., 1999, The Battle of the Ironclads, Arcadia Publishing, Charleston, SC pp. 11ff.

[28] Quarstein, p. 31.

[29] Quarstein, p. 39 ff.

[30] Rippon, pp. 59 ff.

speed problem was now reversed and was solved by installing reduction gears, done for the first time by Charles Parsons in 1910 in the *Vespasian*. Steam turbines with reduction gears became the standard for warships for the next six decades.

The Royal Navy commissioned the HMS *Dreadnought* in 1905. The turbine-driven *Dreadnought* set a new standard for large naval vessels that was soon emulated by the other large navies. Several classes of ships during the WWI era were equipped with turboelectric drives. High-speed turbines were connected to generators, which would then power motors driving low speed propellers. This modernization was effective during the early 20th Century, but limited motor and generator technology resulted in discarding this configuration in favor of the geared turbine drive.

Coal was used up through the early part of the 20th Century as fuel for boilers in reciprocating steam engines as well as steam turbines. However, coal was labor intensive and produced enormous quantities of ash for disposal. Dumping coal ash into the ocean left obvious trails for enemies to see. Smoke from lower grades of coal easily revealed locations of ships. The navies moved to oil as the main source of fuel. Oil was much easier to transfer when refueling ships. It could be pumped aboard, rather then shoveled and allowed the development of underway replenishment, a technique still used by naval forces.

Following WWI, the navies continued to advance in the area of propulsion and overall ship technology. Ships continued to increase operating steam pressures, allowing for more power, as well as a reduction in the actual size of the equipment needed for power. New alloys and materials were used in order to facilitate these new higher pressure and high temperature systems. Shipboard distillation plants and boiler water treatment minimized corrosion and scaling problems. Boiler pressure achieved its peak in the 1200 psi naval steam plants employed throughout WWII and the following decades.

The peak of commercial steam powered ships is the SS *United States,* placed in service in 1952. Capitalizing on the WWII marine engineering for fast U.S. battleships and aircraft carriers, the propulsion system of this ship featured eight 1200 psi boilers with steam turbines that generated more than 240,000 shaft horsepower and drove four propellers. The *United States* was designed for a maximum speed in excess of 38 knots and, on its maiden voyage to England, set the trans-Atlantic speed record with an average speed of over 35 knots.[31] Airlines ended the era of steamships as passenger liners around 1969, when the *United States* was taken out of service. Today, the ships that retain the passenger-carrying role are cruise liners, operating primarily for the entertainment of passengers, rather that transportation.

THE AGE OF STEAM & INTERNAL COMBUSTION

After WWII, gas turbine engines were introduced into propulsion. The first gas turbine-powered U.S. merchant ship was a converted Liberty ship, *John Sargent*. Utilizing a controllable pitch propeller, the ship went into service in 1956. At least one other Liberty ship was similarly converted, but the fuel efficiency, ruggedness, and increased size of available marine diesel engines meant that diesels became the engine of choice for most of the merchant ships that were built post-WWII.

In 1954, the U.S. Navy commissioned the USS *Nautilus*, the world's first nuclear-powered submarine. The *Nautilus* employed a steam turbine driven geared reduction drive to two propellers. Nuclear propulsion was adapted to surface ships and employed in US Navy cruisers and aircraft carriers as well as submarines. The Royal Navy, the French Navy, and the Soviet Navy (now Russian Navy) also built nuclear-powered submarines. The French Navy commissioned a nuclear-powered aircraft carrier, the *Charles de Gaulle* in 2001. Commercially, the American ship SS *Savannah,* completed in 1962 and named for the first steamship to cross an ocean in 1819, the German ship, *Otto Hahn,* a series of nine Russian icebreakers[32] and a Russian container ship, *Sevmorput*, were the only commercial nuclear-powered ships. Of these nuclear commercial ships, only some of the Russian icebreakers and the *Sevmorput* remain operational. Other than for military applications, nuclear propulsion has been relatively inconsequential to commercial ship design. The fact that the typical naval pressurized water reactors can only produce saturated, or slightly superheated, steam is actually a step backward from the superheated, high-pressure systems that were common in oil-fired steam warships.

Over the past several decades, the supremacy of steam propulsion has been reduced by diesel engines and gas turbines as the preferred prime movers. The main advantages of these engines include reliability, ease of maintenance, and low manning requirements. In fact, many merchant ships now have unmanned engine rooms. The primary long-haul ships of today's merchant fleets are container ships, roll-on/roll-off (RORO) ships, tankers and bulk carriers. These ships are generally diesel-powered, some with low-speed, direct-drive, non-reversing diesel engines and others with drive trains that are more responsive and capable of backing. Warships have largely shifted to gas turbines driving controllable pitch propellers via reduction gears and geared diesel drives, or combinations of gas turbines and diesels, but a few classes retain steam turbines. All current nuclear warships and submarines use steam turbines as the prime mover.

At this vantage point, the next era of ship propulsion technology will likely center around advanced electric drive systems with the electrical power being supplied by internal combustion engines and fuel cells. Nuclear power is currently a niche system for naval submarines and aircraft carriers, but lacking political acceptance prevents wider use of this technology. Past and existing naval reactor plants belong to the Age of Steam, but may continue into the era of advanced electric drives with direct energy conversion of nuclear reactions to electricity.

From a ship hull form and materials perspective, naval architects are experimenting with more lightweight and plastic composite materials for both military and

[31] http://members.tripod.com/richard_a14/PagesFacts.htm, 07/18/03.

[32] http://russiaphoto.hypermart.net/nib/index.htm, 07/18/03.

commercial applications. High-speed catamarans and surface effect ships are using water-jet drives.

It is probably safe to state that the era of marine steam is nearing its end. Yet, over the last two hundred years, steam technology provided the optimum power for maritime commerce and naval operations.

INTRODUCTION

The 2003 ASME International Mechanical Engineering Congress will again present the Engineering and Business Management Track, with ten sessions that emphasize entrepreneurial ventures, competitiveness, ethical methodologies, international business relations, and managing new product development. There will be workshops on engineering profit centers, managing intellectual property, partnering to enhance technical capabilities and project management. Business skills needed by newly graduated engineers will be identified and taught. Intercultural business ethics and engineering professionalism will be highlighted. Governmental partnering is also a focus this year.

In these sessions, international organizers and panelists from industry, consultancies, government, nonprofits, and academia will team to offer the broadest multidisciplinary views possible on each of the topics.

All of the sessions in the EBM Track for 2003 are co-sponsored by the Management Division or the Technology and Society Division. ASME Staff were instrumental in organizing the web tool, assisting managers and organizers, and coordinating the necessary communication and planning.

Ted A. Aanstoos, P. E., Track Manager

NOTE: The papers presented herein are those reviewed for publication and submitted by the required deadlines. Other presentations in this technical area were oral presentations only.

Proceedings of IMECE'03
2003 ASME International Mechanical Engineering Congress
Washington, D.C., November 15–21, 2003

IMECE2003-41200

ENTRY LEVEL MECHANICAL ENGINEERING CAD SKILLS ASSESSMENTS AND IMPROVEMENTS

Stephen J. Schoonmaker
Manager, Engineering Systems
Grove Worldwide

ABSTRACT

This paper presents information and discussion with respect to CAD skills for mechanical engineers entering the industrial environment. CAD systems are now an integral part of the functioning of virtually all engineering departments. This paper discusses how critical these systems have become to the practice of mechanical design and engineering. Naturally, then, CAD system skills become a consideration during the hiring of mechanical engineers. In order to deal with this situation, companies have considered or implemented CAD system skills assessments. The author investigates some of these assessments. Based on this investigation, a set of "core" skills or issues is presented which should be part of assessments or general hiring expectations. The core skills are not centered on specific CAD system software, but on general principles and practices that are essential to using these systems effectively.

INTRODUCTION

This paper presents information on CAD skills assessments, particularly with respect to entry level mechanical engineering. First, these skills assessments are reviewed in current industrial practice. Second, a CAD core knowledge base is presented for a potential industry standard assessment. Last, this core knowledge base is discussed with respect to educational issues or risk factors. Hopefully this will lead to improvements in the educational experience of mechanical engineering students as well as improvements in the introduction of entry level mechanical engineers into the workplace.

NOMENCLATURE

1. 3DDo (3-D Documentation) – Model-based annotation (dimensions, notes, etc.) placed on 3-D models intended for distribution and interactive viewing in order to completely replace 2-D drawing documentation.
2. BOM (Bill of Material) – The basic organization of data in manufacturing. It is generally a list of items (parts, materials, etc.) that go together to form a product. There can be a complicated hierarchy of relationships between these items and lists. BOM's are generally revision controlled in a manner similar to drawings, so they are often controlled or managed to some degree within the engineering department.
3. CAD (Computer Aided Design) – The most generic and widely-used term for computer software and systems that automate and enhance the design process. 2-D CAD works with drawings; 3-D CAD works with 3-D models (i.e. part models, assembly models, and associative drawings related to 3-D models).
4. CAE (Computer Aided Engineering) – Computer software and systems that automate and enhance the engineering calculation process. Typical CAE software includes FEA (Finite Element Analysis) for primarily mechanical stress calculations, CFD (Computational Fluid Dynamics) for fluid flow calculations, and "Dynamics software" for dynamic event simulation.
5. CAM (Computer Aided Manufacturing) – Computer software and systems that automate and enhance the production process. Machine tools that are driven via computer software get the basic information on how to produce the parts from the CAM software.
6. CADD – See MCAD.
7. MCCK (Mechanical CAD Core Knowledge) – A "body" of knowledge proposed by the author to capture the most important concepts in the CAD systems for mechanical design (but not

necessarily specific to any particular CAD system).

8. ECO (Engineering Change Order) – A basic business process that controls how engineering information (drawings, BOM's, 3DDo, etc.) are made available to the wider organization (or "enterprise") as well as to suppliers and/or customers.
9. MCAD (Mechanical Computer Automated Design) – Some CAD systems are sufficiently large and sophisticated that they tend to warrant a new designation, i.e. MCAD instead of CAD. However, CAD is still widely used in referring to any of the systems (whether a simple system for a few people or a massive system with enterprise-wide use). Another term that seemed to fail to make the same distinction was CADD ("Computer Aided Design and Drafting")
10. OEM (Original Equipment Manufacturer) – An organization creating an "end user" product (as opposed to just a supplier or vendor). For mechanical design activities, this can be considered companies such as automotive companies, aerospace companies, machine tool companies, heavy equipment companies, etc..

BACKGROUND

The author's perspective on the subject is driven by a variety of experiences within the industrial environment. In terms of background, the author spent 11 years in the development, documentation, verification, and training for Computer-Aided Engineering (CAE) software systems (with various units of the Ingersoll-Rand company); this was followed by 8 years in all aspects of CAD systems deployment and use. A 3-D CAD system deployment project led by the author included training 100's of users for a global enterprise's team-based 3-D CAD system [1]. In general, this project indicated that designers and engineers were not well prepared for the state-of-the-art CAD systems. Although this was largely due to the new nature and capabilities of 3-D CAD systems, there was also a basic inadequacy in understanding the modern CAD system. This inadequacy is likely to continue without an effort to raise the level of appreciation for these systems and the skills needed to use them well.

As further background, the author received his mechanical engineering degree from Worcester Polytechnic Institute in 1984; at that time, CAD was not an integral part of the undergraduate experience (4 seats of ComputerVision were available on a very limited basis, if memory serves). Although CAD is addressed in undergraduate programs now, it seems there may be no standardized approach to it in many cases.

As mentioned in the Nomenclature above, the term CAD is meant to refer to all aspects of Computer-Aided Design. It is not meant to be limited to "detailing", "drafting", or the production of drawings. Although the term MCAD (Mechanical Computer Automated Design) may be more accurate in describing the activities in question, the term CAD is quite "embedded" in the industrial environment and will be used in this paper.

TYPICAL CAD-BASED SKILLS

Table 1. lists some of the CAD-based activities which would be typically performed by entry level engineers in an engineering department (at least for an OEM-class organization). It can be seen that the CAD system is now beyond just the creation of drawings. Indeed, the importance of using and understanding 3-D CAD systems in particular can hardly be underestimated. Given the central importance of these skills, industry is naturally quite interested in the use of skills assessments to gauge the capability of new hires (or perhaps existing employees). The next section takes a look at some of these assessments.

CAD SKILLS ASSESSMENTS "CURRENT STATE"

The author has found that there are a number of currently available schemes and approaches for CAD skills assessments. However, they do not meet the current requirements of mechanical design engineering departments on a meaningful, industry-wide basis.

In terms of commercially available approaches from independent sources, most of these are centered on AutoCAD®. However, this is not the type of CAD program found in the large, mechanical OEM setting (probably because the broad range of activities found in Table 1. are generally not a part of this particular CAD system); it appears that many AutoCAD users are doing architectural or civil engineering projects. The sites listed in Table 2. seem to be serving this non-mechanical market.

There are also CAD skills assessments associated with more OEM-appropriate CAD systems (such as CATIA®, I-DEAS®, Pro/Engineer®, Unigraphics®, etc.). In some cases, these efforts are directed by the CAD system vendor. In other cases, these efforts are directed by independent user's groups or by value-added resellers of the CAD systems. Of course, these assessments are really directed toward the use of the specific CAD system. Although some of these skills are portable between systems (such as understanding what a sketching constraint does), completing a skills assessment in one CAD system does not mean much with respect to another skills assessment for a different CAD system.

An example of a "strong" assessment system from a specific CAD system vendor is CATIA [2]. The vendors associated with this system have advanced assessments to the point of certification. There is a formal approach to the skills, and a high level of recognition of the value of the process. Although CATIA is one of the more prevalent systems among OEM's, it does not have the status of an industry standard (at least not to the degree AutoCAD does within the civil engineering context).

Table 1. Typical CAD-based activities for entry-level mechanical engineers.

CAD-based Entry Level Activity:	Description:
2-D layouts, designs, and calculations	Creating sketches and drawings for new concepts in conjunction with more experienced engineers. These drawings often assist with geometric calculations.
3-D design and part modeling	Entry level engineers are often assigned the task of creating 3-D part models for parts or "details". This may or may not include annotation for manufacturing and processing.
Part-level drawing production (2-D)	In cases where 3-D models are not created, entry level engineers need to be able to create new released drawings. This familiarity is needed in case they perform a drawing approval function.
Part-level drawing production (from 3-D)	In cases where 3-D models are created, then entry level engineers need to be able to create "associative" drawings from them. Although "detailers" may be performing this activity, the 3-D CAD system ability to automate the production of the "geometry" in a drawing has made the use of detailers much less common.
Assembly modeling	In addition to part modeling, entry level engineers can be expected to do assembly modeling (basically a 3-D CAD only task).
Assembly-level drawing production (2-D)	Same as with part drawings discussed above.
Assembly-level drawing production (from 3-D)	Same as with part drawings discussed above.
BOM data	Depending on the organization, sophisticated software systems (such as enterprise resource planning) may be employed to manage the flow of all data in the enterprise. Entry level engineers may be expected to interface the CAD modeling with these systems.
Data Revision Control	The "state" of data within the CAD system (or its related product data management system) may be managed by entry level engineers. This could include researching the interchangeability of potential revisions.
Collaboration	Many organizations now rely on remote access software to also manage the process of designing and analyzing systems with respect to suppliers and customers on a global basis. Entry level engineers would be expected to handle this process as well.
CAD data translations	In conjunction with the collaboration process, CAD data can be expected to be translated between various systems. Entry level engineers often are involved with this process.
CAE interface	The 2-D and 3-D CAD models in the design environment are often used to assist with CAE activities for analyzing designs.
CAM interface	The 2-Dand 3-D CAD models in the design environment are often used to assist with CAM activities for producing physical parts.

Without a greater penetration of the overall industrial environment by CATIA, the author does not feel that CATIA certification is an industry-wide solution for CAD skills assessment. Just the same, the approach used may be a model for an industry-wide system.

Another approach for assessing skills of this type in the industrial setting is to use a "home grown" system. The author has used simple checks to determine if designers understand some essential CAD system concepts (it might be considered a training evaluation exercise rather than a formal assessment, examination, or certification, however). The following are typical examples of questions:

1. Where do you select your input units for a Model File?
2. Do the units you select have any effect on the data in a library?
3. What are the 3 steps to making a feature on a part?
4. What is an "open part"?

Table 2. Examples of AutoCAD-type skills assessments.

Web Site reference:	Description:
http://www.xlquest.com/cadusers.html	The assessment service seems to be called XL Quest. Mentions Mechanical Desktop (3-D), but this may be just 3-D "extensions" for architectural use.
http://www.computersindesign.com/content/training/schassess.html	Seems to be a specialized assessment service provided by a consulting firm.
http://www.cons4arch.com/cad/cad-knowledge-assessment.htm	Seems to be another specialized assessment provided by a consulting firm.
http://www.progressprint.com/	An on-line assessment service.
http://www.synergis-training.com/files/skill/skill_assess.shtml	An AutoCAD® "reseller" offers an assessment called CAD IQ.

Some of these questions are really only relevant for the specific CAD system; however, some of the questions are generic to 3-D modeling. So although this "reality check" approach has some value, it also does not represent a valid solution to the problem of a standardized assessment approach.

Within the academic environment, the author was not able to get meaningful information on the standardization of CAD training and/or skills assessments. The author contacted the Engineering Design Graphics Division of the American Society for Engineering Education, but the author got the impression that this group is not necessarily in favor of using CAD, never mind standardizing on content in the subject. There is also a National Collegiate CADD Conference, but it appears to be focused only on architectural and civil engineering CAD.

The most interesting efforts to standardize CAD training and skills came from some non-profit organizations. These are noted as follows:

1) National Occupational Skill Standards for Computer Aided Drafting and Design (CADD); Foundation for Industrial Modernization, 1994. This appears to be an attempt to provide a significant amount of what the author was originally interested in. However, it is almost 10 years old (i.e. no mention of assembly modeling); apparently is was not well publicized or received [3].
2) American Design Drafting Association (ADDA). This group seems to still be active, but the author has not noticed its presence in terms of mechanical engineering education or practice; it may be generally serving the vocational arena. Although educators (at all levels) may perceive there are still "drafters", "detailers", "designers", etc., the truth is that the latest CAD systems make it rather easy to create a fully detailed drawing based on 3D models (which are almost always made by more highly trained designers or engineers). Thus engineers are quite often taking care of their own drawings. Unless a new 3DDo-based detailing skill arises [5], detailers are likely to be irrelevant at least in the mechanical design activity. Of course, it may still be valuable to get ASME involved with a group such as the ADDA [4].

When investigating the current state of CAD skills assessments, it must be mentioned that there are Human Resources issues to consider. When an employer gets involved with testing of employees or potential employees, it is essential to have a test that is considered fair and reasonable, and this can be hard to prove. If a bargaining unit is representing employees, then this must also be considered. Many companies do not pursue a CAD skills assessment because of these legal issues. Considering these issues, the role of independent groups (such as the ASME) may become very valuable.

PROPOSED MECHANICAL CAD CORE KNOWLEDGE

Looking at the current state of CAD skills assessments, it appears that there is no real standard approach to assist industry in determining if entry level mechanical engineers are capable of doing the kinds of tasks listed in Table 1. Naturally, this is going to be problematic considering the variety of CAD systems currently available. However, the author feels that there is a central or core knowledge base that may be common enough to all the "major" CAD systems that a Mechanical CAD Core Knowledge (MCCK) can be developed.

The list below shows the author's proposal for what could be included in MCCK (generally based on the material found in Schoonmaker [6]). Many available CAD systems would be appropriate for covering these topics. Indeed, if some of these concepts are not found in the software used in an academic setting, then that CAD system may not, in fact, be appropriate for a mechanical engineering degree program. The author would welcome any feedback on this content or its application, and the author would certainly be willing to assist academic programs with this content and its applicability.

MCCK:

1) Computer Hardware
   a) Computer System Basic Components
   b) Networking
   c) Graphics Adapters and the Graphics "Pipeline"
   d) Ergonomics
2) Computer Software
   a) Software Layer paradigm
   b) Memory Management
   c) File Management
3) Basics of Mechanical Drawings
   a) Typical drawing types
   b) Title Block/Rev Block/BOM
   c) Drawing types
   d) Drafting standards
   e) Views
4) 2-D CAD
   a) "Smart Paper" Paradigm
   b) Paper Space and Model Space
   c) Automatic Dimensioning
   d) Views as Graphics Viewports
   e) Appearance - Layers/Filters/Fonts/Weights/Colors
   f) Vector Graphics vs. Bitmaps
   g) Groupings and Instancing
5) 2-D CAD Management
   a) 2-D Data Files
   b) 2-D CAE Integration
   c) 2-D CAM Integration
   d) Plotting
   e) Electronic ECO Processes
   f) Metadata
   g) 2-D Translations
6) 3-D CAD Concepts
   a) Model Terminology
   b) Geometric Terminology
   c) 3-D Computer Graphics Terminology
   d) Modeling Concept 1 - Need To Do 2-D Before 3-D
   e) Modeling Concept 2 - Solid Models Are Really Surface Models
   f) Modeling Concept 3 - Historian - Yes; Sculptor - No
7) 3-D Part Modeling
   a) The Modeling Steps
   b) Types of Features
   c) Sketching
   d) Sketch constraints
   e) Adding new features
   f) Surface issues for new features
   g) Boolean operations
   h) 3-D Geometric Properties
8) 3-D Surface Modeling
   a) Surface Characteristics
   b) Surface modeling operations
   c) Basic part tolerance
   d) Second order continuity
9) 3-D Assembly Modeling
   a) Assembly Structure
   b) Product Structure
   c) Positional Information
   d) Instancing
   e) Assembly constraints
   f) Assembly model calculations
   g) Animations
   h) Part-to-Part associativity
   i) BOM creation
10) 3-D CAD Management
   a) Parent/child relationships and Metadata
   b) Version Management and Revision control
   c) 3-D CAE Integration
   d) 3-D CAM Integration
   e) 3-D data files
   f) 3-D printing and RP
   g) 3DDo

CAD SKILLS ASSESSMENTS "FUTURE STATE"

Given that a standard MCCK might be definable, it stands to reason that it could become the basis for a standardized educational regimen, a CAD skills assessment program, or an examination for certification. The goal of such a program should be to see that the proper foundation for understanding the application of CAD systems has been obtained for a typical entry level mechanical engineer. However, the academic "community" as well as groups such as the ASME would have to get involved in the process. Industry needs this cooperation.

Assuming a certified or at least a presumed foundation of skills is in place, then the employer can more easily deal with the specific CAD system training needed. This MCCK approach would also hopefully allow the academic environment to concentrate more effort on engineering science (by not feeling they have to meet a perceived need for a specific CAD system training).

ACADEMIC RISKS

Of course, there are some risks associated with teaching CAD during the engineering education experience (regardless of a standardized content such as MCCK). For instance, CAD is not as important as engineering science topics. Some programs may decide to have students "pick CAD up" on their own. However, many mechanical engineering programs have long provided "shop"-like classes that really just relate to how manufacturing or design work is done in practice. CAD, it seems, falls into this category.

Of course, as the theoretical aspects of engineering become more and more sophisticated, there is less and less time for these "practicum" activities. However, CAD (as did the old engineering graphics classes) provides an excellent opportunity to deal with concepts such as 2-D and 3-D geometry, spatial visualization, and writing/reporting. Learning mechanical visualization and creating clear engineering documentation is probably considered worthy of a formal training effort, so there should be no reason that learning a CAD system could not be considered part of this effort.

Another academic risk is determining which CAD system to use and/or present. This would seem an impossible task without something like the MCCK, and the author assumes that academic departments do consider these sort of issues in deciding which CAD system to present. Of course, the main considerations in CAD system selection are probably going to be costs, the familiarity or preferences of the instructors or professors, and the tactics of the CAD system sales personnel to influence various levels of decision making. However, in the author's opinion, any CAD system vendor saying that their system is the industry standard is without merit. It depends on the sector of the economy (i.e. aerospace vs. machine tool vs. automotive), the size of the firm (small shop vs. mid-size vs. global enterprise), and the way the CAD is or is not integrated to the overall IT environment (such as enterprise-wide product data management). There is a wide diversity in industry for each of these factors, and investigations of the CAD marketplace by market researchers show all kinds of results based on license revenue, number of seats, new sales, market share, and which "kernel" is used. The fact of the matter is that there is no standard CAD system in mechanical design now, and it will continue to be difficult to decide which CAD system to include in any mechanical engineering undergraduate curriculum.

Another potential risk in the academic environment is whether or not to present material on how the CAD systems actually work. This could include programming in the appropriate computer languages, the 2-D and 3-D graphics transformation algorithms and implementing routines, graphics standards, NURBS basics, shading techniques, etc.. In the author's opinion, this is time well spent. The ability to understand and improve the CAD system performance will be enhanced for the entry level engineer, but then this needs to be balanced against the loss of time for other educational material. Although CAD system workings can involve some sophisticated concepts, it certainly should be within the grasp of 4th year students in a mechanical engineering program.

Finally, for mechanical engineering students, probably the most important aspect of an expanded experience with CAD systems would be integration with other software. With respect to CAM, entry-level mechanical engineers would certainly be expected to be capable of dealing with movement of data from the CAD system to the CAM system (as well as trouble-shooting it).

The situation is more uncertain with respect to CAE. Whether or not a graduate degree is a requirement for doing CAE such as FEA-based stress analysis does not have a consensus within industry. Indeed, many CAD system vendors keep pushing to get designers and engineers (that are generally creating just the 3-D design models) to perform FEA of some kind. It is not clear to the author whether this is necessarily appropriate. However, regardless of whether entry-level engineers would perform CAE, clearly they would still benefit from knowing how their design models may be impacted by the CAE activity. Indeed, in CAD some systems, CAD and CAE (even when presumed to be done by separate types of users) are well integrated.

CONCLUSION

An industry-standard core CAD knowledge base has been proposed. This core knowledge is assumed to be viable for entry-level mechanical engineers for OEM-class employment. If there is a sufficient level of interest, the author is considering updating an older, non-profit national program for use by the ASME or other groups to promote further standardization and improvement of entry-level introduction to the workplace.

REFERENCES

[1] Aziz, S., Schoonmaker S. J., Watson K., 1999, "Leveraging 3-D CAD for Integrated Product Development", ASME Mechanical Engineering Congress, Nashville, TN.

[2] Dassault Systemes, Suresnes Cedex, France. http://www.3dplm.com/en/education/certification.asp

[3] Foundation for Industrial Modernization, 1994, "National Skill Standards for Computer Aided Drafting and Design (CADD)", Washington, DC. http://www.nssb.org/projects/cadstd.html

[4] American Design Drafting Association, Columbia, SC. http://www.adda.org/

[5] Schoonmaker, S.J., 2002, "Productivity Enhancement Investigation for 3-D Documentation (3DDo) versus 2-D Drawings Using I-DEAS Master Series®, PLM World, Orlando, FL.

[6] Schoonmaker, S. J., 2003, *The CAD Guidebook – A Basic Manual for Understanding and Improving Computer-Aided Design*, Marcel-Dekker, Inc., New York.

Proceedings of IMECE'03
2003 ASME International Mechanical Engineering Congress
Washington, D.C., November 15–21, 2003

IMECE2003-41674

Development of An Ideal Project Law

Authors: David C. Hall
Senior Risk Manager
SRS Risk Management Services

Ron Kohl
R. J. Kohl & Associates

Roger Graves
Senior Consultant
Davion Systems Ltd

ABSTRACT

Our research collaboration has begun a project to develop an Ideal Project Law (IPL). What we are trying to accomplish is the development of some equation with a mathematical underpinning that can ***usefully*** relate Functionality, Cost, Schedule and Risk which can then be graphed, and this graph then becomes a decision making and communications aid. For some of us, the current contention is that a possible IPL is analogous to the Ideal Gas Law (IGL), which relates Pressure (P), Volume (V) and Temperature (T) for reasonably well-behaved gases. *(P*V)/T = Constant (where the constant has very much to do with the nature of the gas under discussion).* This is due to our belief that the IPL will represent some form of a relationship (inverse or direct) between Cost (C), Schedule (S), Functionality (F) and Risk (R) much like the IGL represents relationships between P, V and T for a given gas. We also expect that at least some of the factors will change based on the type of project, so we might have to develop a range of factors and constants. The Ideal Gas Law informs us, for example, that if you keep V fixed and increase P, then you can expect T to also increase. So "keep this one factor fixed, and change that other factor and then watch the third factor move one way or the other" relationships can be described via the Ideal Gas Law. We believe that this "push on one factor and see what happens to the other factors" feature of the Ideal Gas Law seems to be very analogous to project and operation Cost, Schedule, Functionality and Risk relationships.

This law, or some such function, is absolutely essential. Having such a Law that is proven valid will introduce (or rather tactfully enforce) some systematic thinking in the project and operational management set-up. Otherwise, no matter how elaborate a case is made for project or operational risks, the risk decisions will be left to the whims, fancies and moods of the key decision-making person(s) in the organization or project - we are working towards reducing this subjectivity in decision-making. We may determine that it is impossible to have a 100% scientific outlook on project or operational factors simply because human beings are involved, but if that is the case, we may be able to be at least 80% scientific about it.

INTRODUCTION

Our Risk Management Research Collaboration[1] has begun a project to develop an Ideal Project Law (IPL). The goal of this development project is ***to develop a simple equation which serves to aid in project management risk-based decision making***. Ideally, such an equation should be as simple as possible and should not contain complicated functions. The whole idea is to develop and verify a quantitative tool for evaluating risk in projects and operations based on known quantities such as cost, schedule and functionality. The major challenge is to find out if one size can fit all, or at least most. Such a verified equation could provide a sense of badly needed unity in diversity for all project and operational management work.

The whole idea is to create an accepted method to instill a sense of risk awareness and prudent risk management mentality in the context of project and operational management. Therefore, unless the equation has an explicit Risk-associated parameter (some prefer a Risk Constant based on many factors and others prefer a Risk Tolerance Constant), such an equation would not be useful. So, the use of such an equation could provide two types of information to a manager: 1) Given a prudent risk (or risk tolerance), which is a CONSTANT, what is the effect a change in functionality would necessarily have to have on the cost and/or schedule to maintain that risk (or risk tolerance) and whether that change of functionality is therefore acceptable/affordable to the management or not; and 2) Given a

[1] Our Risk Management Research Collaboration members are the INCOSE Risk Management Working Group, the PMI Risk Significant Interest Group, the APM Risk Significant Interest Group (UK), the National Quality Association (USA) and over 100 individual members from 14 countries.

change in functionality WITHOUT a change in cost or schedule, how much does that increase the risk to the successful conclusion of the project (or how high does your risk tolerance have to be to agree to this change?).

It is hoped that such an equation, once developed, could be applied across all projects (generic) and not have different constants for different projects. If not, then we could get lost into a myriad number of projects with varying shapes, sizes and configurations. Our challenge is to find and verify a one-size-fits-all equation.

HOW TO START?

Our first thoughts on the matter have come up with something that is analogous to the Ideal Gas Law (IGL), which relates Pressure (P), Volume (V) and Temperature (T) for reasonably 'well-behaved' gases. *(P*V)/T = Constant (where the constant has very much to do with the nature of the gas under discussion).* For some of us, the current contention is that a possible IPL is analogous this law since it will represent some form of a relationship (inverse or direct) between Cost (C), Schedule (S), Functionality (F) and Risk (R) much like the IGL represents relationships between P, V and T for a given gas.

The Ideal Gas Law informs us, for example, that if you keep V fixed and increase P, then you can expect T to also increase. So "keep this one factor fixed, and change that other factor and then watch the third factor move one way or the other" relationships can be described via the Ideal Gas Law. We believe that this "push on one factor and see what happens to the other factors" feature of the Ideal Gas Law seems to be very analogous to project and operation Cost, Schedule, Functionality and Risk relationships.

Assuming as a starting point that these four factors can completely define all types of projects and operations, we realize that at least some of them may have to change based on the type of project or operation, so we might have to develop a range for each factor and constant. Although some of us believe that there should be other factors beyond the Risk (R), Functionality (F), Cost (C) and Schedule (S), the current consensus is that any such dimensions beyond R, C, F and S can be described as derivatives of or characteristics within these four. It is our belief that a given project or operation can be fully described as "provide me a thing that has a set of functionalities (F) for a given price (C) and do so within a certain specified amount of time (S) with a set amount of risk (R) to success". A given operation can be fully described as do this thing which has a set of functionalities (F) for a given period of time (S) for a given cost (C) with a set amount of risk (R) to success. Therefore, we consider that these four factors to the primary factors that describe and drive the success or failure of any project (that is supposed to provide a product) or any operation and that any other factors, characteristics, dimensions, etc., can be derived from or are characteristics of one or more of these four.

If this is the case, then an equation such as $F = (C*S)R$ is likely to fully describe any ideal or realistic project or operation. Therefore, this type of equation could readily be used to determine the risk in any project or operation, or in any phase of a project or operation. This equation could also be used to establish the increase or decrease in risk coming from a change to one or more of the other parameters – F, C or S. Such an equation could then be used to compare a number of proposals in terms of risk. If each proposal has a baseline C and S for a given F, then its risk would be readily apparent.

Another example where project work may be analogous to already established physical laws is Heisenberg's uncertainty principle, which states that uncertainty in momentum times uncertainty in position is a constant. In project work, if you demand a fixed delivery date (zero uncertainty in time), then the uncertainty in other areas increases. Why? If the project hits a roadblock, you must either pour in extra cash to resolve the problem (uncertainty in cost) or deliver with reduced functionality (uncertainty in functionality). On the other hand, if you are prepared to be a little flexible with regard to delivery dates, then you can usually manage to smooth out the fluctuations in cost and functionality. In any "Law of Science", no matter the discipline, there is much certainty and absoluteness in any specific Law, but there is much uncertainty when trying to apply that specific Law in some practical situation or system. This is one of the very essences of the difference between theoretical pursuits and the engineering reality of actually building equipment and structures based on such Laws. But given the historically rich set of "data points" that the Project and Operational Management groups have, it is entirely conceivable that we could develop some equivalent equation (such as the IGL) and be able to identify one (or more) of the equivalent "constants" and perhaps even project-unique constants or other parameters.

The IGL is valuable because it is directly applicable for common circumstances (i.e., "the behavior of all gases tends to that of an ideal gas at low enough pressures, and even air is reasonably approximated as ideal") and it provides a standard around which exceptions can be defined and quantified. Within the stated caveats, the IGL gets by with a linear constant. Under some ideal range project cost and time relationships may be linear and predictable, but projects NEVER operate within that range; anyone who has "crashed" projects knows that cost and time relationships are not linear. Most of us could probably sketch a curve to approximate this relationship and talk conceptually about inflection points and the asymptotic limit, but anyone who could actually quantify this would deserve a Nobel Prize. And it is not predictable, we usually don't know exactly where on that curve a project actually is until we start to push one of the variables and observe the rate at which the others change.

DEVELOPING THE FACTORS

ESTABLISHING HOW TO DETERMINE FUNCTIONALITY

Should we simply use project or operation size as the functionality factor? Or should we add a few of the things normally put into the risk equation to minimize figuring out the risk factor? Cost and Schedule are expected to be single factors, so should Functionality be also? If that is so, what goes into establishing project or operation size - numbers of functions required, complexity of operation, number of people, etc.? It is assumed that Functionality should be an increasing number as complexity of the project or operation goes up. This means that at least the size of the project or operation must be taken into account. But is that the only consideration? Other considerations thought of include use of technology, whether or not to develop software as well as hardware, etc., but most of that should be (may be?) taken into account by the risk factor. So two major questions arise: what is to be contained in the Functionality factor, and what is to be contained in the Risk factor?

In addition to these questions there seems to be another and possibly very serious stumbling block - in what units do you measure the various quantities in the equation? Cost and schedule are obvious but how would you measure functionality in a way that can be made compatible with dollars and time? Listed below are some projects to consider:

a) Developing an new anti-malaria drug
b) Constructing a flood defense system for a coastal town
c) Staging the 2012 Olympic Games
d) Developing a new guided missile for a fighter aircraft
e) Laying an oil pipeline in the Gulf of Mexico

What consistent measure of functionality could be applied to this disparate but perfectly real group of projects? Unless you can establish a valid measure for F you cannot expect to be able to establish risk. The problem is that some of these projects have a series of attributes that collectively add up to the required "functionality".

One method suggested to start the factor development is to use an existing, proven tool for determining functionality, Function Point Analysis. This methodology, developed for the software field, looks at the various ways in which users can interact with a computer system, instead of simply equating software program functionality to mere size, or lines of code. To do this, it divides the system requirements into those that handle the data requirements of the end user, and those that handle the transactional requirements. In our case, we may be able to broaden the scope of this existing methodology to include the functionality of any system, not just software systems. As a start ask, "what does this system do for the end user?" For example, an office building houses and supports office workers, and from this point of view has a one-dimensional functionality. However, if the building combines residential, commercial and office space, it has a much higher functionality because it provides many more functions for the end users.

Note that in the latter case the building itself may not be any larger, or contain more complex systems (such as power, HVAC, elevators, and so on) than a simple office building. However, the requirements that determine the design of the various systems that make up the building will be more complex than in the case of a simple office building. The complexity of the system requirements is intimately connected with, and is probably a direct indicator of, the level of functionality of the system. One definition that makes sense is that functionality represents both the capabilities of the end product of the project (the project end item, or PEI), together with the effort that was necessarily expended in getting there. In the case of a large office building, the total functionality in the PEI is quite large - it houses and supports several thousand office workers, providing all the ancillary services such as HVAC, elevators, fire suppression, telecommunications, and so on. However, the effort expended per unit of functionality is quite low, because little or no development effort is required. One suggestion is that functionality should be defined as the capabilities of the project end item, multiplied by the effort necessarily expended in getting there.

ESTABLISHING HOW TO DETERMINE RISK

There are many suggestions on developing the Risk factor, three of which are outlined in this paper:

1. Include a separate "risk factor" in the IPL.
2. Build risk into "R", the project constant.
3. Use ranges for the main parameters (F, S, and C)

Our preference at this time is to use suggestion 2, that is define the project constant "R" to include risk. That way we can specify what level of risk we expect in this type of "ideal project". So a product innovation project might have a high risk component built into R, whereas building another bridge might have a lower-risk element in its R. What we're trying to do here is to define the expected relationship between F, S and C for "the ideal project of this type", i.e., describing the expected relationship between its parameters, which of course includes risk. We expect different project types to have more inherent risk, so this needs to be in R. So perhaps we can say there is a general IPL for "THE ideal project", defined as F = (S*C)*R where R=1. Then we can define some values of R for different types of projects and operations. We need to define how to derive R, and what its contributing factors are. If we use R for the risk element of the equation, then high risk project/operation types might have R>1, while low-risk project/operation types might have R<1. If this is so, then what else goes into defining R?

OTHER POSSIBLE RELATIONSHIPS AMONG R, F, S AND C

Since we are at an early stage in this work, there are numerous relationships between R, F, S and C being proposed. One of the most complete is described in the following paragraphs. This alternative definition starts from the same fundamental equation (F = k(C*S)), but uses different definitions of the various factors in the equation.

FUNCTIONALITY

Functionality is defined (for this alternative relationship) as the benefits that the project end item (PEI) will bring either now or at some future time to the user(s). The example quoted earlier of the construction of an office building where there is little or no research and development effort required, provides immediate benefits to the users. A project to send men to Mars would, in contrast, require an enormous research and development effort but would provide very limited immediate benefits. However, the reason for such a project would presumably be the potential for very large future benefits to the human race. Note that this definition does not take into account the effort expended in arriving at the PEI. The justification for this is that it is very difficult, if not impossible, to define benefits in absolute quantitative terms and consequently to compare benefits from one project to another (try comparing quantitatively the benefits of developing a new anti-malarial drug and laying an oil pipeline). Since we cannot define benefits in absolute terms, it follows that we cannot define a meaningful general relationship between effort expended to obtain a benefit and the magnitude of that benefit.

We can, however, define a quantitative relationship between benefits within the same project. Consider the case of construction of a road to link two cities, A and B. If access is provided at the end points only, this road provides the functionality for journeys from A to B and B to A (denoted AB and BA respectively). If we now link a third city C, the functionality (the possible journeys) is now AB, BA, AC, BC and CB. Note that we cannot generalize the relationship between the original functionality pair (AB and BA) and the effort to achieve that functionality, since the effort will depend upon external factors such as the distance between the two cities and the difficulty of the terrain, and could be totally different between any two solutions. However, we can certainly derive a quantitative relationship between the effort to link A and B with a road, and the subsequent effort to link C to that road.

This definition of Functionality also makes the assumption that the project is not managed in so inept a fashion that the end item is useless. For example, it is quite possible to spend millions on developing a computer program that would be of no use to anyone. Under the definition used in this alternative consideration, this end item would have zero functionality. However, as will be noted later, this eventuality will be covered under the general heading of risk.

COST

The quantity C in the IPL must be measured in terms of spend rate (d$/dt), otherwise the IPL produces unrealistic predictions. Consider the example of a project with a $10 million overall budget and duration of one year:

F = k x $10M x 1 year

According to the IPL using C in simple cost terms, we could achieve the same functionality if we were to reduce the budget by a factor of ten and increase the duration by the same amount:

F = k x $1M x 10 years

In the first case, we have a spend rate of $10M per year and in the second case we have a spend rate of $0.1M per year, or only 1% of the first case. (There is some argument over this analogy as some of us contend that the risk factor (k) would change considerably between the two cases.) This sounds like perpetual motion – we can achieve anything we want for effectively zero expenditures as long as we do it slowly enough.

Now considering the case where C is measured in terms of spend rate. The IPL in this case says that a spend rate of $10M per year for one year would give the same result as a spend rate of $1M per year for 10 years:

F = k x $10M/year x 1 year = k x $1M/year x 10 years

While this is not yet totally believable, it is more believable than the perpetual motion concept. In this case, the time factors cancel out and F is simply measured in terms of incurred cost, multiplied by whatever units are attached to k. In other words, the IPL now says that if you have a total budget C to spend, then the functionality F you can achieve with it should be independent, in theory, of the rate at which you spend that money. We could then write the IPL as:

$F = k * C$, where C represents the incurred costs.

In a real project or operation, functionality is not usually realized as smoothly as money is spent, but in discrete steps as work is completed. We can therefore say that functionality is a stairstep function of cost. This changes the IPL to now read as:

$F = g(C)$, where g represents a step function

The function g should also contain a time dependency, to account for the fact that in the real world there tends to be an optimum time scale for any project. If the time scale is compressed too much, the costs go up because there are simply too many people working concurrently, who therefore tend to get in each others' way. If the time scale is expanded too much, costs also go up because of such factors as requirements becoming out of date, necessitating rework in the affected areas.

TIME-DEPENDENT AND TIME-INDEPENDENT COSTS

Cost can be divided into time-dependent costs (such as salaries) and time-independent costs (such as equipment and raw material). Consequently, the cost input to any project or operation can be represented by $C_i + h(T)$, where C_i represents the time independent costs and h(T) is some function of time representing the accumulation of time-dependent expenditures. The relationship between cost and functionality could then be described as:

$F = g(Ci + h(T))$

RISK

The effect of risk on any project or operation can be summarized in terms of the cost/time inputs to the project or operation necessary to produce a given functionality, and the actual functionality that results. Risks which result in an increase in the cost and/or time necessary to produce the given functionality can be represented as two risk factors, r1 and r2, that affect time-independent costs and project duration respectively:

$F = g(Ci/1+r1) + h(T/(1+r2))$ where r1 and r2 are positive numbers

R1 represents increases in fixed costs, while r2 represents schedule delays. If both are zero, the equation reduces to the risk-free form. Risks can also result in a reduction in the delivered functionality for a given cost/time input. For example, if cost and time are fixed in an IT project but the project runs into difficulty, the PEI is likely to be delivered without some of the promised capabilities or with a large number of bugs. In either case, the delivered benefits to the users are less than they would have been had the project not run into difficulties. This type of risk can be represented by a third risk factor, r3, as shown:

$F = g(Ci/(1+r1) + h(T/(1+r2))/(1+r3)$

This form of the equation also covers the situation mentioned earlier where inept project management causes a large amount of effort to be expended with little or no tangible benefits as a result.

SIMPLIFICATION TO AN IDEAL FORM

The project law derived in the above paragraphs can be reduced to an ideal case by using a number of simplifying assumptions as follows:

1. The step function g includes a sufficiently large number of small increments that it can be replaced to a first approximation by a simple linear increase in functionality with expenditure (k).
2. The accumulation of costs with time represented by the time-dependent function h(T) is in fact a simple linear function, i.e., time-dependent costs are proportional to time (m).
3. The risk factors r1, r2 and r3 are numerically equal.

This allows development of the suggested alternate form of the IPL:

$F = k(C_i + mT)/(1+r)^2$

WHY DEVELOP AN IPL?

First of all, who are the users of this IPL? As project professionals we could get very sophisticated, but I believe our prime customers are those managers (sometimes characterized as those who "still believe in the tooth fairy") who believe that if they "push" we will "try harder" and our projects will accomplish more with less. If a layman wanted to increase the capacity of a gas tank without changing pressure, volume or temperature, any engineer would say "you can't do that because of the IGL" and it would carry the weight of scientific law, without any need to discuss exceptions or whether or not the gas was "ideal". We need something simple and clear enough that the manager and customer can see that, a) what they are asking for does, in fact, constitute an increase in scope/function/quality/capacity; and b) they can't have that increase without a corresponding increase in the "other factors, cost, schedule or risk", "BECAUSE OF THE IPL" which is universally accepted as having the validity of a scientific law.

This law, or some such function, is absolutely essential. Having such a Law that is proven valid will introduce (or rather tactfully enforce) some systematic thinking in the project and operational management set-up. Otherwise, no matter how elaborate a case is made for project or operational risks, the risk decisions will be left to the whims, fancies and moods of the key decision-making person(s) in the organization or project - we are working towards reducing this subjectivity in decision-making. We may determine that it is impossible to have a 100% scientific outlook on project or operational factors simply because human beings are involved, but if that is the case, we may be able to be at least 80% scientific about it.

CONCLUSIONS

When we studied the Ideal Gas Law in college, the degree of accuracy of the calculation was subject to the degree of accuracy in the inputs. This was usually expressed as 2, 3 or 4 significant digits. Unfortunately, the accuracy of what would be inputs into this "Ideal Project Law", namely the baseline cost and schedule estimates and the precision (or quantifying) of project specifications, are not measured in significant digits but in 10's of percentage points or, in the case of some IT projects, 100's of percentage points or even orders of magnitude. This may introduce enough noise into the equation as to make it impossible to a) prove it to an "unbelieve", or b) estimate an appropriate constant that would allow meaningful calculations to be done over ranges of inputs. On top of the inability to estimate accurately, there is also the tendency of management to misrepresent those estimates for political and/or business reasons.

CONTACTS:

David C. Hall
Secretary, Risk Management Research and Development Collaboration Executive Committee
Senior Risk Manager
SRS Risk Management Services
Phone: 301-641-1530 voice
E-Mail: david.hall@msd.srs.com
www.msd.srs.com

Ron Kohl
R. J. Kohl & Associates
3581 Mar Lu Ridge Road
Jefferson, Md 21755
Phone: 301-874-3509
E-Mail: rjkohl@prodigy.net

Roger Graves
Senior Consultant
Davion Systems Ltd
Phone: 613-258-6090
Fax: 613-258-7884
E-mail: rgraves@davion.com

HOMELAND SECURITY

Proceedings of IMECE'03
2003 ASME International Mechanical Engineering Congress
Washington, D.C., November 15–21, 2003

IMECE2003-42598

A COMMON APPROACH TO RISK ANALYSIS FOR HOMELAND SECURITY DECISION-MAKING

J. Robert (Bob) Sims
Becht Engineering Company
22 Church Street
Liberty Corner, NJ 07938

Bilal M. Ayyub
Center for Technology and Systems Management
University of Maryland
College Park, MD 20742

Kenneth R. Balkey
Westinghouse Electric Company
P.O. Box 355
Pittsburgh, PA 15230-0355

Richard E. (Gene) Feigel
Hartford Steam Boiler Inspection & Insurance Co.
One State St.
Hartford, CT 06102

ABSTRACT

For many years, risk analysis has been used extensively to inform decisions by government and industry. Different methodologies have been employed, resulting in differences in terminology and approach that make it difficult to compare the results of analyses in different fields. Dealing with terrorist threats requires prioritizing the allocation of resources across a broad spectrum of possible targets. Therefore, a common approach is needed to allow comparison of risks. This paper provides a brief outline of an approach that will allow the results of risk analyses performed using current methodologies to be expressed in a common format and terminology to facilitate resource allocation decisions. The results of a risk analysis should never be the only basis for decision-making, but a decision made without employing risk analysis will probably not result in the best outcome.

INTRODUCTION

Since the tragic attacks of September 11, 2001, governments and private industry have made significant efforts to assess the vulnerability of critical assets and systems, and to develop ways to mitigate the risks (NRC 2002, U.S. Department of Homeland Security 2003a and 2003b, and White House 2002). A common framework for performing risk analysis to assess the effectiveness of proposed critical asset protection initiatives is needed to support decisions by government and private industry. This framework should allow the risk in one area to be compared to risks in other areas to support resource allocation decisions.

A methodology for quick preliminary screening of proposals as well as for detailed quantitative analysis is needed. In all cases, the terminology used and the bases for characterizing risk should be the same to allow comparisons across all sectors of the economy.

Many well-established approaches to risk analysis are currently being used by a broad spectrum of public and private organizations. Any common approach must recognize and make maximum use of these existing methods (Ang and Tang 1984, ASME 2002, Ayyub and McCuen 2003, Ayyub 2003, and Balkey et al. 1991, 1992 and 1993).

TERMINOLOGY

In the authors' opinion, the most important step in developing a common approach is to establish common terminology. Clear communication among all homeland security stakeholders- and this includes essentially everyone- is essential to ensure that our limited resources are optimally allocated.

Key risk analysis terms, with some of the issues that should be addressed to develop common, broadly accepted definitions, are discussed below.

- **Consequences** are the outcomes of an undesired event. Consequences include both short and long term outcomes. Property damage/loss consequences and business interruption costs should be directly expressed in monetary units (e.g. dollars). Consequences involving potential loss of life, injury, environmental damage, changes in the quality of life, personal freedom and other intangibles should be quantified to the extent possible as discussed later in this paper.

- **Consequence Scenario** is a verbal description of one path through an event tree. The scenario should include both short and long term consequences.
- **Event Tree** is a graphical presentation of the paths that the consequences of a terrorist action can take. Typically it starts with the triggering event, although it can start before or after. Usually there are several alternative immediate results of the triggering event. Assuming that the event has occurred, the sum of the probabilities of each of the alternatives must equal 1. Each of the immediate results typically has one or more downstream branches. The tree continues to branch until the final consequence is reached. For minor damage, this may be a simple repair followed by returning the facility to service. For major damage or complete facility destruction, the effects may continue for many years. An event tree should reflect the full set of consequences.
- **Mitigation** is a term used to describe actions taken to reduce the probability, consequence or both.
- **Probability** is the chance, or odds, that an adverse consequence will result from an undesired event. The probability must be in a 0 to 1 range. The measure can be quantitative (e.g. 0.01), or qualitative (e.g. low, moderate, high). The probability must be associated with a specific consequence and a defined time frame. In most cases, the probability will change with time, so the period during which the probability is considered valid must be defined. The term likelihood is often used as a synonym for probability, particularly when the measure is qualitative. However, it may be less confusing to consistently use a single term (e.g. probability) to refer to the chance of an adverse event.
- **Qualitative Analysis** uses inputs based primarily on descriptive information. A multidisciplinary team uses judgment and experience to determine the probability and consequence of an adverse event. Formal expert elicitation methods should be used to assure transparency and correct weighting of a qualitative model.
- **Quantitative Analysis** uses numeric data to develop the probability and consequence of an adverse event. The probability of an event and the ensuing consequences are expressed as distributions or single (deterministic) values. The probabilities and consequences are combined using mathematical models.
- **Risk** is the combination of the probability and consequences of an adverse event. When the probability and consequences are expressed numerically, risk is the product of those values. Qualitative expressions of probability and consequence are often combined on a risk matrix (see Figure 1). However, as discussed later in this paper, it is usually desirable to express qualitatively derived probabilities and consequences numerically so risk values can be compared to the results of other analyses. The number of "bins" on each axis of the risk matrix in Figure 1 can be increased or decreased to meet the needs of individual studies. The definitions of probability, consequence and risk have focused only on adverse outcomes. In the financial community, risk is associated with both positive and negative outcomes, but for the purpose of clearly communicating the results of homeland security risk analyses, we believe it preferable to associate risk only with adverse outcomes.
- **Threat Analysis** describes methods used to determine, in the ideal case, the probability of a specific terrorist action at a specific location. As a practical matter, threat analysis will typically provide more general information, such as the probability of a general type of attack at a general class of facilities. However, this information can be used to develop appropriate approximations of the probability of an attack at a specific facility for use in facility specific risk analyses.
- **Triggering Event** is a specific event that sets in motion a series of subsequent events leading to adverse consequences. In a risk analysis, it is the "point of no return" beyond which there will be some level of adverse consequence. It is a useful concept because it provides a focal point, but it is not essential to define a triggering event in all cases. For a site-specific risk analysis, the explosion of a terrorist bomb is an example of a triggering event. The events leading to the explosion (e.g. planning of the attack, obtaining the necessary materials) are not associated with specific consequences for that facility, although they might have implications in other areas.
- **Vulnerability Analysis** studies the probability of a range of consequences (damage) to a specific facility or system as a function of the characteristics of the system.

METHODOLOGY

Although a detailed discussion of risk analysis methods is outside of the scope of this paper we will provide a general overview of key issues. Risk analysis methods can be qualitative or quantitative, or a combination. Most successful risk analyses incorporate both methods. The level of detail employed in different risk analysis methods may vary significantly, but in all cases, the objective should be to determine the probability of each credible triggering event, as well as the probability of each credible resultant consequence scenario. Risk analysis is a dynamic process, because the inputs and databases used are continuously being updated. Also, information developed in a later phase of the process may indicate the need to revisit previous work. Therefore, there is no single "step-by-step" approach. The following discussion of methodology is divided into three parts. The general steps common to most risk analyses are described first. Second, we indicate possible modifications to and application of those steps to qualitative analyses commonly used for preliminary screening. The concluding section offers a few comments on quantitative methods.

General Steps Common to most Risk Analyses.

The following steps are typically a part of most risk analyses done to support homeland security:

1) **Identify the threat.** This is primarily the responsibility of law enforcement and intelligence agencies, working with the private sector.
2) **Determine the probability of a specific triggering event at a specific facility or class of facilities.** This information must come primarily from law enforcement and intelligence agencies. However, much of this information is, very appropriately, classified. It will be necessary to develop a system that provides facility owners and their risk analysts access to the information that they need while keeping it out of the hands of terrorists.
3) **Determine the range of consequences that could credibly result from the postulated triggering event.** Even if the triggering event can be defined precisely, which is unlikely, a range of consequences can credibly result. For example, the damage resulting from an attack on an oil refinery may be limited to the refinery itself, or depending on the plant location and train of events following the triggering event, surrounding facilities and neighborhoods may be damaged. Long-term impacts may include shortages of fuels, chemicals and other products, as well as economic impacts resulting from demands to provide additional protection for other facilities. The most effective approach is usually to construct a series of consequence scenarios that model the events expected to follow the triggering event. These scenarios should include both short and long term impacts on the facility and, ultimately, impacts on the country as a whole. Typically, an event tree is constructed, with probabilities associated with each branch. The product of the probabilities at any point along a path is the overall probability of the event at that point.
4) **Combine the probability and consequences to get the risk.** When the probability and consequence are expressed numerically, risk is the product. Otherwise, a risk matrix (Figure 1) is typically used.
5) **Determine whether the risk is acceptable.** If the risk is sufficiently low, it may be acceptable without further mitigation. Facility owners and regulatory authorities are responsible for establishing risk acceptance criteria.
6) **If the risk is not acceptable, propose one or more mitigation steps and rework the analysis until an acceptable level of risk is achieved.** Benefit-cost analysis is often used to help determine the cost effectiveness of mitigation steps.

Qualitative Methods

Qualitative methods are typically used to prioritize threats and vulnerabilities to establish candidates for more detailed (i.e. quantitative) analysis. In some cases, qualitative methods may also be used for final resource allocation decisions. Qualitative methods usually categorize probabilities and consequences rather than calculating specific numerical values or distributions for each. An example of probability categories is shown in Table 1. Each category covers an order of magnitude range of probabilities. For example, Category C covers the range of probabilities from 0.001 (1 in 1,000) to 0.01 (1 in 100). To be able to compare risk analysis results, probabilities should be developed over a consistent time period. For example, the probability of an attack over the next month may be used in an analysis to develop a short-term mitigation strategy. In all cases, the time frame considered in the analysis should be stated, since the probability and the consequence will change with time. The analysis should be redone before the end of the time frame considered in the initial analysis, or if any of the information used as input to the analysis changes. The analysis of risk requires data that might not be available requiring the use of expert opinions (Ayyub 2002).

Probability Categories

Table 1 shows an example with six probability categories covering six orders of magnitude, from less than 0.00001 (1 in 100,000) to greater than 0.1 (1 in 10). This range is sufficient for most analyses. Although some would wish to have probabilities of catastrophic events lower than 0.00001 per year, it is usually not practical to establish lower probabilities using qualitative analysis approaches. Quantitative approaches should be used where it is necessary to deal with lower probabilities. Most qualitative analyses done today use fewer categories, in many cases as few as three. Typically, quantitative probabilities are not formally associated with the categories. For example, High, Moderate, Low categories are frequently used, but the meaning of each is left to the judgment of the team doing the analysis. To one individual, "high" may mean a 0.5 (1 in 2) probability, while to another, "high" may be 1 in 1000. This leads to inconsistent results and an inability to compare one analysis to another. If each level of a three level matrix covers two orders of magnitude, the "high" category becomes $\geq$ .01 (1 in 100), and the "low" category becomes $<$ 0.0001 (1 in 10,000). This may be adequate for some purposes. Alternatively, the range can be restricted to three of the six categories shown in Table 1. For example, if the "worst case" consequences represents minor property damage and minor injuries, it may be appropriate to use only the top three probability categories from Table 1, together with the bottom three categories from Tables 2 and 3 to construct a 3 x 3 matrix.

Recognizing that the probability determined in a qualitative risk analysis can vary with the approach used, the data and information available, the experience of the experts involved and many other factors, it is not always possible to determine the probability within a single order of magnitude range. However, this problem can be managed by using "what if" approaches to consider how the results will change if the probability is, in fact, higher or lower than the best estimate of probability determined in the analysis.

Consequence Categories

Tables 2 and 3 show examples of consequence categories. The categories in both tables attempt to place consequences into approximate order of magnitude ranges. The tables have been developed specifically to address homeland security concerns for large scale attacks. A similar approach applied to an individual industrial facility may use categories that are as much as three orders of magnitude lower in severity. For example, for a homeland security analysis, the top Safety, Health and Environmental (S/H/E) category (Category I) in Table 2 is assigned to incidents with fatalities numbering in the thousands. For an industrial facility, the top, or catastrophic, category may be assigned to a consequence involving a single fatality. Similarly, the economic consequences considered catastrophic by a small chemical facility may be as low as $10 million rather than the $10 billion shown in Table 3. Broad comparison of risk analysis results can be done even if different tables are used for different applications if the verbal descriptions and dollar ranges are clearly identified.

Qualitative methods typically provide estimates of consequence within an order-of-magnitude. In some cases, it is possible to determine the consequence with more precision than an order of magnitude range. For example, it may be possible to do calculations of consequence costs with an accuracy of ±25% with minimum effort, while still using qualitative, order of magnitude judgments for probability

Quantitative Methods

Obviously, the heart of risk analysis is to determine the appropriate probability and consequence for each threat. Analysts using quantitative methods attempt to calculate probability and consequence using historical failure and incident databases, combined with a formal expert elicitation process to apply the data appropriately to the specific circumstances considered in the analysis. Mathematical models are often used to develop detailed scenarios with associated probabilities. In many cases, the resulting probability and consequence are expressed as single numerical values, although distributions are often used. Simulation methods and what-if scenarios are often employed to develop probability ranges and calculate uncertainties for complex models.

For homeland security applications, game theory may be a useful approach to help manage the complexities associated with defining threats as well as the long-term impacts of terrorist actions.

Determining Risk

Whether qualitative or quantitative methods are used to determine probability and consequence, risk should be expressed in a way that is easy to understand, and that can readily be compared to risks calculated for other sectors using different methods. Ideally, risk should be expressed as a single number. Although there are several difficulties with this approach, developing a single risk number along with qualifying information may have more advantages than disadvantages when used to inform homeland security decision-making, and for facilitating communication among key stakeholders, including the public.

By definition, a quantitative analysis produces numerical values for probability and economic consequence. Therefore, the economic risk can be calculated as the product of those values, typically expressed in currency units (e.g. dollars). As discussed earlier, it is necessary to associate numerical ranges with both probability and consequence for a qualitative analysis. Therefore, an economic risk number can also be calculated from the results of a qualitative analysis using mid point values for each range. Since the order of magnitude ranges plot as a straight line on a log scale, a "log mean" can be used to develop a single number from the range. This approach also facilitates comparing risks, since the scaling produces lines of 'constant risk.'

Although the approach described above is straightforward when the consequences are purely economic, it is necessary to find a way to quantify the non-economic aspects of consequences such as loss of life, injury, short and long term damage to the environment, restrictions to liberties and other consequences that are difficult to quantify. For example, it will be important to find a way to quantify quality of life issues, particularly the value of freedoms that we have traditionally enjoyed.

There have been debates for many years about whether it is appropriate to assign a value to human life and limb. The issue is complex, and will not be discussed in detail in this paper. However, when deciding among many potential courses of action to protect our infrastructure and our homeland itself, it is necessary to weigh the potential for fatalities against the potential for loss of freedom and our way of life. If minimizing fatalities is our only goal, we could build fortresses around most portions of our infrastructure, and provide armies of armed guards, resulting in a significant reduction in our quality of life, with only a small reduction in the probability of fatality for the average citizen.

It is more sensible to develop a basis for assigning values to life, limb and even certain freedoms in order to provide a basis for calculating a risk in one area that can be compared to the risks in others. Each mitigation action considered will have pros and cons. Assigning a value to each consequence of the proposed action will allow our resources to be allocated in the most effective way

Assigning values to "difficult to quantify" consequences is a public policy issue, which must be addressed through the political process. The approach that has arguably been the most successful is to conduct well-structured surveys focusing on the public's willingness to pay for an array of realistic (not just conceivable) alternatives. This approach has a long history, and promotes public acceptance by developing a broad consensus. However, even if all of the consequences can be expressed in dollars, allowing a single risk value to be determined, risk analysis results should include a discussion of the components of the risk, including a description of all of the “difficult to quantify” consequences and the probability that each will occur. This information will give decision makers a better basis for making the tough choices that they will inevitably face.

Risk Acceptance

Assuming that risk can be quantified in dollars, it is usually appropriate to evaluate potential mitigation steps using a benefit-cost analysis. However, when all cost-effective mitigation has been applied, there will still be some residual risk. Since risk analysis, and particularly benefit-cost analysis, should not be the only basis for a decision, it is necessary at some point to decide whether the residual risk is acceptable. When there is strong competition for resources, as there will be in the homeland security area for many years to come, allocating resources to the area where they will be the most effective will still leave some significant risks unmitigated. Allocating more resources (e.g. increasing the homeland security budget) is an option that can be evaluated by extending risk analysis to evaluate the change in risks in other areas that would result from the reallocation.

The risk matrix in Figure 1 shows an example of risk acceptance, by designating combinations of probability and consequence as either “High”, “Medium” or “Low”. If a risk matrix is used to help determine the acceptability of risk, it is important that it contain a consequence category that is low enough that even if the probability is close to 1.0, the risk will be acceptable. Also, a probability category must exist that, when multiplied by the highest consequence to be considered in the analysis, will result in an acceptable risk. These are the upper left and lower right hand boxes on Figure 1. A straight diagonal line drawn between these boxes is an iso-risk (constant risk) line. The boxes along this line should represent an acceptable risk if the end points do. Diagonal lines to the left and below the iso-risk line represent a lower risk and should also be acceptable. The designations on these boxes should be used as examples only. Actual criteria for risk acceptance should be a matter of public policy.

ACKNOWLEDGMENTS

The Authors would like to acknowledge the contributions and support of the Risk Analysis Team of the ASME Critical Assets Protection Initiative.

REFERENCES

Ang, A. H-S., Tang, W. H., 1990. “Probability Concepts in Engineering Planning and Design,” Vol. II Decision, Risk, and Reliability, John Wiley & Sons, NY.

ASME, 2002, “Standard For Probabilistic Risk Assessment For Nuclear Power Plant Applications,” ASME RA-S-2002, New York, NY.

Ayyub, B. M., and McCuen, R. H., 2003. Probability, Statistics and Reliability for Engineers and Scientists, Chapman-Hall/CRC Press, FL.

Ayyub, B.M., 2001, Elicitation of Expert Opinions for Uncertainty and Risks, CRC Press, FL.

Ayyub, B.M., 2003, Risk Analysis in Engineering and Economics, Chapman & Hall/CRC Press. FL.

Balkey, K.R., Ayyub, B.M., Vic Chapman, O.J., Gore, B.F., Harris, D.O., Karydas, D., Simonen, F.A., and Smith, H., 1991. "Risk-Based Inspection - Development of Guidelines, Volume 1 - General Document," CRTD - Vol. 20-1, The American Society of Mechanical Engineers, also NUREG/GR - 0005 Vol. 1 by Nuclear Regulatory Commission, Washington, DC.

Balkey, K.R., Ayyub, B.M., and Chapman, O.J.V., 1992. "International Risk-Based Inspection Technology Development," Winter Annual Meeting of the American Society of Mechanical Engineers, Anaheim, CA, November 8-13, 1992, 92-WA/SAF-6.

Balkey, K.R., Ayyub, B.M., Vic Chapman, O.J., Gore, B.F., Harris, D.O., Phillips, J.H., Krishnan, F.A., Karydas, D., Simonen, F.A., Smith, H., and Vo, T.V., 1993. "Risk-Based Inspection - Development of Guidelines, Volume 2 - Part 1 Light Water Reactor Nuclear Power Plant Components," CRTD - Vol. 20-2, The American Society of Mechanical Engineers, also NUREG/GR - 0005 Vol. 2 - Part 1 by Nuclear Regulatory Commission, Washington, DC.

Bush, S. H. and Sims, J. R. 1999. “A Comparison of Criteria for Acceptance of Risk”, PVRC Final Draft Report on Project 99-IP-01. Sponsored by the ASME Foundation and ASME Codes and Standards.

National Research Council’s (NRC), 2002. Making the Nation Safer, The Role of Science and Technology in Countering Terrorism, Washington, DC.

U.S. Department of Homeland Security, 2003a. The National Strategy for The Physical Protection of Critical Infrastructures and Key Assets.

U.S. Department of Homeland Security, 2003b. The National Strategy to Secure Cyberspace.

White House, Office of the President of the United States, 2002. National Strategy for Homeland Security.

Figure 1
Example of a Risk matrix

Probability Category	VI	V	IV	III	II	I
A	L	M	M	H	H	H
B	L	L	M	M	H	H
C	L	L	L	M	M	H
D	L	L	L	L	M	M
E	L	L	L	L	L	M
F	L	L	L	L	L	L
	Consequence Category					

H = High Risk
M = Moderate Risk
L = Low Risk

Table 1
Example of Probability Categories

Category	Description	Annual Probability Range
A	Very Likely	≥ 0.1 (1 in 10)
B	Possible	≥ 0.01 (1 in 100) but < 0.1
C	Unlikely	≥ 0.001 (1 in 1,000) but < 0.01
D	Highly Unlikely	≥ 0.0001 (1 in 10,000) but < 0.001
E	Not Credible	≥ 0.00001 (1 in 100,000) but < 0.0001
F	Practically Impossible	< 0.00001 (1 in 100,000)

Table 2
Example of S/H/E Consequence Categories

Category	Description	Examples
I	Catastrophic	≥ 1,000 fatalities; major long term environmental impact
II	Major	≥ 100 but < 1,000 fatalities; major short term environmental impact
III	Serious	≥ 10 but < 100 fatalities; significant environmental impact
IV	Significant	< 10 fatalities; short term environmental impact
V	Minor	Injuries only; minimal environmental impact
VI	None	No significant consequence

Table 3
Example of Economic Consequence Categories

Category	Description	Examples
I	Catastrophic	≥ $10,000,000,000
II	Major	≥ $1,000,000,000 but < $10,000,000,000
III	Serious	≥ $100,000,000 but < $1,000,000,000
IV	Significant	≥ $10,000,000 but < $100,000,000
V	Minor	≥ $1,000,000 but < $10,000,000
VI	None	< $1,000,000

Proceedings of IMECE'03
2003 ASME International Mechanical Engineering Congress
Washington, D.C., November 15–21, 2003

IMECE2003-42612

MANAGING & PROTECTING INFRASTRUCTURE ASSETS

Daniel C. Rees, SCIENTECH, Inc.

Kenneth I. Rubin, PhD., PA Consulting

KEYWORDS

Infrastructure Security, Asset Based, Vulnerability Assessment, Risk Assessment, Risk / Cost Analysis, Security, Homeland Security

ABSTRACT

The events of September 11th focused renewed attention on protection of our nation's critical infrastructure. Utilities across the nation have an increased awareness of risks and are recognizing the potential vulnerability of their physical assets, and also the assets embodied in their employees, their knowledge base, their information technology and their customers. Utilities must now grapple with the possibility that their infrastructure assets may be targets of direct physical threats — or serve as conduits for indirect physical threats.

As the concern for protecting our nation's infrastructure intensifies, each utility is being asked to reassess its ability to provide safe and reliable services to customers and communities as a whole. However, improvements to protection of utility assets must be performed with constraints of limited capital and operating budgets.

Security threats from terrorist and related events are relatively new to the utility industry, so standard industry-wide protocols are just now being developed. Serious security practices have evolved in some discrete areas, such as high-risk government buildings, nuclear power plants, and airline terminals. Utility infrastructure physical assets are typically dispersed, so, standard approaches to security (developed for enterprises with highly centralized assets, such as nuclear weapons production facilities) are difficult to apply. Managers must then face a balancing act between demands for security and the resources needed to enact and finance those actions.

This paper describes the Vulnerability Self Assessment (VSAT™) methodology and software that provides a structured, cost-effective approach for utilities to assess their vulnerabilities and to establish a risk-based methodology for making necessary changes. The VSAT™ methodology groups utility assets into the classes of People (utility staff), Physical Plant, Knowledge Base, Information Technology Platform, and Customers.

The methodology and software are flexible, customizable, and user friendly. VSAT™ software is equally applicable to deliberately caused or natural disasters. In addition to a library of prototypical assets, included in the software application are threat and countermeasure libraries. As users proceed through self-assessments, VSAT™ automatically documents the analysis process during each step. VSAT™ helps users identify critical asset(s) and potential single points of failure (SPFs). The VSAT™ process culminates in a series of risk-reduction-cost reports that presents findings in clear and concise ways. This is important, because the goal is business continuity and, at the end of the day, VSAT™ provides solutions that enable utilities to mitigate risks of business interruptions at least cost.

INTRODUCTION

The events of September 11th focused renewed attention on protection of our nation's critical infrastructure. Utilities across the nation have an increased awareness of risks and are recognizing the potential vulnerability of their physical assets, and also of the assets embodied in their employees, their knowledge base, their information technology and their customers. Utilities must now grapple with the possibility that their infrastructure assets may be targets of direct physical threats — or serve as conduits for indirect physical threats.

Utilities must assure their governing boards, customers, and the public that they are crisis-resilient; have a thorough understanding of their vulnerabilities and risks of business failures associated with these vulnerabilities; that they can respond to crises with practiced, effective plans and a well-trained staff; and that they can recover from a crisis with minimal interruption of services to their customers.

Most utilities are finding that this is easier said than done. There are several complications:

- The water and wastewater utility industry (and other parts of our infrastructure) have only recently begun to consider security from terrorists and related events, and consequently, industry standard protocols are not yet well developed;
- Traditional support organizations, with which utilities are used to working on such global issues, do not have genuine security offerings or have only recently begun to develop them as the entire industry gears up for this line of work;
- Because of the highly dispersed nature of water and wastewater assets, standard approaches to security developed for enterprises with highly centralized assets don't fit very well;
- Organizations must respond to the public and political demand for security, which given the last point, would be inordinately expensive if unbounded by practical budget considerations and developed in the absence of systematic, risk-based analyses; and
- Utilities have limited budgets for making security improvements and thus need a technical approach that supports their decision making to assure they have achieved the greatest improvement with a finite budget. A risk-based approach is needed but most utility personnel are not trained in risk assessment techniques.

The VSAT™ (Vulnerability Self-Assessment Tool) methodology and security planning software, which is rapidly becoming an industry standard, has been developed to address these complications. VSAT™ is provided free of charge to all public and private water and wastewater utilities and to associated local, state, and federal organizations.

The methodology and software are flexible, customizable, and user friendly. VSAT™ is equally applicable to deliberately caused or natural disasters. In addition to a library of prototypical assets, included in the software application are threat and countermeasure libraries. VSAT™ prompts users to document assumptions and decisions at each step. It helps user s identify critical asset(s) and single points of failures (SPF). Where risks of failure are unacceptable, VSAT™ leads users through optional security improvement plans balancing risk reduction against cost to improve. VSAT™ guides users through the following 11-element process:

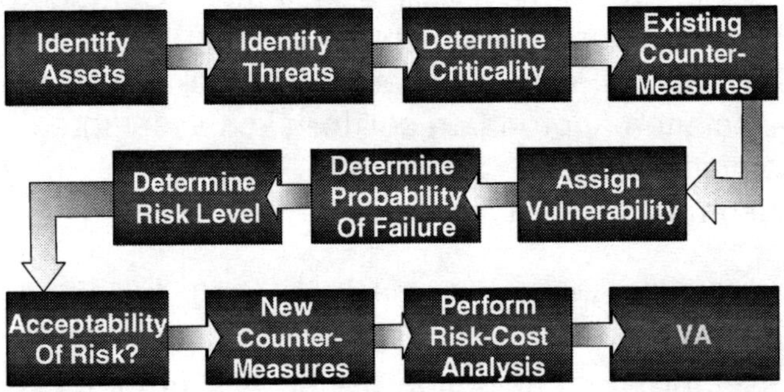

This process compels a thorough evaluation of all the potential risks associated with identified threats (chemical and biological contamination scenarios will be among the threats considered) against all enterprise assets, with the last block representing completion of the Vulnerability Assessment (VA). Where risks are unacceptable, VSAT™ provides a decision process to select new counter-measures that reduce risk in priority order according to risk reduction and cost criteria.

VSAT™ TECHNICAL BASIS

VSAT™ is based on qualitative risk assessment techniques and supports baseline vulnerability analysis as well as security improvement analysis that helps managers choose new countermeasures to reduce vulnerability. Finally, and perhaps most importantly, VSAT™ assists in the management decision process by enabling managers to trade-off cost for risk.

VSAT™ was specifically designed to assist utilities, many of whom have no background or experience in risk assessment, to help them address all the tasks necessary to complete a Vulnerability Assessment (VA) for Wastewater Utilities as well as meeting the six basic elements that EPA requires in a Water System VA (which may become a requirement for Wastewater Systems and other industry sectors in the future). In addition, VSAT™ imposes the rigor necessary to ensure that vulnerability assessments are comprehensive.

The key attributes of the VSAT™ methodology are described below.

VSAT™ is Asset Based and addresses widely dispersed and inter-related assets.

Typically, security risk assessments are performed on facilities that are well bounded and located in a finite geographical area, e.g., D.O.E. Weapons Complex and commercial nuclear power plants. In the case of water and wastewater utilities, this not the case as they have multiple locations of source water that may be located many miles from their treatment plants, have a large number of dispersed pumping stations, as well as multiple customers served by a complicated distribution network that do not yet have extensive security measures employed. Thus, traditional "barrier analysis" methods do not support the needs of many utilities.

In addition, the classic security risk assessment approach is typically applied to the physical assets only. While physical assets are certainly important, they are only a portion of the assets that a utility must protect in order to stay in operation. Thus, VSAT™ uniquely groups utility assets into five distinct classes needed to support utility operations:

- People (utility staff),
- Physical Plant,
- Knowledge Base,
- Information Technology Platform, and
- Customers

VSAT™ comes with standard asset templates for water supply and wastewater systems, that can also be adapted for other Sectors (e.g., Electric Utility, Gas Utility, Telecommunications Sectors). These are typically not available in other methods and these asset templates allow for easy initialization of the analysis.

Qualitative risk based method that does not require formal training or education in risk assessment.

VSAT™ was developed with the premise that the users would not need to have formal training or education in risk assessment. This is important because on the one hand, most of the utilities have little background in use of risk assessment techniques and on the other hand, they realize a need to have a risk-based approach to minimizing their vulnerabilities. VSAT™ uses well documented qualitative risk assessment techniques to support a thorough evaluation of risks associated with facilities from threats (man-made external, man-made internal, or natural disasters) that utilities need to protect against. What is unique in the VSAT™ Methodology is the approach that makes potentially complicated risk assessment into a straightforward analysis. The core of VSAT™ relies upon the standard risk equation of Risk as the product of consequence times likelihood:

$$\text{Risk} = \text{Consequences} * \text{Likelihood}$$

The consequences (termed "criticality" in VSAT™) are determined by the use of either a decision tree approach or by using "Criticality Evaluation Tables". An example of a criticality table is shown below:

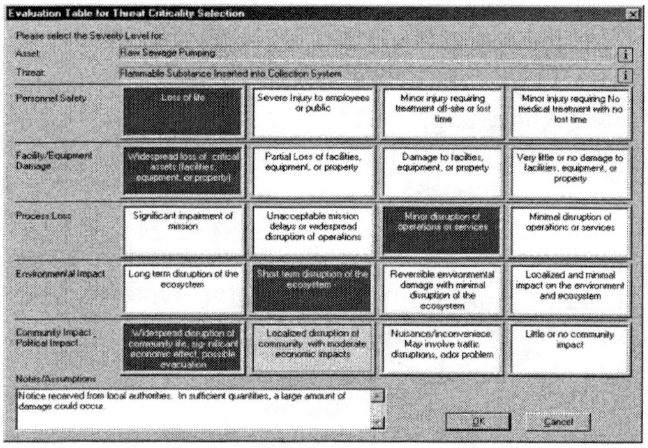

Note that there is a separate criticality table for each asset type described above, and that each criticality table has up to five different attributes for criticality (note that the user, if desired, can modify the attributes and their definitions). In the criticality table example shown above for physical assets, the five attributes included are Personnel Safety, Facility/Equipment Damage, Process Loss, Environmental Impact, and Community-Political Impact. The utility analyst selects the appropriate cell for each asset/threat combination and documents in the Notes / Assumptions area the basis for the decision.

The likelihood aspect of risk in VSAT is termed "Vulnerability" and is determined for each asset/threat combination by the use of either a decision tree approach or by the use of "Vulnerability Evaluation Tables". Vulnerability in VSAT is defined as a function of probability of failure times and probability of occurrence

$$\text{Vulnerability} \propto P_{(\text{failure})} * P_{(\text{occurrence})}$$

In this evaluation, the utility analyst selects both a probability of failure of the asset/threat combination (given existing countermeasures), as well as the probability of occurrence of the event. This joint probability distribution then provides the user with the vulnerability of the asset/threat combination. As with the criticality evaluation, the VSAT software provides a means of documenting the basis for the decision.

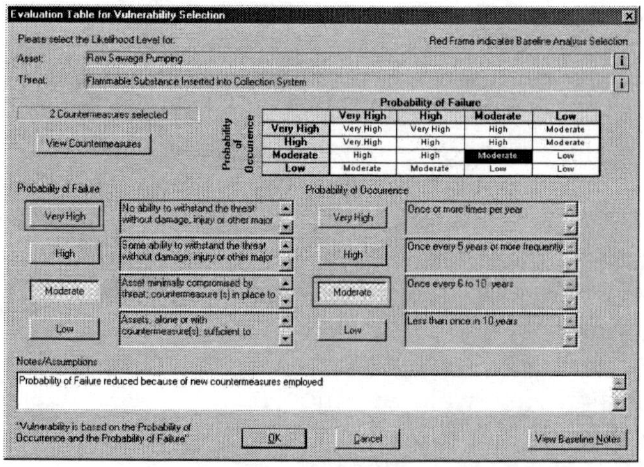

Based on an assessment of criticality and vulnerability, each asset is assigned into one of 16 risk categories using the VSAT™ risk matrix.

The standard VSAT™ risk matrix of is a "four by four" with four levels of criticality and four levels of vulnerability. Note that the user can define the risk matrix from a "three by three" up to a "five by five" depending on the desired level of granularity of results. VSAT™'s risk matrix displays the number of asset-threat combinations in each cell as a convenient way to track the overall risk profile of the facility. The risk matrix from the VSAT™ software is shown in the figure below. Note that all asset categories can be displayed as well as only an individual asset category. The results for individual sectors or all sectors can also be selected.

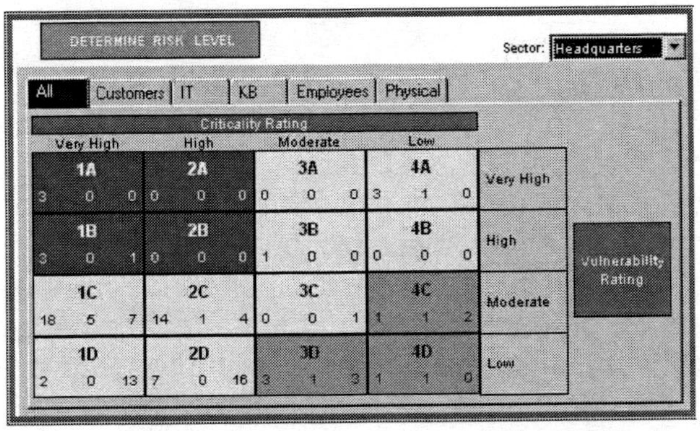

At the end of this analysis for each asset/threat combination, risk levels will have been assigned to all selected assets. This results in a current or "baseline risk profile", which defines where the high, medium and low asset/threat combinations are in the utility operation. This in itself is very useful information for the utility, and is typical of the ending point in most qualitative risk assessment approaches. However, it raises the issue as to what can be done to reduce the risk. In VSAT, once the baseline level of risk is determined, the next step is to determine if risk can be reduced by lowering criticality (by adding redundancy) and/or reducing vulnerability (by adding physical protection, intrusion detection devises, new policies or procedures, etc.) – in general, both are termed countermeasures.

Potential risk reduction is determined by re-assessing the criticality and vulnerability of each asset/threat combination to determine whether, and the extent to which, new countermeasures would reduce risks for each of the high-and medium-risk asset/threat combinations, VSAT™ includes a "countermeasure library" that describes potential countermeasures, their technology basis, the pluses and minuses, as well as capital and operating expenses. If multiple countermeasures are applied to a single asset/threat combination the relative benefit of each countermeasure to the criticality or vulnerability reduction is specified.

Application of redundancy and/or countermeasure may change the risk level (which matrix cell) of the analyzed asset / threat combination. This is obviously a goal of performing an improvement analysis. For each movement between risk matrix cells, VSAT™ calculates the risk reduction units (RRUs) resulting from the criticality and/or vulnerability reduction. The figure below shows the application of countermeasures that reduce one asset threat combination from matrix cell 1A to 4A and another that is reduced from 3A to 4B.

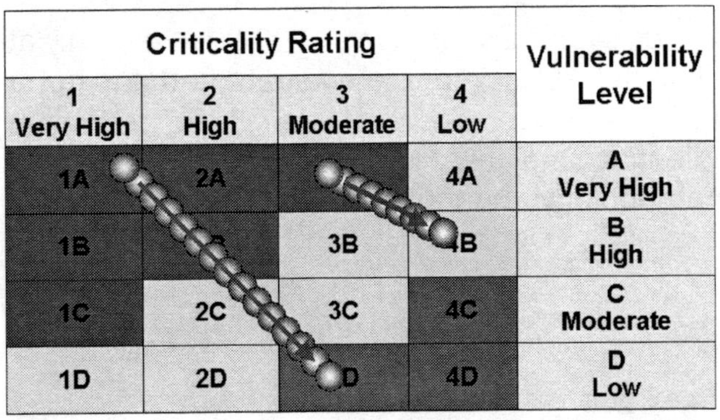

Each cell in the matrix is assigned a risk value, ranging from 100 (very high criticality and very high vulnerability) down to 35 (low criticality and low vulnerability). The values reflect change from risk level color (e.g., red to yellow), the horizontal movement and vertical movement. The lowest value of 35 reflects that there is always some risk level. Thus when an asset/threat combination changes it's location from one risk cell to another, as shown above, there is a change in both the criticality and vulnerability due to the set of countermeasures that have been applied. This delineation of individual countermeasure benefits is also unique to the VSAT approach and is possible because of the application of the RRU concept. RRUs result from determination of the vector values (criticality and/or vulnerability) and calculating the change to both criticality and vulnerability for each countermeasure. This then forms the basis for countermeasure selection based upon it's RRU potential as well as cost, which is shown in VSAT™ as $/RRU.

VSAT™ Cost Risk Evaluation Capabilities

VSAT™ is then used to select the most cost-effective countermeasure profile. As each countermeasure is reviewed, it is "checked" in VSAT™ dialog box and the cost is added to VSAT™'s Capital Costs window and the "Risk Reduction Units" associated with each countermeasure are accumulated. At the same time, the overall risk profile is updated to show the number of asset/threat combinations that are in the High, Medium, and Low risk categories.

A utility may have 1000 asset/threat combinations or more to evaluate and it is not a trivial exercise to evaluate their reduction potential and cost risk attributes without a methodology and tool such as VSAT™.

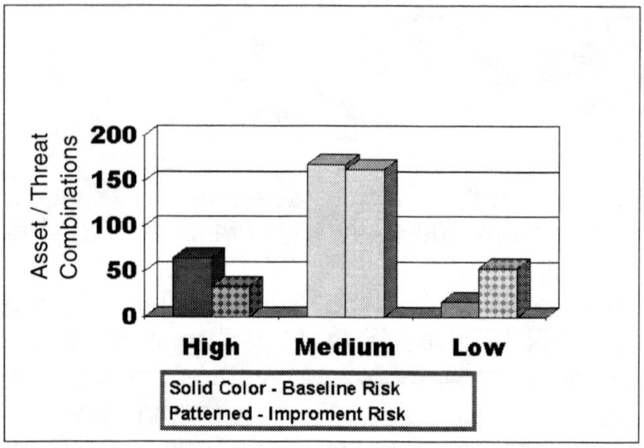

VSAT™ compiles cost and risk profiles for "Scenarios" that describe various combinations of countermeasures. Note that the cost/risk analysis can be based on capital costs, annualized costs, total RRUS, those with the best dollar per RRU ($/RRU), or other specified measures. An example of this scenario analysis is shown below.

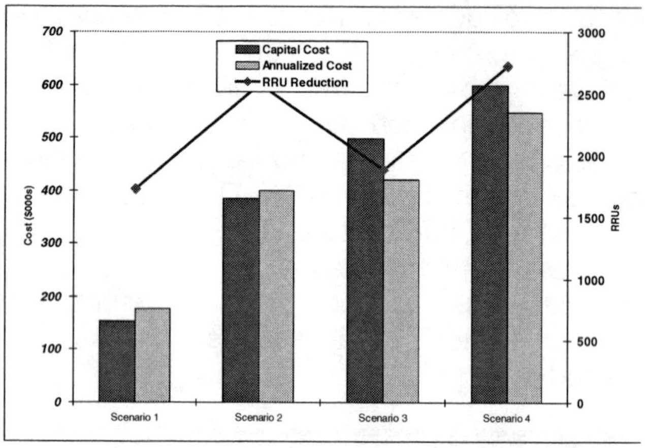

Prepare Plan of Action and Security Report

The user then develops an Assessment Report and Security Plan using the scenarios developed to determine the improvements and information about available budgets. VSAT™ has more than 25 standard reports that can be used to prepare analysis, brief management, and communicate among the security team or with outside parties, as appropriate.

SUMMARY OF VSAT™ METHODOLOGY

The VSAT™ methodology and software offers several distinct advantages over traditional qualitative risk paper and pencil approaches:

- An asset-based approach that is scalable from very small to very large facilities,
- Accommodates the full range of utility assets, not just physical assets,
- Has standardized asset templates to allow utilities to quickly begin their analysis,
- Has built in "libraries" of threats and countermeasures,
- Provides a means for risk assessment by personnel who do not have an extensive risk assessment background,
- Provides a method to evaluate potential improvements analysis on a quantitative basis,

- Is easily updated as assets, threats, or countermeasures change; and
- Documents all decisions and assumptions, enabling managers to re-create their thought processes leading to decisions.

SUMMARY AND CONCLUSIONS

While VSAT™ was originally developed for the water and wastewater industry, it has been adapted for use in public buildings, airports, electric utilities, and gas utilities. VSAT's approach to risk assessment and its functionality is widely applicable to any enterprise with substantial physical, IT, and /or employee assets, About 5,000 copies of VSAT™ software are in use today, available free of charge to public and private water and wastewater utilities and associated local, state and federal agencies and related organizations.

VSAT™ development began in January, 2002 and has had two major updates. As part of the development of the software, provisions were made to update users with new information on threats, countermeasures and on-line references, as well as to anonymously share information on threats and countermeasures employed. Secure, web-based information sharing will help new users in the water and wastewater industry complete their assessments with the latest in industry thinking.

Acknowledgments

VSAT™ was developed by PA Consulting and SCIENTECH, Inc. under contract to the Association of Metropolitan Sewerage Agencies (AMSA), with funding in part from the U.S. Environmental Protection Agency.

References

"Asset Based Vulnerability Checklist for Wastewater Utilities", prepared for the Association of Metropolitan Sewerage Agencies, December, 2001.

Proceedings of IMECE'03
2003 ASME International Mechanical Engineering Congress
Washington, D.C., November 15–21, 2003

IMECE2003-42695

A Cyber Security Self-Assessment Method for Critical Infrastructure Protection

Clifford S. Glantz
Pacific Northwest National Laboratory

Robert B. Bass
Pacific Northwest National Laboratory

James R. Cash
Pacific Northwest National Laboratory

Garill A. Coles
Pacific Northwest National Laboratory

Andrea J. Currie
Pacific Northwest National Laboratory

David J. Gower
Battelle Columbus Operations

Jeffrey J. Heilman
Pacific Northwest National Laboratory

Matthew D. Lammers
Battelle Columbus Operations

Jeffrey L. Thomas
Battelle Columbus Operations

Keywords: cyber security; risk assessment; risk management; critical infrastructure

ABSTRACT

Pacific Northwest National Laboratory is working with federal- and private-sector representatives to develop a cyber security self-assessment method (the Method) for use in the electric power industry. The Method will assist facility owners in identifying and characterizing cyber vulnerabilities, potential adverse consequences, cyber security risk levels, and cost-effective protection and mitigation measures.

The Method supports a comprehensive, expeditious assessment that can be integrated efficiently into normal business practices. Personnel in information systems, plant operations, and security all would contribute to the assessment.

The risk assessment stage of the Method evaluates both external and internal threats. In the risk management stage, potential protection and mitigation measures are evaluated based on risk reduction; the costs of purchasing, installing, operating, and maintaining new protection and mitigation measures; and changes in the cost of doing business that may be associated with the implementation of new measures.

INTRODUCTION

Digital systems are being used to operate key processes, provide monitoring information, plan future activities, design equipment, perform testing, maintain quality assurance, track maintenance, schedule work, and for many other functions. Many digital assets are linked to computer networks that may extend across a facility and, in many cases, may be connected to large and diverse corporate networks that have thousands or tens of thousands of users. At the same time as the use and connectivity of digital systems is expanding, the cyber threat environment is growing more hostile. New domestic and international adversaries are emerging, and new tools are appearing that can be used by adversaries to exploit vulnerable systems. As a result of these developments, cyber security risks are increasing, and there is a growing need to address these risks in a systematic manner [1].

In recognition of the growing importance of cyber security, the U.S. Department of Energy (DOE), U.S. Nuclear Regulatory Commission (NRC), and private industry are working to implement new cyber security self-assessment methods. In this paper, we describe one such method being developed by Pacific Northwest National Laboratory (PNNL) with input from government agencies and private industry to assist power plants in assessing and managing cyber security risks. The Method is designed for use by an in-house cyber security assessment team that would work with the owners and administrators of digital systems and facility decision-makers. The Method provides a structured approach that will assist a facility's assessment team in examining its critical digital assets, assessing the consequences to the plant of a successful exploitation of each critical digital asset, evaluating the vulnerabilities associated with these assets, evaluating the effectiveness of existing protection and mitigation measures, estimating risks, and evaluating cost-effectiveness of new potential protection and mitigation measures. The Method addresses concerns about safety, physical security, emergency preparedness, continuity of power, and plant efficiency. The Method builds upon the work of Alberts and Dorofee [2,3], Stoneburner et al. [4], and Swanson [5].

NOMENCLATURE

Critical Digital Asset: A digital device or system that plays a role in the operation or maintenance of a and can impact the proper functioning of that critical system. Critical systems include safety, plant security, emergency preparedness, and operational control systems.

Digital Asset: A digital device or system that is directly or indirectly connected to a critical digital asset.

Digital Device: A component whose operational function is dependent on the programmed execution of an internal, electronic, digital processor.

Digital System: One or more digital devices combined to perform programmed functions as a unit.

Risk: "A combination of the probability of an adverse event and the nature and severity of the event" [6]. In the Method, risk is defined as the combination of the susceptibility of a critical digital asset to cyber exploitation and consequences to the plant from that exploitation.

Susceptibility: A relative measure of the cumulative vulnerability to exploitation for a critical digital asset. It is an indication of the security of a critical digital asset based on the identified vulnerabilities, the severity of the vulnerabilities, and the effectiveness of existing protection and mitigation measures to reduce or eliminate the vulnerabilities.

Vulnerability: A weakness in the physical or electronic configuration of a digital asset that could allow an action that compromises the cyber security of the asset.

OVERVIEW OF THE METHOD

The Method consists of the major components depicted in Fig. 1. The first is to organize an in-house, multidisciplinary team to conduct the cyber security self-assessment. Three broad technical areas should be adequately represented on the assessment team:

- *information technology*, covering the areas of cyber security, software development and application, computer system administration, and computer networking

- *power plant operations and design*, including knowledge of overall facility operations, plant technical specifications, and plant dynamics – Staff representing this technical area must be able to trace the impact of a vulnerability or series of vulnerabilities in a critical digital asset (or connected digital asset) outward through plant subsystems and systems

- *physical and operational security*, including in-depth knowledge of the plant's physical and operational security programs.

Specialized *in-depth cyber security skills* may be required to perform the electronic validation testing and optional scanning activities. It is acceptable to use corporate-level cyber security staff or an independent cyber security organization to conduct electronic testing. A self-assessment team might consist of a

- cyber security specialist
- plant systems engineer
- plant operator
- computer systems expert (familiar with both the plant's engineering and business computer systems)
- computer networking specialist
- plant security specialist.

Formation of the assessment team is followed by the six stage process indicated in Fig. 1. Stage 1 of the Method is focused on gathering information on plant-wide cyber security policies, procedures, and practices. This information includes plant-wide requirements for the physical security of cyber resources (e.g., locks, alarms, surveillance) and digital systems (e.g., requirements for the configuration of firewalls, restrictions on the use of wireless communications). It also includes organizational practices (e.g., cyber security training requirements for system administrators, restrictions on access to computer systems, requirements for testing computer systems).

The second stage in the Method focuses on identifying the critical digital assets to be assessed. Critical digital assets are defined as those digital devices or computer systems that have an important function in plant safety, security, emergency preparedness, and operations. Assets that directly control critical plant functions and those that provide data used to design, operate, calibrate, support, or maintain critical plant functions are included in this definition. Interconnected digital devices and computer systems that perform a critical function often are classified as a single critical digital asset.

Stage 2 also includes a characterization of the consequences to the plant if an asset were exploited. This characterization involves determining the consequences to the critical functions performed by the critical digital asset and the resulting impacts to the plant subsystems and systems supported by the asset. Consequences are evaluated separately for impacts to the

- confidentiality of information
- availability of the critical digital asset
- integrity of the critical digital asset's function or data.

Stages 1 and 2 are largely independent of each other. For this reason, the order in which they are conducted can be reversed, or the assessment team may choose to conduct work on both these stages in parallel.

Stage 3 concentrates on gathering information on the selected critical digital assets through a tabletop review and walk-down inspection of those assets. The objective of this stage is to characterize the connectivity, vulnerabilities, and cyber security countermeasures associated with each critical digital asset and any directly or indirectly connected digital assets. This stage is the most important and time-consuming component of the Method.

Stage 3 begins with identifying each asset's owner(s) and system administrator(s). It continues with acquiring up-to-date information on the asset's function, components, locations, connectivity, vulnerabilities, and existing protection and mitigation measures. In addition to information on the designated critical digital asset, information on all digital assets directly or indirectly connected to the critical digital asset also should be gathered.

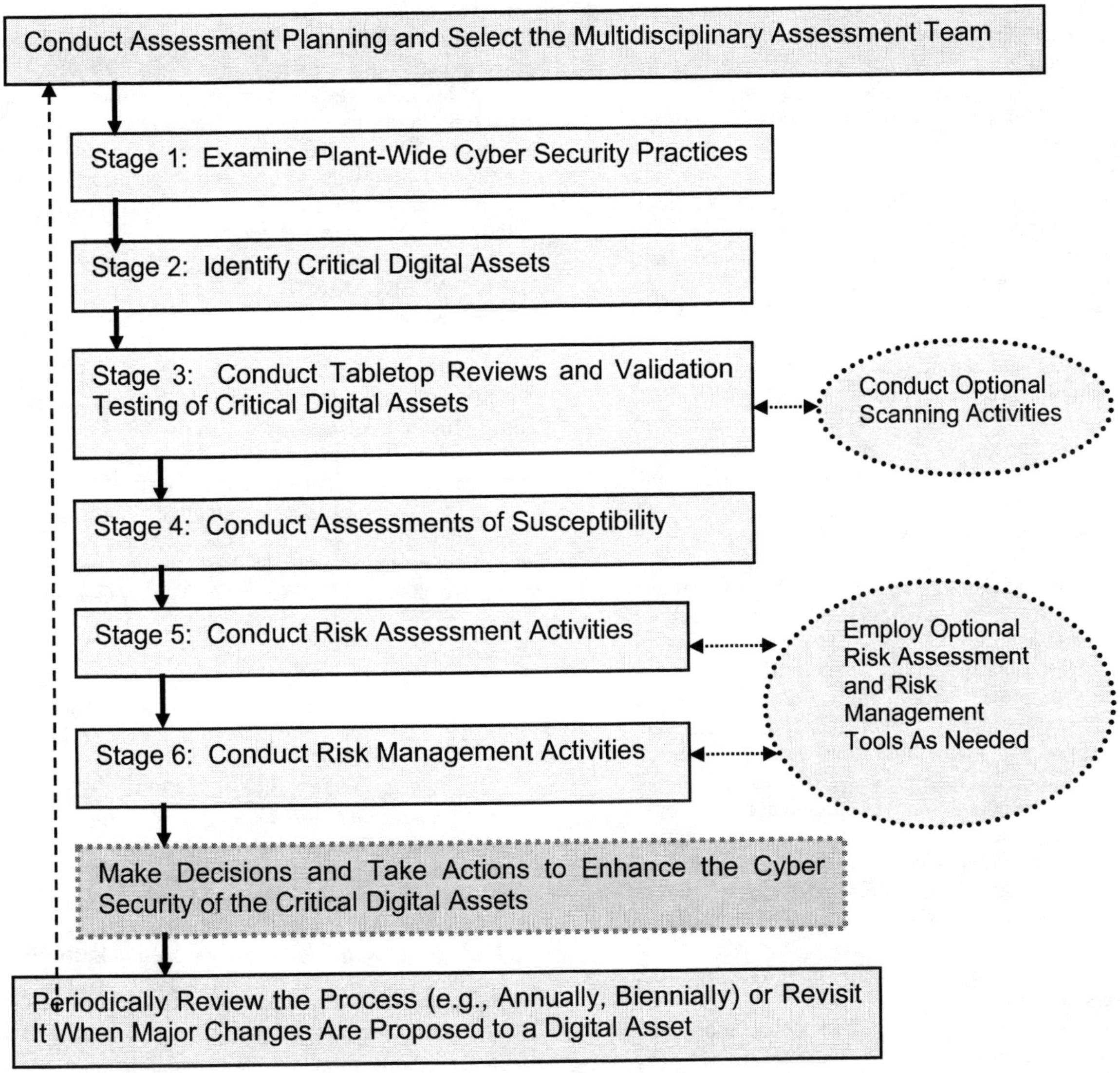

Figure 1. Simplified Flowchart for the Cyber Security Self-Assessment Method

Interview sessions with the owners and system administrators of critical digital assets often are an important source of information. This information should be confirmed by a walk-down inspection of the critical digital assets—including computers, other digital devices, and communication lines. The walk-down also can reveal vulnerabilities that were not captured in the documentation or interviews.

Stage 3 also involves logical testing of the critical digital assets to validate findings and detect other potential vulnerabilities. Logical testing involves an "inside-out" approach that starts at the critical digital asset and moves outward to identify all connectivity paths to and from the device. Each path is evaluated to determine its endpoints, accessibility, and security. This is contrast to the "outside-in" approach of penetration testing that starts from the hacker's perspective and attempts to find a path inward to a critical digital asset.

The Method also allows for optional electronic scanning. At many facilities, the level of expertise required to perform scanning or penetration testing would not be available in-house. Corporate-level or contractor support may be required to help support this optional portion of the investigation.

In Stage 4 of the Method, assessments of susceptibility are conducted. The results of the tabletop reviews and the validation testing are analyzed to determine the susceptibility to cyber exploitation of a critical digital asset or each individual connectivity pathway leading to the critical digital asset. Susceptibility is a relative measure of the cumulative vulnerability that a critical digital asset has to being exploited. It is an indication of the security of a critical digital asset based on the identified vulnerabilities, the severity of the vulnerabilities, and the effectiveness of existing protection and mitigation measures in place to reduce or eliminate the vulnerabilities.

In Stage 5 of the Method, a measure of cyber security risk is generated for each critical digital asset by considering the cumulative susceptibility of a critical digital asset and the consequences that would occur from an exploitation event. This measure allows for the ranking of risk levels for all of the critical digital assets.

In Stage 6 of the Method, risk management activities are performed to evaluate the cost-effectiveness of potential protection and mitigation measures. Risk management involves reviewing ideas for potential new protection and mitigation measures acquired during interviews with system owners and administrators. It also involves having the assessment team identify other new potential protection and mitigation measures (including cross-cutting measures that can benefit multiple critical digital assets). Detailed information is gathered on the cost and effectiveness of potential protection and mitigation options. An objective of this stage is to estimate the cost to purchase, install, and operate the protection or mitigation measure under consideration as well as estimate the cost of any productivity impacts associated with the measure.

The next step in the risk management stage is to evaluate the risk reduction to all critical digital assets as a result of successfully implementing the protection or mitigation measure. The cumulative costs associated with each individual or set of protection and mitigation measures are compared with the cumulative risk reductions achieved. Results are compared with those from other potential protection and mitigation measures. After assessing the results of the cost-benefit analysis, the assessment team should propose a course of action. This may involve implementing one or more of the protective actions that have been assessed, or it may involve electing to use available resources to address other, more pressing issues. Careful documentation of the assessment process (and all other stages of the Method) is important so that information can be reviewed by both decision-makers and future assessment teams.

If the risk assessment or management process outlined above does not adequately address all risk issues, the assessment team has the option of using additional risk assessment or management tools to further the analysis process. This may involve the use of tools that the plant has used in the past or the introduction of new quantitative tools into the decision-making process. It should be emphasized that all facilities are encouraged to supplement the approach presented in this Method with any additional procedures and tools they believe would help them to further enhance effective decision-making in the cyber security arena.

TESTING THE METHOD

The Method was tested at four power plants and refined based on feedback from the assessment teams.

CONCLUSIONS

PNNL has developed a cyber security self-assessment method that will be used by power plants to assess their cyber security program. The Method is designed to guide assessment teams in the identification of cyber security vulnerabilities and the potential consequences to their facility in the event vulnerabilities are exploited. The Method provides risk assessment and management components to assist the assessment teams in ranking their cyber security risks and in decision-making on the adoption of cost-effective protection and mitigation measures to reduce risks.

The Method views cyber security from an operational standpoint and focuses on the overall impact of the exploitation of a critical digital asset on the safe and efficient operation of the plant (as opposed to an information systems perspective that would focus on the impact of a cyber attack on the operation of the affected digital asset).

ACKNOWLEDGMENTS

The authors acknowledge Guy Landine (Southern California Edison) and Brad Yeates (Southern Company) for their contributions to the development of the Method. Members of the Cyber Security Task force of the Nuclear Energy Institute also participated in the development and review of the Method. Eric Lee, Chris Graham, Matt Chiramal, James Buchanan, and Al Tariff of the U.S. Nuclear Regulatory Commission supported and guided the development of the Method. The authors also acknowledge the assistance and support of the U.S. Department of Energy and the North American Electric Reliability Council.

REFERENCES

1. Office of the President. 1998. The Clinton Administration's Policy on Critical Infrastructure Protection: Presidential Decision Directive 63. White Paper, May 22, 1998.
2. Alberts, C. J., and Dorofee, A. J., 2001, *OCTAVE Method Implementation Guide Version 2.0 – Volume 1: Introduction,* Carnegie Mellon Software Engineering Institute, Pittsburgh, Pennsylvania.
3. Alberts, C. J., and Dorofee, A. J., 2001, *OCTAVE Method Implementation Guide Version 2.0 – Volume 2: Preliminary Activities,* Carnegie Mellon Software Engineering Institute, Pittsburgh, Pennsylvania.
4. Stoneburner, G., Goguen, A., and Feringa, A., 2001, *Risk Management Guide for Information Technology Systems*, Special Publication 800-30, National Institute of Standards and Technology, Gaithersburg, Maryland.
5. Swanson, M., 2001, *Security Self-Assessment Guide for Information Technology Systems,* Special Publication 800-26, National Institute of Standards and Technology, Gaithersburg, Maryland.
6. Presidential/Congressional Commission on Risk Assessment and Risk Management.,1997, *Risk Assessment and Risk Management in Regulatory Decision*-Making, Final Report, Volume 2, March 27, 1997, http://www.riskworld.com/Nreports/1997/risk-rpt/volume2/html/v2epaa.htm (May 26, 2003).

Proceedings of IMECE'03
2003 ASME International Mechanical Engineering Congress
Washington, D.C., November 15–21, 2003

IMECE2003-42742

ANALYTICAL APPROACHES TO ADDRESS HOMELAND SECURITY ISSUES

Mr. Gregory M. Holter
Pacific Northwest National Laboratory[1]
Environmental Technology Division
P.O. Box 999
Richland, Washington, 99352, USA
Phone (509)375-3967 fax (509)375-5921
gm.holter@pnl.gov

Mr. Jonathan Young
Pacific Northwest National Laboratory[a]
Environmental Technology Division
P.O. Box 999
Richland, Washington, 99352, USA
Phone (509)375-4510 fax (509)372-4995
jonathan.young@pnl.gov

ABSTRACT

Homeland security concerns arising since September 11, 2001, have captured national attention and sparked a number of responses at all levels of government. As events have unfolded and the nature of the situation has become better understood within the US, the need for effective planning and response has resulted in the identification of significant analytical challenges. These challenges relate to a number of different needs, including the following:

- estimating the probability and the potential impact of various threats,
- identifying the need for and effectiveness of specific counter-measures, and
- assessing the combined results of interacting activities and events.

Analytical approaches traditionally used for safety engineering and risk analysis, coupled with analytical approaches borrowed from other systems analysis disciplines, can be usefully adapted to help meet these challenges.

This paper identifies and discusses several illustrative examples of the analytical challenges currently being faced with respect to homeland security. Linkages are then examined between these specific challenges and traditional analytical approaches from a variety of disciplines, including safety engineering and risk analysis. Since effective cooperation among responsible agencies and organizations has been identified as an issue of concern and is essential to achieve an effective homeland security strategy and response capability, issues relating to multiple interacting activities are specifically highlighted.

INTRODUCTION

The events of September 11, 2001, shocked the United States into recognizing the full scope of an evolving set of homeland security concerns. Although we had experienced previous terrorist actions against US interests, both at home and abroad, the reaction to 9/11 abruptly changed our perspective and thrust us into a new situation.

It became clear that, as a nation, we would have to make some significant changes in our approach to the internal security of the nation. This awareness led President George W. Bush to establish the new Department of Homeland Security (DHS) to provide a single federal entity with responsibility for homeland security in the US. DHS, in turn, is responsible to coordinate with all other organizations with responsibilities related to homeland security, at all levels of government.

According to information published on their website, the first priority of DHS is to protect the nation against further terrorist attacks. Component agencies of DHS will analyze threats and intelligence, guard our borders and airports, protect our critical infrastructure, and coordinate the response of our nation for future emergencies. Implicit in this is the need for effective cooperation and coordination with other agencies such as the Nuclear Regulatory Commission (NRC) and the Department of Energy (DOE), law enforcement agencies at all levels (including state, county, and local), and first responders.

To support its mission, DHS will apply science and technology (S&T). According to the DHS website, "The

[1] Pacific Northwest National Laboratory is operated by Battelle for the U.S. Department of Energy

Department's research and technology focus will seek to develop capabilities to detect and deter attacks on our information systems and critical infrastructures. The S&T Directorate will develop a national research and development enterprise to support homeland defense. This research and development effort will be driven by a constant examination of the nation's vulnerabilities, repeated testing of our security systems, and a thorough evaluation of the threats and weaknesses in the system."

To ensure that these new homeland security efforts are effective, and to help minimize the resulting impact on resources, analyses need to be conducted to make sure that we are meaningfully addressing the appropriate problems, in proper priority, with responses that can be effective and integrated into a successful homeland security structure. Conducting the needed analyses presents some interesting challenges. Many of these analytical challenges can be addressed by adapting analytical approaches developed and traditionally used for safety engineering and risk analysis, in conjunction with approaches borrowed from other systems engineering disciplines.

Researchers at the Pacific Northwest National Laboratory (PNNL), along with those at many other institutions within the US, are looking at various approaches to meeting these analytical needs. This paper provides a brief overview of some selected challenges and some potential responses to these challenges, illustrating how existing approaches can be adapted to these new challenges. The paper is not intended to be comprehensive in identifying the full range of current efforts, nor is it intended to investigate any particular approach in great depth. It is intended to point out how existing approaches can make a significant contribution to homeland security in the US.

HOMELAND SECURITY ANALYTICAL CHALLENGES

As engineers, we have all been trained to employ various types of analyses to help solve societal problems. While those analyses may not provide all of the information needed to address a particular set of problems, they are helpful in developing recommendations, testing ideas, narrowing options, making decisions, and prioritizing actions.

Some key analytical needs associated with understanding and responding to homeland security issues include:

- estimating the probability and potential impact of various threats,
- identifying the need for and effectiveness of specific counter-measures,
- assessing the combined results of interacting activities and events.

These needs are similar to analytical needs associated with a number of existing engineering focus areas.

On their website, DHS notes that "One important DHS priority will be to examine the vulnerabilities found in security systems. The emphasis will be on catastrophic terrorism - threats to the security of our homeland that could result in large-scale loss of life as well as triggering major economic repercussions. The focus will be on both evolutionary improvements to current capabilities and the development of revolutionary new capabilities."

As John Parachini of the Center for Nonproliferation Studies at the Monterey Institute of International Studies indicated in his testimony before the House Subcommittee on National Security, Veterans Affairs, and International Relations (2000), terrorism represents a very fluid and elusive security threat, significantly increasing the challenge of crafting effective countermeasures. Assessment of various types of identified threats is more straightforward than identifying possible terrorist threats and their impacts. We don't fully understand terrorists' objectives and thought processes, and much of the impact of terrorism is intended to be psychological. Systematic approaches such as applied in the fields of safety engineering, risk analysis, and systems engineering will help provide a useful framework for such work.

The identification of the need for counter-measures is based, at least in part, on the potential impacts associated with the various threats. Analyses need to include consideration of the potential effectiveness of candidate counter-measures, so that we can maximize the value of our actions.

There are a number of different organizations with responsibilities in regard to homeland security, with their own individual missions and mandates. Thus, we need to consider the interfaces between these organizations and the resulting inter-organizational issues. These issues include the affect of combined actions of multiple organizations, as well as the need for effectively sharing information and integrating results.

A recent example of inter-organizational issues is shown by a set of NRC measures to address terrorism concerns relating to nuclear power plants. The NRC determined that they could increase the safety of US nuclear power stations from possible terrorist attacks by removing public access to information on the locations of these stations on the NRC web page. Subsequently, they removed nuclear power station location information from their internet websites. At the same time, they requested the Federal Aviation Administration (FAA) to establish a 10-mile exclusion zone around each nuclear power station to reduce their vulnerability to possible airborne attack. This resulted in the FAA publishing new aviation maps including these exclusion zones. Unfortunately, these 10-miles exclusion circles, being perfectly centered on each plant, provide a precise location for each nuclear power station. Therefore, the effects of the second countermeasure obviated the intended objective of the whole approach. (Incidentally, relatively detailed information on the locations of nuclear power stations also remained available to the public on the internet at other websites than the NRC's, such as that of the DOE Energy Information Administration [EIA].).

ADAPTING ANALYTICAL APPROACHES

Analytical approaches that were originally developed to address a specific set of problems can often be adapted to new problem sets, thus taking advantage of established capabilities. This strategy has often been used in engineering. There are a number of existing analytical approaches that can be adapted to help facilitate investigations into homeland security issues, and some of these approaches are discussed in this section.

In "The National Strategy for the Physical Protection of Critical Infrastructures and Key Assets" (2003), one section addresses the need for the use of modeling, simulation, and analysis activities. The Strategy notes that these methods can help to prioritize activities and investments, facilitate planning and decision support by helping to map the complex interrelationships among the different elements involved in the risk environment, and provide for real-time decision support to help mitigate an event.

Fault tree/event tree and PRA approaches should be readily adaptable to homeland security issues. Although these approaches are usually applied to the evaluation of more tangible risks, they are quite flexible and can be adapted to assess risks associated with homeland security issues. In looking at homeland security issues, there are many examples of chains of events that could result in an impact that we would want to mitigate. We can use these approaches to analyze the probabilities at each stage in the chain, and also to evaluate countermeasures that might influence those probabilities.

As Dr. Woo (2002) observes in his paper on quantitative terrorism risk assessment, the classic definition of risk is that it is a product of hazard and vulnerability. Assessments of hazards resulting from terrorist acts can be carried out similarly to hazard assessments for natural or other man-made hazards. On the other hand, identifying the probability of specific hazards resulting from terrorist activities, i.e., the vulnerability to specific terrorist acts, can be considerably more complex and difficult.

One approach that would provide significant benefits is the ranking of potential hazards by relative probability. This does not provide for identifying the absolute probability of events but still allows decisions to be made and priorities to be established based on the relative likelihood of different events. Stungis and Schori (2003), for example, have developed a decisional model of the process that a terrorist organization may use to prioritize and select potential terrorist targets. They note that it is important to be able to sort through the many possible terror targets and prioritize them by the probable severity of their individual impacts, in order to maximize our ability to protect US assets. They indicate that they have implemented a computerized version of their model, which not only allows for this prioritization but also provides the capability to conduct sensitivity analyses. The authors have some concerns with the specific model proposed by Stungis and Schori, particularly in that they include potential impact as part of the motivation for terrorist actions and thus part of the likelihood term, so if one calculates risks as hazard times vulnerability with their model, potential impact is included twice, which could significantly skew results. However, the concept proposed by Stungis and Schori remains valid.

Decision analysis techniques such as the use of decision trees or decision matrices can also help provide a systematic basis for assessing threats and evaluating possible actions. These techniques could be used to evaluate decision processes of terrorist groups, to try to better predict their actions. They can also be used to develop decision bases for response actions prior to the time they are needed, so that the decision process can be accelerated and made more effective at the time that specific responses are required. Furthermore, decision analysis techniques can also be particularly useful in the context of identifying the types of information needed to make the required decisions as events unfold, so that planning and communication protocols can be put in place to make sure such information will be more readily available.

The use of N^2 (N-squared) diagrams, applied in systems engineering to evaluate interfaces, is one of several techniques that can be adapted to analyzing combined results of interacting activities or events. Interacting events are of interest in evaluating the impact of simultaneous threat actions, and also in considering the combined results of related countermeasures. Further discussion of the evaluation of the combined effects of countermeasures is included below.

OBTAINING AND SHARING INPUT INFORMATION

The National Strategy for the Physical Protection of Critical Infrastructures and Key Assets notes that "Information is a crucial tool in fighting terrorism, and getting the right information to the right party at the right time is a top priority." The Report of the National Commission on Terrorism (2000) goes on to say "The Commission has identified significant obstacles to the collection and distribution of reliable information on terrorism to analysts and policymakers."

There are several types of challenges in obtaining and sharing the needed input information to carry out meaningful analyses, including:

- Difficulties in *identifying what specific data is required* to support the desired analyses,

- Difficulties in *obtaining reliable and applicable data* on which to base the desired analyses, and
- Inter-organizational difficulties in *collecting, distributing, and sharing some forms of information* that can provide important input for analyses.

There are a number of potential difficulties in identifying the necessary data on which to base meaningful analyses of homeland security issues. This is partly because of the elusiveness of the threat and partly because of the difficulty in predicting the full chain of events and resulting ramifications that can occur following a major terrorist action. Jessica Ruvinsky (2003) refers to the difficulties in identifying what she calls "residual consequences," and notes that ripple effects can turn a tragedy into a disaster. Careful scoping of the analyses will help limit the needs for such elusive data, but this trade-off must be weighed against the possibility of reducing the value of the analytical results by this scope limitation.

We also need to consider the potential difficulties in obtaining reliable or applicable (useful) data on which to base analyses. As noted by Ruvinsky, information relating to terrorism comes in from a wide range of different sources, and the information is variably reliable. We need to remain aware of the quality of the data we seek for our analyses, and to consider the impact of this on the reliability or usefulness of our results.

It is also imperative to consider the inter-organizational issues associated with collecting, distributing, and sharing the information needed to feed analyses of homeland security issues. As noted by Steven Miller (2003), enormous amounts of raw intelligence were already being collected prior to 9/11, and this has presumably increased substantially since then. But collection is only the first step in using this type of data, which must also be sifted, analyzed, integrated with other information and assessment, and then communicated effectively to those in a position to use the resulting integrated set of information. This implies the need for a systematic understanding of which organizations and functions need what data, mechanisms for breaking down the jurisdictional boundaries between organizations, and improved mechanisms for communicating the necessary data in a timely fashion.

The National Strategy for the Physical Protection of Critical Infrastructures and Key Assets suggests that overcoming the obstacles relating to information sharing entails the following elements:

- "Identifying what is to be accomplished by exchanging security-related information;
- Defining the type of information that must be shared to accomplish that purpose;
- Determining how and when to share and safeguard critical security information most properly;
- Deciding who the appropriate recipients of such information will be;
- Assigning responsibility for analyzing information and determining the threat implications; and
- Assigning responsibility for appropriate action once that information has been analyzed and the threat implications are clear."

This sort of systematic approach will go a long way toward improving the flow of key information.

EFFECTIVELY SHARING/INTEGRATING RESULTS

A number of reports coming out following 9/11 addressed the issue of the need for effective cooperation and coordination among the agencies and organizations responsible for protecting homeland security within the US. This need had been previously recognized, but the events of 9/11 and the response to these events underscored this issue.

Difficulties in effectively coordinating and integrating the actions of various organizations and agencies can significantly reduce the combined effectiveness of their actions. The potential for adverse impacts often increases in instances where the implementing agency is not the agency developing the objectives or the one with facilities and resources to protect. The potential for adverse impacts is also increased in areas where several agencies have overlapping jurisdictions.

An example of such adverse interactions, discussed previously, involved efforts by the NRC to limit availability of information on nuclear power station locations. This example showed that, although the objectives of the individual measures were appropriate, the combined implementation of these measures resulted in serious impacts on their usefulness.

A key challenge is to provide a method that allows for consistent, systematic evaluation of combinations of countermeasures to determine if their combined effects meet the objective(s) of each measure or whether they result in an adverse interaction (e.g., defeat an objective). Such a method should also support optimization of the combined effects of the applicable countermeasures.

Among other initiatives at PNNL relating to homeland security, researchers are investigating an approach to address concerns about the combined effectiveness of countermeasures, based on adaptation of analytical tools that have been traditionally used in related technical fields (e.g., safety analysis, systems engineering, reliability analysis). The approach characterizes the objectives, the features, and the impacts of each countermeasure and then uses this information to evaluate the combinatorial impacts of multiple related countermeasures. This approach is being designed to provide a picture of the net impact of the set of countermeasures and their implementation on the set of individual countermeasure objectives. The method seeks to provide a systematic way to examine each combination of objectives, features, and effects for every identified countermeasure. It is based on the concepts used in the development of the systems engineering N^2 diagram (described previously).

The approach being developed is represented schematically in Figure 1. Note that, in order to be effective, this approach requires, at a minimum, the effective sharing of information regarding the countermeasures being considered for implementation; challenges relating to sharing this type of information among different organizations and agencies were discussed previously.

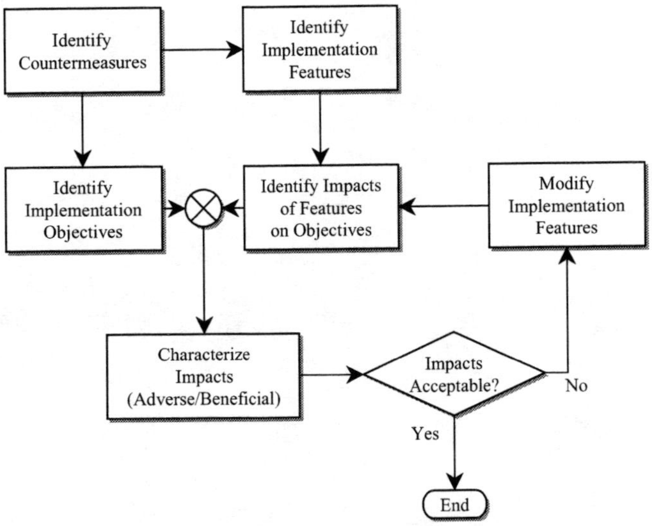

Figure 1 Schematic for Assessing Adverse Interactions

A modified N^2 diagram provides the basis for the characterization of impacts from the various countermeasures on each other, represented in the lower left block of the figure. N^2 diagrams are described in many publications addressing systems engineering, including the NASA Systems Engineering Handbook (1995). A simple example of an N^2 diagram is shown in Figure 2.

Blocks representing different interacting functions or processes are shown on the diagonal (shaded in this example). Outputs from each block move in horizontal direction (either left or right), and inputs to each move in vertical direction (either up or down). This type of diagram is one approach that allows explicit representation of how outputs from each block can serve as input to each of the other blocks, in a pair-wise representation.

F1	F1→F2	F1→F3	F1→...	F1→FN
F2→F1	F2	F2→F3	F2→...	F2→FN
F3→F1	F3→F2	F3	F3→...	F3→FN
...→F1	...→F2	...→F3	...	...→FN
FN→F1	FN→F2	FN→F3	FN→...	FN

Figure 2 Example N^2 Diagram

The modified N^2 diagram allows the outputs from each countermeasure to be defined as inputs into each of the other countermeasures, providing a formalized mechanism for pair-wise evaluation of impacts. If unacceptable impacts are identified based on this pair-wise evaluation, specific implementation features of selected countermeasures can be modified and the analysis repeated, until an acceptable set of interactions results.

Given that agencies or organizations are sharing responsibilities for implementation of countermeasures, they should be motivated to cooperate in terms of providing appropriate information to allow for such analyses. Nevertheless, there are many policy and other constraints on information sharing that will need to be overcome, and the challenges associated with these changes should not be underestimated.

CONCLUSIONS

Although there are significant challenges in providing meaningful analyses to aid effective planning and response to homeland security issues, there are a number of currently available and well understood analysis techniques that can be usefully adapted and applied. There is still a significant challenge associated with obtaining useful and representative data on which to base these analyses. Even so, there are a number of precedents for the development of data relating to phenomena that are poorly understood (for example, to provide input to PRA exercises relating to nuclear power incidents that have never been experienced).

Another important challenge is associated with the effective sharing and integration of analytical results among the involved organizations and agencies. However, the authors believe that there are approaches available that can significantly improve our ability to respond to this challenge.

The authors believe that the intelligent adaptation of current analytical approaches can provide the basis for considerably increased understanding of homeland security issues and the usefulness of proposed responses. The fundamental basis for such intelligent adaptation is a clear understanding of the specific information needs associated with understanding homeland security threats, planning for the countering of these threats, and responding effectively to actual threats when they occur.

ACKNOWLEDGMENTS

The authors wish to thank the many colleagues at PNNL and other organizations with whom they have discussed ideas relating to this subject area.

REFERENCES

Miller, Steven E., March 2003, "After the 9/11 Disaster: Washington's Struggle to Improve Homeland Security," in Axxess, no. 2, 8-11, Stockholm, Sweden.

National Aeronautics and Space Administration, June 1995, *NASA Systems Engineering Handbook*, SP-6105.

National Commission on Terrorism, 2000, *Countering the Changing Threat of International Terrorism.*

Parachini, John V., July 26, 2000, "Combating Terrorism: Assessing Threats, Risk Management, and Establishing Priorities," Testimony before the House Subcommittee on National Security, Veterans Affairs, and International Relations.

Ruvinsky, Jessica, March 12, 2003, "Post-9/11, Scholar Advocates Objective Risk Assessment," in Stanford Report, Stanford University.

Stungis, George E. and Thomas R. Schori, March 2003, "A Terrorist Target Selection and Prioritization Model," in The Journal of Homeland Security, Analytic Services Inc.

The White House, February 2003, "The National Strategy for the Physical Protection of Critical Infrastructures and Key Assets." Washington, DC.

U.S. Department of Homeland Security, 2003, http://www.dhs.gov , Washington, DC.

Woo, Dr. Gordon, 2002, "Quantitative Terrorism Risk Assessment," Risk Management Solutions Ltd.

AUTHOR INDEX

PROCEEDINGS OF ENGINEERING/TECHNOLOGY MANAGEMENT — 2003:
Safety Engineering and Risk Analysis
Technology and Society
Engineering Business Management
Homeland Security

Book Number: I00697